UNDERSTANDING SCIENCE

PROMETHEUS BOOKS
Buffalo, New York

Published 1992 by Prometheus Books

 Inquiries should be addressed to Prometheus Books, 59 John Glenn Drive, Buffalo, New York, 14228-2197, 716-837-2475 / FAX 718-835-6901

96 95 94 93 92 5 4 3 2 1

Library of Congress Cataloging-in-Publication Data

Strahler, Arthur Newell, 1918-
Understanding Science; an introduction to concepts and issues / by Arthur N. Strahler
423 p. cm.
Includes bibliographical references and index.
ISBN 0-87975-724-8 (cloth: acid-free)
1. Science. I. Title.
Q158.5.S73 1992
501--dc20 92-6704
CIP

Printed in the United States of America on acid-free paper.

Credits and permissions:

Excerpts from DICTIONARY OF PHILOSOPHY by Peter A. Angeles. Copyright © 1981 by Peter A. Angeles. Reprinted by permission of HarperCollins Publishers.

Chapters 1 and 13. Quoted passsages by Bernard G. Campbell. Reprinted with permission from: Campbell, Bernard G. *Human Evolution: An Introduction to Man's Adaptations*. 3rd Edition. (New York: Aldine de Gruyter) Copyright © 1985 Bernard Campbell.

Chapters 2 and 10. Exercepts from *An Introduction to Logic and the Scientific Method* by Morris R. Cohen and Ernest Nagel, copyright 1934 by Harcourt Brace Jovanovich, Inc. and renewed by Ernest Nagel, reprinted by permission of the publisher.

Chapter 10. Quoted passages by Tobias Dantzig. Reprinted with the permission of Macmillan Publishing Company from NUMBER: The Language of Science, Fourth Edition by Tobias Dantzig. Copyright 1930, 1933, 1939, 1954 by The Macmillan Company. Copyrights renewed © 1958, 1961 by Anna C. Dantzig, 1966 by Henry P. Dantzig and George B. Dantzig, and 1982 by Mildred B. Dantzig.

Cover design by Kathleen Sarazin.

Designer, illustrator, compositor: Arthur N. Strahler

To the memory of
Harriet Newell Brittan,
who taught me
Skepticism
and the
Joys of Learning

PREFACE

My goal in this book is to make the philosophy of science accessible and intelligible to science students, their teachers, and just about any person with a liberal college education who would like to learn something about the subject. One might suppose that with a large number of philosophers of science installed in our colleges and universities, where they offer lecture courses in their field, at least a few of them would already accomplished what I have set out to do and would have done a vastly better job of it. To date, we have no such general work specially written for science majors.

Perhaps the obstacle has been one of two different perspectives. Until quite recently, science philosophers seem to have been preoccupied with prescriptive analysis—the "oughts," as they see them—whereas practicing research scientists are preoccupied with getting their job done—the "is"es of the business. But even as both groups have now converged their respective sight lines to a common focal point, the philosophers continue to discourse in a different tongue, almost unintelligible to the scientist. As a self-appointed translator, using science as my native language, I have labored in the interest of my science colleagues, giving them the benefit of my untold hours of agonizing over what the philosophers mean by their alien vocabularies and involuted phraseologies. Linguistic reduction has its perils, and doubtless I have misled my readers in many places, but not intentionally. Fortunately, some philosophers who are obviously good teachers have published lucid statements on particular subjects, and I have quoted them freely with both satisfaction and gratitude.

My book comes in two parts. Part One is an overview of the many facets of science philosophy, keeping things as simple as possible and emphasizing basic attitudes and realistic goals. What science is not may be just as important as what it is. For many readers, the first six chapters of Part One will cover most of what they wish or need to learn about the workings of science, but I hope they will browse the remainder of the book and delve into some topics that may look interesting.

Part Two attempts to deal seriously with criteria of demarcation of all the major separate knowledge fields. I have followed closely the writings of philosopher Mario Bunge on general ontology and epistemology. A possible innovation is my classification of logic and pure mathematics as an ideational field. Science cannot be understood in isolation and there is much to be learned about the nature of science by examining its interactions with certain of the ideational fields, especially religion, ethics, and the sociopolitical ideologies. Keeping these fields clearly separated in content and intent from science is more important than ever before as Western society seeks increasingly to use and abuse its science and its scientists for ends of dubious propriety.

Some of the content of this book derives from my 1987 work, *Science & Earth History—The Evolution/Creation Controversy*, published by Prometheus Books. The chapters on philosophy of science selected from that work have been extensively revised and augmented by new paragraphs and sections. An important improvement is a new chapter titled "The New Philosophy of Science." Part Two contains much new writing on the fields of knowledge and the criteria of their demarcation.

ACKNOWLEDGMENTS

I am indebted to the following persons who reviewed chapters of my manuscript: Harold I. Brown (Chapter 5), Mark A Melton (Chapter 10), and John R. Armstrong (Chapters 11 and 14). Their fields of special competence are (in the same order) philosophy of science, mathematical statistics and geology, theology and geology. Because several chapters in Part One are taken in large part from *Science & Earth History*, my indebtedness for extensive manuscript review of that work must again include both Mark A. Melton and John R. Armstrong, along with Ronnie J. Hastings (physics) and Peter A. Angeles (philosophy). Among the authors who gave permission to quote extensively from their published works, I am especially grateful to Mario Bunge (philosophy) and Bernard G. Campbell (anthropology).

Steven L. Mitchell, Editorial Director of Prometheus Books, gave valuable guidance in developmental stages of the manuscript as well as careful supervision through the production stages. As the word-processing keyboarder who functioned as designer, compositor, and illustrator through the final stage of camera-ready pages, I take full responsibility for whatever strange glitches you may discover between the covers of the book.

ABOUT THE AUTHOR

Arthur N. Strahler holds the Ph.D. degree in geology from Columbia University. He was appointed to the Columbia University Graduate Faculty of Pure Science in 1941, serving as Professor of Geomorphology from 1958 to 1967 and as Chairman of the Department of Geology from 1958 to 1962. A Fellow of the Geological Society of America, his published research has dealt with processes and forms of fluvial erosion, morphometry and statistical analysis of landforms, and systems-theory applied to geomorphology and geohydrology. He is the author or coauthor of textbooks on physical geology, the earth sciences, physical geography, and environmental science.

Abbreviated Contents

Expanded Contents

PART ONE Science—Once Over Lightly

PART ONE

SCIENCE—ONCE OVER LIGHTLY

Introduction

Philosophy of science is often viewed by practicing scientists as foreign territory—a *terra incognita* filled with strange and threatening notions described in esoteric and incomprehensible language. Philosophy of science holds the role of general overseer of all scientific knowledge; always looking over the scientist's shoulder and asking probing questions. Many scientists resent this intrusion, which can be embarrassing in the extreme, particularly when the philosophers discuss among themselves the activities of the scientists, pointing out misconceptions and weaknesses in the generally accepted methods of scientific practice.

When I began to write about the philosophy of science I envisioned my prospective group of readers as mostly nonscientists—outsiders beyond the walls of the science establishment. Then, as I progressed into the more abstruse problems of the philosophy of science—bizarre subjects such as "reductionism in biology"—it gradually dawned on me that very few research scientists and probably not very many college instructors of science have ever given much thought to the nature of science. I quickly became aware of how little I knew of the philosophy of science, despite some dabblings in that area early in my career. Mostly, it was cut-and-dried positivistic stuff with lots of lip service to testing, corroboration, and falsification. Operational rules for actually doing scientific research were relatively simple and, once the requirements of the science community were understood, we could get on with the really important job, which was to get some graduate students, attract adequate contract funding, and keep the output of journal articles coming.

Science philosopher John Ziman has expressed some sadness that most practicing scientists exhibit no fervor in philosophical analysis of what they are actually doing. He nevertheless shows tolerance for their dereliction and takes a pragmatic view of the situation, for he writes:

> One can be zealous for Science, and a splendidly successful research worker, without pretending to a clear and certain notion of what Science really is. In practice it does not seem to matter. Perhaps this is healthy. A deep interest in theology is not welcome in the average

> churchgoer, and the ordinary taxpayer should not really concern himself about the nature of sovereignty or the merits of bicameral legislatures. Even though Church and State depend, in the end, upon such abstract matters, we may reasonably leave them to the experts if all goes smoothly. The average scientist will say that he knows from experience and common sense what he is doing, and so long as he is not striking too deeply into the foundations of knowledge he is content to leave the highly technical discussion of the nature of Science to those self-appointed authorities the Philosophers of Science. A rough and ready conventional wisdom will see him through. (1980, pp. 38-39)[1]

I was pleased to find that Professor Ziman had some more to add, taking a harder line with the "rough and ready" scientists:

> Yet in a way this neglect of—even scorn for—the Philosophy of Science by professional scientists is strange. They are, after all, engaged in a very difficult, rather abstract, highly intellectual activity and need all the guidance they can get from general theory. We may agree that the general principles may not in practice be very helpful, but we might have thought that at least they would be taught to young scientists in training, just as the medical students are taught Physiology and budding administrators were once encouraged to acquaint themselves with Plato's *Republic*. When the student graduates and goes into a laboratory, how will he know what to do to make scientific discoveries if he has not been taught the distinction between a scientific theory and a non-scientific one? Making all allowances for the initial prejudice of scientists against speculative philosophy, and for the outmoded assumption that certain general ideas would communicate themselves to the ucated and cultured man without specific instruction, I find this an odd and significant phenomenon.[1]

Perhaps some of us who came on the science scene a half-century ago would have paid more attention to the philosophy of science if it had seemed to relate to the natural and historical sciences in which we chose to work. Instead, we were turned off by philosophers who dealt almost exclusively with formal science and pure physics. The worst part of it was that these masters of the abstruse did not even recognize biology and geology as being science in the first place. I well remember a futile exchange of letters with Ernest Nagel, a leading philosopher of science on the same campus, exploring some ideas on the nature of historical science. Upon rereading those letters, I see now that he had no idea what I was trying to say and I, in turn, found nothing of relevance or value in his response. Things have changed in recent years. Contemporary philosophers such as Philip Kitcher (1982) are talking about things in the range of my experience, which is in part geology but includes the history of the planet and the evolution of its life forms. Perhaps with the emergence of a philosophical analysis of historical science, more young professionals in the natural sciences will get interested in the nature of their science.

The modern scientific view of the universe can be described as *naturalistic*, using an adjective that has its historical roots far back in philosophy, explaining all phenomena by strictly natural processes—as distinct from explanations invoking supernatural forces. I could have just as easily used *mechanistic* as the adjective, but that is a harsh word, suggesting the actions of a machine and the work of an inventor. Another choice would have been *materialistic*, but for most persons that adjective carries a negative association in terms of moral and religious values. The naturalistic view is that the particular universe we observe came into existence and has operated through all time and in all its parts without the impetus or guidance of any supernatural agency. The naturalistic view is espoused by science as its fundamental assumption.

We have only to look back a few centuries to find a time when naturalistic science was practically nonexistent in Western Europe. With the fall of the Roman Empire in the fifth century A.D. and the onset of the Dark Ages, whatever science had been accumulated by Greek scholars—among them Thales, Democritus, Hippocrates, Plato, Aristotle, Aristarchus, and Ptolemy—was put securely under wraps by the Christian Church. During that dark medieval period, science survived in the hands of the Moslems, who had translated many of the Greek scientific writings and proceeded to add a few ideas of their own.

The Catholic Church had its own version of how the universe, its contents, and its governing laws came into existence. Christian theology had an answer for every question about the physical universe: it was created by God, along with the natural laws that sustain it and govern all change. An omniscient and omnipotent God, acting with a divine purpose, designed everything that is or has been. God's work was and is perfect, and humans were forbidden to question or demean it.

On the other hand, investigation and description of nature was permitted and even encouraged to disclose the marvelous works of God, and in so doing, to glorify His name. Under the name of "natural theology," this license to seek new knowledge of nature was precisely what was needed to permit a resurgence of science, budding during the Renaissance of learning in the fourteenth and fifteenth centuries and blossoming in the scientific revolution of the sixteenth and seventeenth centuries. The Protestant Reformation made no real difference in what was happening, because Protestants retained the same doctrine of universal divine creation and espoused a similar natural theology. Theologians actually encouraged the pursuit of natural theology in the belief that it would assist in revealing the true nature of God

(Polkinghorne 1989, p. 7). They took their warrant from the Apostle Paul, who wrote that God has shown himself to humans through the creation of the world, ". . . being understood by the things that are made" (Romans 1:20).

Natural theology enabled devout Christians, and even members of the clergy itself, freely to investigate nature. Thus Carolus Linnaeus, a Swedish botanist and taxonomist, was free to establish in 1735-37 the modern system of classification of plants and animals. For him, it was an exposition of God's wondrous creation. Of course, the Bible contained the true story of creation, and sooner or later natural theology was to come up with a description of some aspect of nature that was in direct conflict with the dogma of Christian theology. That had already happened in 1543, when Copernicus brought forth his heliocentric theory that the earth and planets move in circular orbits about the sun. (The idea had been suggested by Aristarchus in the third century B.C., but had been lost from sight during the Dark Ages.) The heliocentric heresy persisted and was strengthened by the astronomical findings of Tycho Brahe, the formulation of laws of planetary motion by Johannes Kepler, and finally by direct observational data put forth by Galileo in 1610. Evidence finally forced the theologians to shift their support from the established dogma of a geocentric universe to the new heliocentric model, but it did not weaken their conviction that God was the creator. What we see here was simply a correction of a human error in having held a mistaken concept of God's work. The "tail" of science had begun to "wag" the dogma of theology, and was to continue to do so with increasing strength.

During the mid-1600s science gained strength through the formation of the Royal Society in England and the Academy of Sciences in France. The former included at the time of its formation Robert Boyle and Robert Hooke, now great names in the history of physics. Boyle was deeply religious, as were others within the society, and was able to reconcile scientific research with religion through natural theology (Faul and Faul 1983, p. 41). Isaac Newton joined the group somewhat later, after he had discovered the law of gravitation and the spectral composition of white light. Following the publication of his *Principia* in 1687, Newton turned to studies of alchemy and theology. This may seem a strange mixture of interest—science, pseudoscience, and religion—but it was evidently considered quite acceptable at the time.

This early history of Western science cannot help but suggest that Christian theology provided the trunk from which sprouted and grew the great branch of naturalistic science. Perhaps a better analogy is that theology gave birth to science, which became a freeswimming organism growing to gigantic proportions. Parent and offspring have ever since stood in varying levels of conflict. We shall pursue this conflict in a later chapter on the criteria of demarcation between science and religion. Before that point, however, it is more directly useful to take our cue from

Newton's interest in alchemy, which we would now clearly recognize as false science, or pseudoscience.

One of our early objectives will be to investigate both science and pseudoscience from various angles in order to find some reliable criteria whereby one can be distinguished from the other. This task occupies Chapters 7 and 8. There is much to be gained by such a comparative analysis beyond establishing a set of distinguishing criteria. Perhaps the most important gain is an in-depth understanding of the nature of science. Years ago, one of our university students, applying for admission to graduate school after a term of overseas service in the Peace Corps in a remote third-world country, was asked what he had learned from that experience. His reply: "I learned what the United States is really like." Recently, I read the same statement made by another Peace Corps returnee. Perhaps it has been made innumerable times and is now just a cliché: We learn most about ourselves by comparing our state with another state that is very different.

Part One furnishes an overview of science and the scientific method. Some of the topics included are covered only briefly and will need reinforcement in later chapters. For many readers, however, this part of the book will suffice as an introduction to the philosophy of science. You may want to browse through the remainder of the book, selecting sections of personal interest such as logic and mathematics, ethics and morality, science and religion, or science versus "creation science."

Credit

CHAPTER 1
What Is Science?

The first six chapters offer an overview of the philosophy of science—once over lightly—so that you can gain a general idea of what science is all about. There are positive features to be considered as well as misconceptions to be straightened out. Many of my academic colleagues in science and philosophy could do the job more authoritatively than I; many of them have already done so. But, after reading their presentations of the subject in both popular and academic styles, I'm not sure they have actually represented science to the average literate adult in a way that makes clear essential differences between science and the various other forms of knowledge on which it constantly impinges. What's more, many accounts of science as an activity seem to be telling us what scientists are supposed to be doing, rather than how they actually do things.

One difficulty is that the giants in science and philosophy are prone to restrict the scope of science rather severely, usually to physical science, and even as narrowly as to nuclear and quantum physics alone. We must, on the other hand, cover a large area of natural science quite far removed from the ideal behavior of matter on a subatomic scale. Besides dealing with the origin and physical evolution of the universe—a field that does indeed rest in large part on principles of theoretical physics—we must include the geological and biological evolution of our own planet Earth over a time span of billions of years. Here we will find extremely complex aggregations of matter that have long and involved histories of continuous development. Scientists who investigate these historical areas of knowledge need to adopt specialized views of science. Be prepared for some unfamiliar notions of what science is and how it works.

Science as a Method of Obtaining Knowledge

Let us begin by thinking of science as a method of obtaining knowledge. Science philosopher Ernest Nagel, for many years a member of the Columbia University faculty, has shown how modern science originated in common sense knowledge and how it has "grown out of the practical concerns of daily living." He gives examples:

> . . . geometry out of measuring and surveying fields, mechanics out of problems raised by the architectural and military arts, biology out

> of problems of human health and animal husbandry, chemistry out of problems raised by metallurgical and dyeing industries, economics out of problems of household and political management, and so on. (Nagel 1961, p. 3)

Granted this historical continuity between common sense methods and genuine scientific methods, it would be a mistake, says Nagel, to conclude that "the sciences are simply 'organized' or 'classified' common sense" (p. 3). As one point, he notes, common-sense methods of doing things were rarely accompanied by any explanation. This we would expect of a strictly trial-and-error process of learning. Where explanations were offered, they were typically far off the mark. For example, it was thought that the value of foxglove as a heart stimulant could be explained by the heartlike outline of the foxglove flower (p. 4). Nagel conjectures: "It is the desire for explanations that are at once systematic and controllable by factual evidence that generates science" (p. 4). He concludes that organization and classification of knowledge through explanatory principles defines the distinctive goal of science.

Let's back up to square-one and attempt to reach a working definition of the scientific method. We start with a simple declarative statement, or proposition, and enhance it in stages in response to a set of queries. (Terms in parentheses are suggested alternatives or synonyms for the immediately preceding terms.)

- Science is the acquisition of knowledge in one of several ways (processes, systems).

  Q: Knowledge of what?

- Science is the aquisition of knowledge *of the real world* (external reality).

  Q: What is the quality of that knowledge?

- Science is the acquisition of *reliable* knowledge.

  Q: How reliable is it? Is it infallible?

- Science is the acquisition of reliable *but not infallible* knowledge of the real world.

  Q: Where does explanation fit into the picture?

- Science is the acquisition of reliable but not infallible knowledge of the real world, including explanations of the phenomena.

That's only a start, because all kinds of additional questions and qualifications come to mind and must be addressed, and that takes a lot of careful analysis. We will attend to these matters a bit later. So much for now about science as a method of obtaining knowledge.

Science as a Category of Knowledge

Philosophers recognize many categories of knowledge, and scientific knowledge is only one of them. But what is knowledge? Having asked that question, we immediately confront a concept so elusive that about all we can do at this point is to find an excuse to get out of it and on with something more tangible. Situations like this are often handled by making a circle of words—it's called a tautology—in which, as you move around the circle, each word means the same thing as the word before it and the word that follows it. Let's try this with the term "knowledge." Knowledge of something is acquired by knowing; to know is to have cognition; cognition is a condition of awareness; awareness is the knowledge of something.

A major preoccupation of philosophers from classical times to the present is to break out of such neat tautological traps; they make a market in it under the name of *epistemology*. Epistemology deals with such questions as: What is knowledge? How is knowledge acquired? What are the kinds or fields of knowledge? A few pages further on in this chapter we will firmly tackle the first two questions. For the moment you may be excused for simply saying: "I know what it is to know, so tell me something I don't already know."

Science philosopher Mario Bunge of McGill University thinks of the various kinds or classes of human knowledge as *cognitive fields* (1984, pp. 37-38). This substitution of words may not seem like a significant accomplishment, but it does lead to a first attempt to organize and classify the varieties of human knowledge. Here is how Professor Bunge gets started on that organization:

> We shall characterize a science, as well as a pseudoscience, as a cognitive field, genuine or fake. A *cognitive field* may be characterized as a sector of human activity aiming at gaining, diffusing, or utilizing knowledge of some kind, whether this knowledge be true or false. There are hundreds of cognitive fields in contemporary culture: logic and theology, mathematics and numerology, astronomy and astrology, chemistry and alchemy, psychology and parapsychology, social science and humanistic sociology, and so on.[1]

Professor Bunge has not told us what knowledge is, but he has told us something new and important, something substantial. Notice the pairing of the cognitive fields he has listed. The first member of each pair belongs to science; the second to non-science, including pseudoscience, or false science. Not that all my readers will agree with the professor on the things he has put in the pseudoscience group, for if you are a believer in astrology or parapsychology, you may feel outraged.

Nevertheless, a salient point has been made here. We can think of "mountains" of knowledge, separated into two mountain ranges, between which lies a great gulf. So there are two great classes of cognitive fields. One, Bunge says, consists of *belief fields,* in which the knowledge rests on belief—belief in something that cannot be observed to exist physically. He cites religions and political ideologies as examples. He also puts pseudoscience in with the belief fields, much to the discomfiture of the theologians! The other great class, Bunge says, consists of the *research fields*, in which knowledge rests solely on observation of the real world. He puts science—both basic and applied varieties—in this category along with the humanities. Bunge gives us one distinguishing feature that clearly separates the two classes: "Whereas a research field changes all the time as a result of research, a belief field changes, if at all, as a result of controversy, brute force, or revelation" (p. 38).

Already, we are well launched on our climb up the mountain of science because we have been given a valuable piece of information about our mountain and others of the same chain: it is in continuous change, so we must select our route carefully. An old trail shown on the map may be washed out in places. Others who climb after us may not find our trail passable and will choose another path that looks better. Those who like things neat and easy may want to climb a mountain in the other range, where well-worn trails are permanent and secure, amply fitted with strong handholds and guide ropes.

Perhaps we have jumped ahead a bit too fast and far, but two purposes have been served. First, we have put all forms of knowledge into two distinct compartments, or boxes. Humans seem to gain a sense of satisfaction in classifying things into compartments with solid barriers, particularly when each item fits only one compartment. Second, we have sown the seeds of conflict, and conflict guarantees interest and purpose. Human adherents to the knowledge in one compartment distrust or hate those subscribing to the kind of knowledge in the other one—at least some of them do. Some persons, of course, divide their own brains into two compartments, one for each class of knowledge, and seem to get along fairly well. George Orwell's word for this strategy is "doublethink." In any case, there is more than enough hostility to support several major ideological wars.

In Part Two, we go into much greater depth and detail on the fields of knowledge, redefining the two great classes and establishing criteria of demarcation between them, and among knowledge fields in each class.

The Scope of Science

If science is a fund or body of collected knowledge, we need first to decide what scope that knowledge covers or deals with. Let us switch from "knowledge" (that which is known) to a synonym, "information"; it is a word that seems better to remove the knowledge that resides in the brain to an external location where it is more readily accessible. We can then observe that science consists of a body of information that can be stored as well as freely communicated by humans. Information entering one person's brain, analyzed and temporarily stored there, can be transferred to the brain of another human only by use of symbolic statements in oral, written, or pictorial form. In such forms, information can also be available in a general storage pool.

If science is a fund of information, what is the nature of this information? About what does science inform us? The question takes us into philosophy because our reply is very sweeping: Science gathers, processes, classifies, analyzes, and stores information on anything and everything observable in the universe. We like to think that science deals only with that which is real in the sense that it is observable and identifiable as either *mass* or *energy*.

We need to review briefly at this point just what is meant by "observable universe." First, we will use the term *matter* to include both mass and energy. This is feasible because mass can be converted into energy and vice versa. Almost all literate persons are at least vaguely familiar with the famous equation revealed by Albert Einstein: $E = mc^2$ (Energy is equal to mass times the speed of light squared.) This equation specifies exactly the relationship between mass and energy. Next, we note that the universe consists of matter in a framework of *space* and *time*. For most practical purposes in science, space is three-dimensional. Physicists often find it necessary to combine space and time into a single term, *spacetime* (Davies 1984, p. 151). Thus, we can summarize by saying that the physical universe consists of matter in spacetime. It is very important to add the concept that the structures and states of matter change in space with the passage of time; i.e., that the physical universe is not a static or changeless universe. These relationships between mass and energy apply to both classical physics (Newtonian mechanics) and to the newer relativistic mechanics, which we might wish to designate as the "modern" physics (Atkins 1972, p. 200). The same analysis of the universe in terms of matter and spacetime can be extended into quantum mechanics and the grand unified theories (GUTs) that account for all mass and energy in terms of the four fundamental forces and the kinds of leptons and quarks that make up the universe (Davies 1984, Chaps. 5, 6).

We can use our matter/spacetime description of the physical universe to arrive at a criterion for distinguishing empirical science from at least several fields of knowledge in the belief class. Matter and spacetime consist of *measurable components*, meaning that scientists have the appropriate scales and instruments by means of which observed values can be referred to standard values. In mechanics, these components are usually designated as "dimensions." Mass is a single fundamental dimension in itself (symbol M). Space can be reduced to a product of the fundamental dimension length (symbol L). Thinking of space as volume we realize that the measure of the volume of a cube is the product of length, width, and height, which boils down to length-cubed (L^3). Area (length times width) is expressed as length-squared (L^2); distance along a line is simply length (L). Time, like mass, is a single fundamental dimension (symbol T). Every term used in mechanics can be defined in terms of some combination of M, L, and T. These combinations are all products of M, L, and T. (Addition is meaningless.) Take velocity, for example, described in words as distance per unit time; it reads as L/T. Then take acceleration, defined as velocity per unit time; it reads L/T^2. Follow with energy, defined as the product of mass and acceleration: it reads $M \times L/T^2$.

We can try out the dimensional system on the objects of study in certain belief fields of knowledge. Suppose that we start with ethics and morality. Take, for example, the nouns "goodness" and "wrongness." These express human values. Can science define them and measure them in terms of mass, length, or time? Surely not, because they lack the qualities of physical objects; i.e., they have no physical dimension. Try this out on the field of aesthetics, using the noun "beauty." Agreed that beauty lies in the eye of the beholder, can you assign it one of the three physical dimensions, or a product of them? Test the supernatural realm of religion in the same way, keeping in mind that by definition its contents have no natural (physical) properties. You can visualize an angel, even paint a picture of one, but can you assign mass, length, or time to an angel? Mass and length do not exist in the supernatural, but does time exist there? Angels, of course, endure forever, and so for them time has no meaning in any physical sense. As food for thought, consider the following proposition: One criterion of distinction between the knowledge classes of research fields and belief fields (explained earlier in this chapter) is that the former consists entirely of objects (things) possessing mass, length, or time, whereas the latter is devoid of any such dimensional properties.

Science with content as described above, encompassing all matter and spacetime, is usually called *empirical science*; its content consists of experience with that physical universe, i.e., with knowledge derived by human observation of that universe. The word "empirical" simply means "relying on experience or observation alone" or "based on observation and experience" (*Webster's Ninth New Collegiate Dictionary*).

This brings us to an area of knowledge often called *formal science*; it consists of logic and pure mathematics. Are these fields separate from empirical science? Logic and to a considerable degree pure mathematics are claimed by philosophy as its concerns. Both have no need for examining facts of nature based on observation. Instead, they construct hypothetical or theoretical situations (axioms and a priori propositions) and work from these to conclusions by logical deduction. As Marx Wartofsky puts it, these are "linguistic systems of deductive inference" (1968 p. 99). Clearly formal logic and pure mathematics do not belong in the same knowledge field as empirical science, but they progress by a process of deductive research and deserve a place in the class of research fields. The kinship is obviously close, because empirical science must follow rules of sound logic and must make use of mathematics.

The scope of empirical science is incredibly broad, poking its nose into every nook and cranny of the universe and into almost every facet of our lives. The fields within empirical science are commonly listed as follows: the *natural sciences* include three basic groups: (1) The *physical sciences*, physics and chemistry, deal with the structure of matter and the nature of energy. The experimental method is particularly important in the physical sciences, and much of its research is conducted entirely within laboratories where closely controlled experiments can be conducted. (2) A group of complex inorganic sciences that includes astronomy, geology, oceanography, and meteorology. These require observation of nature outside the laboratory. (3) The *biological sciences*, traditionally divided into zoology and botany, deal with living cells and their aggregations into all life forms. A special feature of the second and third groups is that they must deal with the history of events happening over vast spans of time, such as the history of our solar system and the evolution of life on earth. Set apart from the natural sciences are the *social sciences* that investigate the ways in which the human race organizes itself, behaves, and functions individually and in groups. They include economics, cultural anthropology, psychology, and sociology. (In Chapter 4 we will discuss in greater detail the classification of branches of empirical science.)

Are there any areas of human thought and action off limits to science? For reasons we have already given, empirical science cannot partake of the arts, ethics, or religion, for the products of these fields cannot be evaluated by empirical science. The arts create new, imaginative structures of ideas and sensory experiences that appeal to the emotions. Aesthetic judgments, involving as they do matters of taste, preference, and cultural conditioning, cannot be challenged by science. In ethics, moral values that are chosen as good or right are not subject to challenge or verification by science. Religious tenets, which usually involve belief in supernatural entities, are also beyond the limits of scientific appraisal.

We need to emphasize that supernatural forces, if they can be said to exist, cannot be observed, measured, or recorded by the procedures of science—that's simply what the word "supernatural" means. There can be no limit to the kinds and shapes of supernatural forces and forms the human mind is capable of conjuring up "from nowhere." Scientists therefore have no alternative but to ignore claims of the existence of supernatural forces and causes. This exclusion is a basic position that must be stoutly adhered to by scientists or their entire system of evaluating and processing information will collapse.

Mind and Body

Before turning to the question of how knowledge is obtained, we need to resolve a longstanding philosophical argument about human mental processes, that of the mind/body relationship. Science considers mental activity to be a function of the human brain. *Mind*, as we use the word here, is the functional and subjective aspect of the living brain (Campbell 1985, p. 330). Or as biologist Gunther Stent puts it, mind is the "epiphenomenon" of the brain (1975, p. 1055), meaning that mind is a secondary phenomenon that accompanies and is controlled by the brain. Science must take that position on the mind/body relationship.

At the risk of greatly oversimplifying the issues, consider two rather straighforward alternatives: (a) Mind and body are one and the same basic form of reality; (b) mind and body occupy two separate and distinct fields of reality. In philosophy the study of the nature of reality is known as *ontology*. The first alternative is said to represent a form of *monism*; it is a *monistic ontology*. In the monistic view, "mind" is in the same physical realm with "body," so both are within the purview of science. Under the second alternative (explained below), a *dualism* exists; it represents a *dualistic ontology*.

The monistic view held by science can also be labeled *mechanistic materialism* because it assumes that the functioning of the brain can be explained by the same scientific principles used to explain the sensory activities of the eye and ear, or the contractions of a muscle, or any other organic function. The billions of interconnected nerve cells that make up the working brain carry out all mental activity, including thinking, imagining, remembering, and emoting. "Mental activity," says anthropologist Bernard Campbell, "means, broadly, the neuronal mechanisms that operate between stimulus and response" (1985, pp. 338, 340). Thus, images produced within the brain are also physical phenomena and can be treated by scientists along with any other activities of the brain. This materialistic view of mind as physiologically one with body is today the universal view of scientists.

Perhaps the most widely celebrated version of mind/body dualism was that of René Descartes (ca. 1640). His total package, Cartesian philosophy, regarded mind (mental states or events) as being in a nonphysical realm, a totally separate realm from that of the physical or material world. Beyond saying that "mind" is an unsolved mystery, Descartes did little to enlighten his listeners on the nature of mind. He considered mind so completely removed from the physical realm that its activation requires the intervention of a Deity, acting through a third agency, the "soul." This suggests that Descartes did not regard "mind" as supernatural in the same sense that God is supernatural.

A common dualistic view of more general character, usually associated with a religious dogma or theology is simply that mind is a spiritual substance, created and implanted at birth by God. In this view, "mind" has essentially the same meaning as "soul" and carries with it the idea that the mind is immortal and can leave the body to dwell elsewhere (in another body, in heaven, or in hell). These religious views of mind are obviously inadmissible to science, and scientists simply pay no attention to them. There are, however, nonreligious versions of the mind-brain dualism (dichotomy) which postulate that mind differs from body in important fundamental ways. Included are such ideas as that mind is private and unobservable, or that mind is intangible, nonspatial (occupies no space) and nonlocatable. These ideas perhaps border on empirical science, but involve great difficulties in terms of possible scientific examination. (See Peter Angeles 1981, pp. 172-76.)

The dualistic mind/body concept has recently resurfaced, this time at the suggestion of a scientist of high reputation: George Wald, a 1967 Nobel Laureate. As reported in *Science News* (Thomsen 1983), Wald, in discussing his recent studies of the visual systems of frogs as compared with higher animal forms, suggests that there is something he calls "consciousness" existing "outside the parameters of space and time." As Wald describes it, consciousness pervades the universe as a kind of supernatural "force" capable of directing the development of the material universe and even of having guided the course of cosmology. To find a reputable scientist proposing a theory of supernatural force is disturbing to the community of scientists. If the realm of matter and energy with which scientists work is being influenced or guided by a supernatural force, science will be incapable of explaining the information it has collected; it will be unable to make predictions about what will happen in the future, and its explanations of what has happened in the past may be inadequate or incomplete.

Empiricism, Experience, and Realism

The modern monistic scientific view of a mechanistic (or naturalistic) materialism has its historical origins in the writings of Western philosophers of the late seventeenth and early eighteenth centuries. Two whose ideas we briefly review here are John Locke (1632-1704) and David Hume (1711-1776). Their views go under the name of philosophical *empiricism*.

Locke, an English philosopher, came to the conclusion (ca. 1690) that all human knowledge (with the possible exception of formal logic and pure mathematics) derives initially from *experience*, the input the brain receives through the senses (Russell 1945, p. 609). This process Locke described as *sensation*. The human infant "comes into the world with a mental *tabula rasa*, a blank tablet, upon which experience records impressions; there is no such thing as innate knowledge" (Ferm 1936, p. 447). In addition to sensation, Locke recognized that some of our ideas can also formed by "reflection." (In Chapter 9, we call this *ideation*.) Of the two kinds of "ideas," experience is primary or basic because we cannot form any other kind of knowledge without using the content of prior experience.

Locke's form of empiricism is sometimes labeled "representative realism." *Realism*, as used by philosophers in the field of epistemology (the study of knowledge), ". . . is the doctrine that says that there is a world-out-there (independent of or prior to our knowledge of it) which we somehow grasp in the knowledge relation" (Ferm 1936, p. 449). Realism is an essential tenet of science, as we shall explain in some detail in Chapter 9. Philosopher/mathematician Bertrand Russell describes Locke's contribution as having countered a then prevalent belief in innate (a priori) knowledge and observes: "Locke's thorough-going empiricism was therefore a bold innovation" (1945, p. 610).

Locke's realism was soon challenged by another leading philosopher, Bishop George Berkeley, who proposed instead (ca. 1710) that "the only real existents are minds and their ideas; the world exists only as it exists in the mind" (Ferm, p. 452). This view, called *subjective idealism*, led Berkeley into a definition of knowledge that is meaningless or irrelevant to science, and we need follow it no further.

Scotch philospher Hume is our second great figure in in the emerging field of philosophy of science. He came after Berkeley and extended Locke's empiricism in his 1748 essay, *An Enquiry Concerning Human Understanding*. Hume's writing is difficult for most of us to grasp and is perhaps best restated in our vernacular. Like Locke, he held that ". . . the source and validity of human knowledge lie in sense-experience" (Ferm, p. 455).

What is perhaps now most important to all scientists is Hume's analysis of the process of deriving scientific generalizations or laws that link causes to their effects. Humans have always noticed that in nature there are many sequences of two events, one following the other in time, and that these temporally paired (serial) events are repeated over and over again. For example, a nearby lightning stroke (event A) is always followed closely by a loud thunderclap (event B). Before science (in the guise of Ben Franklin) gained the information necessary to explain this common phenomenon, the cause-effect linkage could only be inferred through long experience. But Hume argued that no cause-effect law can be logically inferred directly from such observation. Bertrand Russell states Hume's argument as follows: "The inference is not determined by reason, since that would require us to assume the uniformity of nature, which itself is not necessary, but only inferred from experience" (1945, p. 665). Hume went on to say that we come to believe that events A and B are always connected, despite that belief being groundless in terms of logic. We will develop this concept in Chapter 2 in the analysis of laws of science. Beliefs such as these, based on experience, serve us well in daily life situations. You just don't stay alive very long if you fail to respond to these beliefs. This "natural realism" or "commonsense realism" is today seen as a genetic endowment based in evolutionary natural selection (see "Sociobiology" in Chapter 13).

With this historical vignette as background, we turn next to a modern scientific analysis of the mental process of gaining the "experience" that Locke and Hume put in place as a form of reality.

Perception and Conceptual Thought

If we agree that the human brain is the organ within which all mental activity takes place, it is worthwhile to follow the pathways of information into and out of the brain and the processing and integration of that information. This process comes under the general heading of *cognition*, the act or process of knowing.

We must elect an arbitrary starting point; let it be the sensory input from environment to the brain—from the outside to the inside. Most systems descriptions begin with the input from external sources. This input information is sensory; in the simplest case it requires direct use of the human sensory mechanisms—visual, auditory, olfactory, or tactile. The sensing process itself needs no further elaboration here. The process is continuous and entirely unconscious, often with no conscious purpose. What happens to the sensory stimuli when they reach the brain is quite another matter, for here a selective process is at work, as well as an integrative process making use of stored, previously processed information. What we actually observe through sensory perception from the outside, according to Professor Marx Wartofsky, a specialist in the philosophy of science, "is largely a function of intent and context, and depends to a great extent on frame of mind, attention, and what we know to look for" (Wartofsky 1968, p. 101). In other words, observing or seeing is a guided process.

Anthropologist Bernard Campbell analyzes the processing of sensory data as follows (1985, p. 330). The brain's observation, which is its own mental image of the external environment, can be called a *percept;* it is based on two kinds of information: (a) input from the senses, and (b) memory of previous experience. The two forms are unconsciously and continuously combined as long as the input of sensory information continues.

Perception, then, is the formulation of percepts; it is a very different mental activity from the intaking of raw sensory data upon which it is partially based. Perception is an activity common to many animal species, for it requires only the sensory apparatus and a certain amount of memory. Nevertheless, it is the basic mechanism for the development of a much more complex brain function found in humans, that of formulation of a *concept,* or *conceptualization.* A concept "is an abstraction from the particular to the class" (Campbell 1985, p. 332).[2] The abstraction is not a conscious activity, but it forms the material used in *conceptual thought,* or thinking. Thinking uses imagination, the ability to conceive of acts, events, or artifacts that have not yet been realized. A clear distinction between percept and concept is of greatest importance in the classification of fields of knowledge.

Is conceptual thought a unique capability of the human species? Campbell considers this to be so, for he states that humans alone can have conceptual thoughts about objects not concurrently visible to the thinker. In other mammals, especially other primates, there is evidence for a classification of experience that can be called "unconscious conceptualization." Evolutionary biologist Theodosius Dobzhansky uses the term *self-awareness* for essentially the same phenomenon and extends this to a special and unique form in humans, called death awareness (1977, p. 453). Conceptual thought increased in power in humans along with the increase in size and surface of area of brain and was accompanied by the invention and growth of language (Campbell 1985, p. 333).[2]

Getting back to perception of the environment, the perceived image, or percept, is completely private within the brain of the perceiver. It is a form of individual knowledge, but of no value in advancing science unless it can be communicated to other brains. This communication can only be made by use of language. Perception, says Wartofsky, is formulated in language: "To perceive something seems to come to saying to oneself or to someone else, 'This is a so-and-so, or such-and-such.' So intimately tied to the framework of language is our perception that our identification of things and properties of things in the language may in effect influence what we see and fail to see" (1968 , p. 104).

A percept transformed to words (language) can be called an *observation.* To distinguish the noun "observation" from the verb form, we should perhaps call it an *observation statement.* In terms of formal logic, an observation statement is expressed as a claim that "some proposition, P, is true." The object of an observation statement is "what's out there in the environment." It states or implies that something exists out there; it is thus an *existential statement.* The observation statement is perhaps the most commonplace kind of statement made by humans. The observation state-

[2]Reprinted with permission from: Campbell, Bernard G. *Human Evolution: An Introduction to Man's Adaptations*. 3rd Edition. (New York: Aldine de Gruyter) Copyright © 1985 Bernard Campbell.

ment is also the unit building block of the structure that is science. To meet the special needs of science, those that are *scientific statements* must conform to a special standard of quality, both in the manner in which they are arrived at and in the language by which they are transmitted.

Science and Language

Scientific data, laws, and theories, consisting of observation statements or indirectly formulated conceptual statements, are communicated by language; they are also stored in the form of language. Language, according to Professor Campbell, exerted a strong positive force in the evolutionary development of conscious conceptualization in humans. He writes:

> But the conscious concept did not appear unaided. It seems probable that it was finally evoked by the use of symbols, gestures perhaps, but more often words, which make up language. The concept "bird" could be brought into full consciousness only by its identification with the word symbol "BIRD." The word symbol was the twin of the conscious concept, and it seems probable that they were born together and grew together. (1985, p. 333)[2]

You might say that language enables the "private" concepts that originate in one human mind to "go public," to become exposed to the light of close examination by anyone who chooses to do so. Transferral of private observation statements and percepts to the public domain requires statement in words and sentences or, in the case of mathematical language, by special symbols. Those special symbols are, however, definable in words. Indeed, they must be expressible in words. We may also wish to include drawings and other forms of graphic expression as ways to transmit and store information. A drawing or map made by a scientist using visual observation of a subject can be taken as a kind of scientific observation statement, but words will ultimately be required to express the content of that drawing or map in a useful or meaningful way.

An important operational rule is that each word in a scientific statement must carry exactly the same meaning to all scientists, at least to all who practice in a given field or area of science. This rule requires that all words be precisely defined. Scientists must be very fussy about definitions, even if that seems painful to others. For each branch of science, one can usually find a glossary that attempts to define all terms found in published works dealing with that branch. Most glossaries are published through a scientific society that is strongly representative of the particular branch of science. If a term is used in two or more meanings, the glossary and any good dictonary will make a special point of explaining each.

But now we come to a real problem. A particular word must be defined by using a set of different words. A nice word like "albedo" probably means nothing to most persons, even though it is a fairly short word and looks simple. It might be a good calling name for a dog! The words we use to define "albedo" must be familiar to most persons with a reasonably good education and trained to use a dictionary. To define "albedo," a scientist needs to use these words: ratio, electromagnetic, radiation, reflection, incident. They, too, are words of science, but they are widely used in science and most scientists do not have to search for their meanings. Persons who are not scientists may need to take each term by itself and ferret out its meaning in everyday, commonsense language. Only then can a person who uses "albedo" be absolutely sure that he or she is communicating the following message to another person: Albedo is the ratio of the amount of electromagnetic radiation reflected by a body to the amount incident to it (American Meteorological Society, *Glossary of Meteorology* 1959, p. 21). The same definition can be written in mathematical symbols, but this also requires that each symbol be defined in words.

What we see here is a kind of hierarchy of language frameworks. At the top is the special scientific framework needed for a particular branch o science. Below it lies a much more general language framework that is common ground for all science. At the very base is a framework of words and meanings common to most literate humans in the daily affairs and activities of living. The process of expressing one framework in terms of the more general one that lies below it is known to philosophers of science as *linguistic reduction*; it proceeds from the special to the more general. The reduction of scientific language to common language (commonsense meanings) is what gives science its credibility. What we have here in practice is "a chain of so-called coordinating definitions or reduction sentences, which lead from the complex theoretical formulation to the basic predicates" (Wartofsky 1968, p. 116).

Scientists set up their own theoretical framework of observational procedures and the language by which observations are communicated and stored within their own circle. They agree on the framework—indeed, they must agree on it—although they may disagree on the interpretations of the observational statements and concepts themselves. By linguistic reduction, the language of that small, elite circle can be brought to broader and simpler statements capable of being more widely understood.

Professor Wartofsky gives a rather useful analogy to explain what linguistic reduction means and how it works (1968, pp. 119-20). Consider our monetary system. The scientists' abstruse language system can be likened to a banknote of large denomination printed on a piece of paper of very little intrinsic worth. What gives the banknote value is that the holder can take it to a bank and redeem it in "coin of the realm," accepted by everyone in payment of a debt. Wartofsky says: "This notion that theoretical terms in science, like 'atom' or 'magnetic field' . . . may be reduced . . . or translated and turned into into basic coin leads us finally to the outcome of the argument for reduction to basic predicates." The value of the basic coin, in turn, rests in the strength of "the whole system that underwrites the exchange—the monetary system and the public agreement that upholds it" (p. 120). At this point, one can perhaps think of some appropriate words of advice for the scientific community: Keep one foot on the ground at all times. If the public doesn't understand what you are saying, they will lose confidence in you. It does look as if science stays alive at the pleasure of the commonsense persons who make up the bulk of human society, and not the other way around. A measure of humility on the part of scientists seems called for, in any case.

What Are the "Facts" of Science?

Can science give us the truth? Try looking up the word "fact" in a dictionary. You will probably find that a fact is something that is "actual" (i.e., an "actuality"). Try cross-checking to see what "actual" means and you will learn that something that is actual is a "fact" (or is "factual"). This is a neat little circle. *Webster's Ninth New Collegiate Dictionary* (1985) tries to extricate itself with its final definition of fact as "a piece of information presented as having objective reality." One of the main concerns of philosophers over the centuries is to ponder the nature of objective reality. "Objective" simply means that your mind is supposed to go outside itself and perceive some material thing or event as it "really is" (i.e., "that which is factual"). Again, we go into a circular maneuver ending in self-defeat, but the experience is chastening. Such reasoning is *tautological*, and we have created a *tautology*. Every first-year student of philosophy gets a good dose of the painful strivings of philosophers to get to the very heart of the problem of reality. To confuse the situation further, let me introduce another word with which we can play the circles game: truth. "Truth" is that which is a fact, which is an actuality, which is a truth, which is a fact . . . *ad infinitum*. When is an observation statement a fact? When is it true?

I suggest that in our search for insight into the nature of science we set some strict limits to how we use the words "fact," "actuality," and "truth." Let us vow never again to say "Scientists discover the truth." Observation statements that are the building blocks of science and of all knowledge within the research fields are designed only to minimize the probability of failing to make a true statement. Let us admit that the human mind or brain will never be privy to truth in science, but rather agree that the special kind of observation statement that is a *scientific statement*, despite being put forward as being a true proposition, actually contains a certain probability of being in error. This is a concept that needs to be developed in depth, for it is the very essence of empirical science and what has come to be called the *scientific method*.

Closely linked with what the previous paragraph says is another statement we should vow never again to say—or even to think in the privacy of our minds: "Scientists believe that . . . (such-and-such is the case or is what happened)." *Webster's Ninth New Collegiate Dictionary* says "to believe" is "to have a firm conviction as to the reality or goodness of something" and also "to have a firm religious faith." "Belief" is defined as "conviction of the truth of some statement or reality of a fact, especially when well grounded." Shouldn't we admit "belief" to the circles game? That would give us four players: fact, actuality, truth, and belief. I propose instead that we leave belief in the realm of religion where it belongs, along with all other nonempirical concepts such as questions of ethics and morality, which may be at least partly religious in origin. A scientist is free, of course, to believe in God and to be a religious person, but forbidden to express belief in the absolute truth of any of the scientific observation statements that make up the body of science. There always is the possibility, no matter how small it may be, that a scientific statement will be shown to be false.

In everyday life scientists, like everyone else, treat many phenomena as if they were facts—the "facts of life." We would be foolish indeed to take a chance that at the precise moment we step off a forty-story building the phenomenon of gravity will be inactivated. Yet a scientist would also be foolish to assert flatly that the gravitational attraction between the same two masses separated by a certain distance is, always was, and forever will be exactly the same.

I can anticipate several of my readers coming up with a challenge to my flat assertion that there is no place in science for such concepts as a fact, an actuality, or a truth. You will put it to me that some events are facts, pure and simple. Consider an act of terrorism committed before the eyes of hundreds or thousands of persons and documented by motion pictures and television. Take the case of the 1981 attempt upon the life of Pope John Paul II in the crowded courtyard of the Vatican. How can I claim that there is some doubt that a particular individual fired bullets that struck the pope? I yield to the obvious. Hundreds of times a day, we observe and react to events that are factual beyond reasonable question. Descriptions of such events can be called *statements of fact* because we find no alternative statement acceptable for serious consideration.

I would like to limit the definition of a "scientific statement" to one that is subject to a finite probability, no matter how small, of being in error when it is asserted to be correct. To use the word "error" implies that there exists an alternative statement (or several alternatives) that can be advanced for serious consideration. Take, for example, the assassination of John F. Kennedy in Dallas on November 22, 1963. That the president received a fatal wound is a statement of fact, but there is some shadow of doubt that Lee Harvey Oswald's rifle was the one and only rifle to fire a bullet aimed at the president. To state that Oswald was indeed the only person to fire a weapon can be classed as a scientific statement because there is a certain degree of probability that the statement is false. There is also an alternative statement to complement it: namely, that a second assassin fired from a position on the grassy knoll. Each of these statements seems to have carried a substantial probability of being in error; for otherwise the lengthy deliberations of a national commission of inquiry would not have received widespread interest and support.

Taking a closer look at statements of fact, we can recognize that they take much the same form as rather straightforward propositions of existence. Mostly, they consist of a noun or pronoun as the subject, a verb, and a predicate that is commonly an adjective or adverb. I shall use examples from elementary mineralogy and geology:

- The streak of hematite is red.
- The crystal form of aragonite is different from that of calcite.
- A major earthquake struck Los Angeles on April 7, 1962.
- A landslide dammed the Madison River, forming a lake.

For the most part, a statement of fact describes a tangible substance, a structure, or a unique event as perceived by the senses and, therefore, subject to being recorded independently by use of commonplace devices, such as the camera or microphone. The statement can be substantiated by a multiplicity of observers and by reference to permanent records, such as photographs, sound tapes, or video tapes. In each case, the suggestion that the statement may be in error is not spontaneously forthcoming (although an observer can be accused of lying). We would not think of countering the first statement by another reading: "The streak of hematite is blue." We would not think of negating the second statement thus: "The crystal form of aragonite is not different from that of calcite."

To summarize, a statement of fact is an observation statement accepted pragmatically and operationally as being true (a) because to substitute a predicate of opposite meaning would be absurd (would violate common sense) and (b) because the statement is subject to verification by many observers and by mechanical means of documentation, with no expressed alternative statements (i.e., no dissent). One feature of the statement of fact is that, by itself, it is not very interesting. Interest arises when the statement leads to anticipation of some consequence that is of interest. In the courtroom, statements of fact lead to the interesting possibility that the accused will be convicted and punished. In science, statements of fact are grist for the mill of conceptualization, where the scientist's interest is concentrated.

Scientific activity is the investigation of that which is to some degree unknown or uncertain. If this were not so, science would be highly uninteresting, to say the least. It is the element of something unknown or uncertain that attracts and intrigues the human mind, as we know from the lure and excitement we feel as a murder mystery begins to unfold. And so, of course, the investigation of a homicide may also be a bona fide scientific investigation—forensic science, you might call it.

Scientific Observations

Scientific statements can include informational packages ranging from the extremely simple to the highly complex. Those statements based on observation are initially arrived at through the same general inductive process we have already assigned to formulation of statements of fact. The simplest class of scientific statements consists of *singular observations*, meaning that they are unique observations. The singular observation may describe an attribute of some object or substance or an event with a fixed place in space and time. Observations commonly consist of single measurements of mass, length, time, or temperature. The following will serve as examples:

- Meteorite A-345 weighs 5.32 kilograms.
- Elapsed time of fall of the boulder was 2.09 seconds.
- Maximum recorded acceleration during the earthquake measured 0.92 g.
- Crystallization was first observed at a temperature of 303 C.

Simple as they seem, measurements are subject to errors: those committed by the observer, those induced by the equipment used and those induced by variations in the environment. Consequently, statistical theory is applied to the distribution and magnitude of the errors. It is customary to add an estimate of the range of error that can be expected when a measurement is repeated many times. This "probable error" is a statement to the effect that, if we perform the measurement repeatedly (or many different observers each make one measurement), a certain percentage of the observed values will, on the average, fall within a specified range. There is also a statement of the probability that a single observation will fall outside a stated range. That probability diminishes rapidly beyond the stated range but never falls to zero. This is why we must never refer to the single measurement as a "fact." Generally, we refer to collections of such observations as *raw data*. With proper precautions based on long experience, scientists may judge the data to be "good," "sound," or "reliable" and proceed with an investigation as if they were dealing with facts. You might put it this way: "There is safety in numbers." With large numbers of repeated measurements carried out by many observers, using many different sets of apparatus, the probability of error has been reduced to an acceptable level.

From what I have just stated, it looks as if scientists are professional gamblers, highly skilled in estimating the odds of winning or losing!

Indeed, they must be familiar with the "laws of chance" and the mathematics that goes into those "laws."

At the risk of seeming to harp endlessly on a favorite theme, I have more to say about the use of "fact" and "truth" in scientific writing. When we say "this is a fact" and "this is true" we are speaking in absolutes. One cannot say "this is an almost-fact" or "this is an almost-truth." On the other hand, as with any concept of the absolute, one can indicate an approach to it, thus: "This seems close to being the truth." "This is almost certain to be a fact."

If we glibly say that "science is made up of facts" and "science contains the truth," we have already trapped ourselves in an untenable position, for we have also said implicitly that scientific knowledge, once put in place, cannot be discarded or replaced. Lacking that capability for self-correction and internal improvement, scientific knowledge could only grow in bulk, like a brick wall to which we can add more bricks, but replace or rearrange none of those already set in mortar. This is why, I suggest, the words "fact" and "truth" belong to formal science (logic and mathematics) but not to empirical science, which starts with observation of nature and, therefore, does not deal in unmodifiable premises. In this sense empirical science is open-ended at both ends.

Of course, scientists and philosophers of science will continue to talk freely of "facts" and their use in research. The more perceptive ones simply change the common meaning of "fact" to one that is not absolute. I discovered this dodge in the writing of a prominent philosopher of science, Thomas S. Kuhn, where he speaks of determining facts with "greater precision" (1962, p. 25). In connection with examples of measurements of such values as positions of stars and boiling points—of physical constants, that is—Kuhn refers to "attempts to increase the accuracy and scope with which these facts are known." Clearly, he implies that "fact" is not an absolute term and that a statement of fact should not be equated to a statement of truth. By suitably adjusting the meaning of the word "fact" he has avoided what would be an oxymoron (a proposition that contains a self-contradiction); in this case, "a fact is a nonfact."

Induction and Deduction in Science

Scientific statements that attempt to describe and explain phenomena or generalize about repeated observations are said to be *inductive*; i.e., they use the mental process of *induction*. The primary information is sensory, flowing from the object to the brain, where it is made into percepts. Perhaps this will be easy to remember because the first syllable of "induction" is "in"—that which first flows into the scientist's brain.

The *method of induction* in science has a special meaning, that of deriving a general statement, or law, from a set of repeated statements that are all in agreement. This set might be a collection of statements made by a single observer or many observers in a large number of different places and at different times. We can illustrate with a case dear to the hearts of philosophers. Suppose that for a large number of observations of swans, all of the swans are found to be black. Can we then safely say: All swans are black? That would be a general statement or law—also called a *universal statement.* The logicians say "No, you can't do that." Your evidence is only that some swans are black, and that's as far as you can go. Perhaps the next swan you encountered would turn out to be white—and there goes your law down the drain. We will say more about this argument in Chapter 2, on the subject of laws of science. There, we will find that laws cannot be thought of as being universally true statements. So induction might seem to have some strict limitations as a scientific method.

Contrasting with the inductive method is its counterpart in the "formal sciences" of logic and mathematics. Both of those disciplines deal in absolutes, using such words as "prove or disprove," "correct or incorrect," "verify or falsify." Keep in mind that formal logic starts with a premise generated inside the brain, rather than being formulated directly from that which is perceived from the outside. The investigation typically begins with such statements as "Suppose that we take two things, A and B . . ." or "Assume that two variable quantities, x and y, are related so that . . . ." The initial propositions, or premises, are defined in absolute terms and manipulated in strictly regulated fashion. The end result is a decision that a subsequent proposition derived from the first is either correct or in error. Thus, formal science can be said to be completely rigorous. This method of gaining knowledge is described as *deductive*; it is the *deductive method.*

Science also makes use of the deductive method, as we shall explain on later pages, but it does so in a somewhat different context or frame of reference. The scientist looks at the general statement or law previously derived by the inductive method, takes it tentatively to be valid, and from it deduces some other propositions that predict what should be the case or should be uncovered by further investigation. The idea of deduction is contained in the words "if, then," expanded to read "*If* proposition *a* is true, *then* proposition *b* must also be true." Deduction in science consists of the prediction of what *should* be found *if* we look for it. Sherlock Holmes used this method repeatedly in searching for new evidence to back up (or undermine) his suspicion as to who might be the culprit.

The accompanying diagram, Figure 1.1, puts together the inductive and deductive methods into a single flow pattern, shown by arrows. The usual starting point is with induction, but you can see that the deduction that follows it only feeds back into to more observation, which in turn can set off more induction. In almost any important and complex scientific investigation, the scientist is recycling this knowledge continually, using continuous automatic feedback to strengthen his or her tentative conclusions. Of course, the process may weaken rather than strengthen those conclusions.

Unfortunately, the absolutes of rigorous formal science tend to spill over into the thinking and expression of the empirical sciences. Philosophers of science are especially prone to blur the distinction between formal science and empirical science. For example, Sir Karl Popper, whose admonitions about evaluating scientific statements we will refer to in later pages, talks about when to declare a scientific statement to be false (Popper 1959). We will find it necessary to transform Popper's logic-language into the language of uncertainty.

Science Defined

We have been discussing methods used by scientists to obtain and process observational data. You may have heard it said that science is actually nothing more than a special method for obtaining information. Should we define science as a special kind of body of information, as suggested on earlier pages, or as the method itself? Perhaps science is both of these things. I propose definitions for each.

Scientific knowledge: the best picture of the real world that humans can devise, given the present state of our collective investigative capability. By

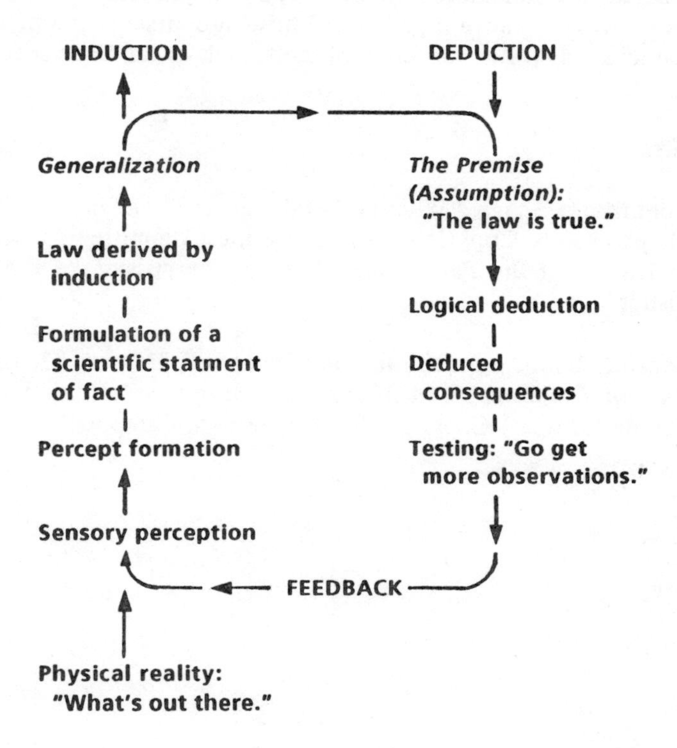

Figure 1.1 The induction/deduction feedback cycle.

"best" we mean (a) the fullest and most complete *description* of what we observe, (b) the most satisfactory *explanation* of what is observed in terms of interrelatedness to other phenomena and to basic or universal laws, and (c) description and explanation that carry the greatest probability of being a true picture of the real world. (A definition of explanation is given in Chapter 2.) Scientific knowledge represents the harvest of human endeavor; it is an artifact and its makers are fallible. Therefore, scientific knowledge is imperfect and must be continually restudied, modified, and corrected; it will never achieve static perfection.

Scientific method: the method or system by which scientific knowledge is secured. It is designed to minimize the commission of observational errors and mistakes of interpretation. The method uses a complex system of checks and balances to offset many expressions of human weakness, including self-deception, narrowness of vision, defective logic, and selfish motivation.

To these terms, perhaps we should add a third. *Scientific community:* the collection of humans applying the scientific method. As we show in Chapter 8, the scientific community has a rather distinct set of characteristics as a society in itself. In other words, there is a sociology of science to be considered in developing a full understanding of how scientists work. Believe it or not, scientists are humans and behave much n the same as any other collection of individuals of the same species.

Credits

1. From Mario Bunge, "What Is Pseudoscience?" *Skeptical Inquirer*, vol. 9, no. 1, pp. 36-46. Copyright © 1984 by the Committee for the Scientific Investigation of the Paranormal. Used by permission of the author and publisher.

2. Reprinted with permission from: Campbell, Bernard G. *Human Evolution: An Introduction to Man's Adaptations*. 3rd Edition. (New York: Aldine de Gruyter) Copyright © 1985 Bernard Campbell.

CHAPTER 2

Laws, Explanations, Theories, Hypotheses

We now turn our attention to major parts of the content of science. Laws of science were mentioned in our description of induction and deduction as methods of science. Laws are the first major class of scientific knowledge we will examine. Analysis of laws raises the question of scientific explanation. Once that topic is explored, we will be in good shape to move into the high clouds of philosophy of science, where the lordly theories of science dwell. Gifted scientists make great theories, which in turn confer greatness on the scientists who invent them. Join this select group and your name will be carved in stone in the friezes of public buildings and your tombstone will attract an unceasing stream of pilgrims.

Laws of Science

Everyone has heard of "laws of science," which are passed off as eternal truths about the real world. In the role of unforgiving tyrants, the laws of science dictate exactly what will happen as energy is transformed and matter is moved about or changed in various ways. There is a law of gravity that tells how rapidly an object falling in a vacuum will accelerate its speed. Laws of motion tell us what happens when one object strikes another. A law of frictional resistance limits the top speed of an automobile.

There has been endless discussion among scientists and philosophers as to what constitutes a "natural law" or "law of nature." One idea that seems to provide some common ground is that of *regularities* in nature; meaning only that the changes and processes taking place in nature show characteristic patterns endlessly repeated. Furthermore, we are told that in such cases of regularity there lies a rather mysterious implication that the law is stating a physical necessity; i.e., that what the law states must always be the case in nature (Nagel 1961, p. 51). Perhaps the notion of necessity has its origins in natural theology, which assumed that laws were created by God, who deemed them necessary to sustain an orderly universe without his continual supervision.

Induction and Laws of Science

We can profit by going back in history to David Hume, in whose time natural theology was dominant and the regularities of nature were simply accepted without requiring any analysis or question as to their logical validity. In Chapter 1, on the topic of deriving scientific generalizations or laws from observations of nature, we found Hume protesting that no cause-effect law can logically be inferred directly from experience (observation). Such an inference is not logically correct, he claims, because it requires the assumption of uniformity in nature—an assumption that is itself inferred from that same observation.

Philosopher Antony Flew, in his chapter on Hume in D. J. O'Connor's 1964 history of western philosophy, describes in some detail Hume's classic argument, or demonstration, showing "the impossibility of deducing universal laws from any evidence which can be provided by experience" (Flew 1964, p. 260). The argument runs as follows. Consider these two propositions: (a) The sun will not rise tomorrow; (b) The sun will rise tomorrow. Both predictions cannot hold empirically true; it's a case of either (a) or (b). In everyday life, any person we judge to be sane and rational selects (a) without hesitation because it follows our collective experience. What we are asserting is: Because it has always happened that way in the past, it will always happen that way in the future. Rephrasing in the manner of classical (Aristotelian) logic this becomes: (a) All known x's are (so-and-so), therefore (b) All x's are (so-and-so). In Chapter 1 we used swans to illustrate: (a) All known swans are black, therefore, (b) all swans are black. Formal arguments of this form are known as *syllogisms*, but the one we have stated is false. Whereas (b) is a *universal proposition* ("*All* x's are . . ."), in contrast, (a) is a *particular proposition* ("*Some* x's are. . ."). A universal proposition cannot be derived directly from a particular proposition.

If we were to try to patch up the syllogism to make it valid, we would need to insert between (a) and (b) a second universal proposition (c) reading "All phenomena in nature are perfectly regular." Our syllogism now reads: "If (a) and (c), then (b)." Flew points out that this attempt to escape from Hume's "dilemma" fails because the assertion of "uniformity in nature" is a purely *a priori* proposition that is not capable of being demonstrated (p. 261).

Hume's demonstration led him to a position of extreme skepticism, bordering on despair, as to the possibility of humans being able to certify as logically correct the process of gaining knowledge of the external world. On the other hand, Hume shared the commonsense conclusion that humans do gain reliable knowledge of the world about them by observation, for he commented: "None but a fool or a madman will ever pretend to dispute the authority of experience or to reject that great guide to human life" (Hume 1748, p. 37).

Philosopher Bertrand Russell also analyzes Hume's demonstration in his 1945 history of western philosophy, fully supporting Hume's argument. But both he and Flew offer an alternative, which is the concept of probability: If many (some) are, it is highly probable that all will be (Flew 1964, p. 261). Russell describes the probability concept thus: ". . . not indeed as giving certainty, but as giving sufficient probability for practical purposes. If this principle is not true, every attempt to arrive at general scientific laws from particular observations is fallacious, and Hume's skepticism is inescapable for an empiricist" (1945, p. 674). The principle of probability is to be our guide throughout this book.

Kinds of Laws

Philosophers have identified several varieties of natural laws. Perhaps the simplest statement that a law might take is that a particular object or substance always possesses a specified property or quality; for example: pure liquid water is always transparent., or pure sodium chloride always crystallizes in the cubic system. These statements are highly specific, rather than general, in what they cover, and we need not consider them in further discussion.

One important class of laws consists of causal laws, which tell how a particular cause is always followed by a specified effect. Good examples abound in chemistry. When you apply a drop of dilute hydrochloric acid to the clean surface of the mineral calcite (calcium carbonate) a reaction immediately follows, causing an effervesence (bubbling) in the drop of acid. Another and perhaps more familiar phenomenon of this type is the reaction in the common lead/acid battery used in automobiles. When the two battery terminals are connected by a conducting wire, reaction of sulfuric acid with lead of the plates immersed in the acid generates a flow of electrons (current) through the wire. There is only the one direction of natural change in these examples, so the cause is easily separated from the effect. These examples are quite specific, but some general cause-and-effect laws of acid reactions can easily be formulated to include them and innumerable others. Yet another type of law—the statistical law—concerns the probability of a certain event occurring by chance alone during a series of trials, such as the rolling of dice, the tossing of coins, or the drawing of lottery tickets.

Here, we shall concentrate on a fourth type or class of laws because it is most typically associated with physical science. Science philosopher Ernest Nagel describes this type as a law that "asserts a relation of functional dependence between two or more variable magnitudes associated with stated properties or processes" (1961, p. 77). Laws we cited in the opening paragraph are of this type. Formidable as Nagel's definition may seem, it is easily clarified in simple steps. First, he requires that two variable properties or qualities be mathematically related to each other. (The equations of analytical geometry illustrate this class.) If *x* and *y* are the two related variables, then *y* changes as *x* changes, or vice versa. For example, in Charles's Law of behavior of an ideal gas, pressure (P) varies with temperature (T). The relationship is of the direct proportional kind, so that when P increases, T increases (if the volume is held constant). This statement can just as easily be turned around to read that when T increases, P increases. A process of simultaneous change of the two related variables is involved, but you cannot assign to one of them the role of a cause and to the other the role of effect. This is obvious because an experimenter can either increase the pressure or raise the temperature. We can conclude that Charles' Law is not a causal law. We are dealing here with *functional dependence laws.* Two other well known examples are the following:

> Newton's first law of motion: Force (F) equals mass (m) times acceleration (a). $[F = m \times a]$
>
> Newton's law of universal gravitation: Force of attraction (F) is directly proportional to the product of the two masses (m_1 and m_2) and inversely proportional to the square of the separating distance (R). $[F \propto (m_1 \times m_2)/R^2]$

In mechanical laws such as Newton's, three fundamental classes of physical terms, known as "dimensions," are represented in each equation: mass, length, and time (see Chapter 1). Length appears in the distance term, R. Length is also present in the definition of acceleration, which is rate of change of velocity, while velocity in turn is defined as distance (length) per unit time. It is by virtue of possessing one or more of these dimensions that each term can be observed and measured, and thus can be described as being empirical in knowledge content.

The use of the word "law" for these scientific statements about universal relationships has been criticized, and perhaps rightly so, because we also use "law" to mean an order or directive set up by humans (or by God). It has been pointed out by geologist James H. Shea that the often-used metaphorical phrase "laws that govern or control nature" is an anachronistic concept (1982, p. 458). Its usage probably derives from the early period of modern science in which even the most independent scientists believed that God created the universe and set up immutable laws of behavior of mass and energy. It seems more fitting today to recognize self-imposed laws that govern or control the scientist who practices science. The word "law" is firmly entrenched in science and we will continue to make use of it.

To summarize, a *law of science* is a form of scientific statement with the following special attributes. First, it is a very general statement in the sense that it is applicable over a wide range of time and space; it is a statement that applies over and over, countless times, in countless situations. In formal logic, this is referred to as a *universal statement.* Second, the statement of the law, once formulated, does not vary with repetition. Third and perhaps most important, the statement enjoys an extremely small probability of being in error because it has been tested in application countless times by innumerable investigators without having once failed its tests. Wherever and whenever the law has been applied, it has successfully predicted the outcome of the events it explains or the experiment set up to test it. Nevertheless, as we have already pointed out in Chapter 1, because laws are derived by induction, they cannot be said to carry the absolute truth. All that can be safely said about a law is that it carries a very high level of confidence, or reliability, such that we are willing to gamble that it will work each time we rely on it. Every human, throughout every minute, hour, and day, places full confidence in the laws of science—it's a simple matter of life or death, safety or injury, success or failure.

Yet another characteristic feature of laws of science is their interrelatedness and interdependence, as explained by philosopher John Hospers:

> The laws of science are not viewed in independence of one another. Together they form a vast body or system of laws, with each law fitting into a system including many other laws, each mutually reinforcing the others. The laws that scientists are most loath to abandon are those that form such an integral part of a system of laws that the abandonment of the one law would require the abandonment or alteration of a large number of other laws in the system. Thus an observation that directly confirms one law indirectly confirms a group of laws, because of the interconnection of the laws in a system. . . . Whether or not something is called a law, then, depends to a large extent on how deeply embedded it is in a wider system of laws. (1980, p. 110)

Do laws explain how nature works? We will discuss at some length in the next section what is meant by "scientific explanation." Each law mentioned above describes a direct relationship between two (or more) physical quantities, but does not tell us why that relationship should be so. Take the law of gravitation, for example. It tells the relationship between any two masses in respect to the magnitude of the force that mutually attracts them. Until recently, however, physicists have not been able to answer the question: Why do masses attract each other in the first place? To answer this question, physicists have invented a particle called the graviton; it serves to transport the gravitational influence between the two masses. Unhappily, gravitons are impossible to detect, because the force of gravitation is so weak (Davies 1984, pp. 95-96). So we conclude that a functional dependence law of science does not contain any self-explanation, nor is there any logical necessity that it should do so. On the other hand, the law of gravitation can be used to provide an explanation (or part of the explanation) of an observed phenomenon (a particular case) involving two or more masses.

In summary, we reemphasize that laws of science are not absolute truths, because there is no absolute certainty that a law as now stated has always applied in the past and will apply through all future time, or that it applies in other realms of space than that in which it has been tested. After all, records of the distant past are largely obliterated and irretrievable, while knowledge of the future is patently unavailable. The concept of unavailability of absolute truth in empirical science is put this way by science philosopher Herbert Feigl:

> The knowledge claimed in the natural and the social sciences is a matter of successive approximations and of increasing degrees of confirmation. Warranted assertibility or probability is all that we can conceivably secure in the sciences that deal with the facts of experience. It is empirical science, thus conceived as an unending quest (its truth-claims to be held only "until further notice"), which is under consideration here. Science in this sense differs only in degree from the knowledge accumulated throughout the ages by sound and common sense. (1953, p. 10)

Philosopher John Ziman says it this way:

> Our experience both as individual scientists and historically, is that we only arrive at partial and incomplete truths; we never achieve the precision and finality that seem required by the definition. Thus, nothing we do in the laboratory or study is "really" scientific, however honestly we may aspire to the ideal. . . . Many philosophers have now sadly come to the conclusion that there is no ultimate procedure which will wring the last drops of uncertainty from what scientists call their knowledge. (1980, p. 38)[1]

The Role of Explanation in Science

Most carefully composed definitions of science include its two basic aims: to describe and to explain. (A third aim often cited is to predict.) We have already delved into the nature of scientific description and how it makes use of special language and observational techniques. Now we need to devote some more attention to the process of explanation in science. What do we mean when we say that science explains the things it observes?

We will show later in this chapter that a set of scientific statements can consist of the description of something observed to exist or occur, but for which no explanation has yet been provided. Nevertheless, most descriptive statements evoke inquiring responses: Tell me more. Fill in the details. How did it come about? What forces caused it to happen? How is it controlled?

Paleontologist George Gaylord Simpson gives a rather straightforward and comprehensible analysis of scientific explanation that applies to both physical and natural sciences (1963, pp. 33-35). He thinks of kinds of explanations in terms of kinds of questions we might want to ask. First is the question "How do things work?" It applies to knowledge of how processes operate in nature. Specific questions might be phrased as follows: How do streams erode valleys? How do animals digest food? Processes of physics and chemistry are involved, and the explanation consists of relating appropriate processes from those disciplines to the specific natural environment or setting under discussion. In explaining how a stream erodes a valley, we need to call on laws of mechanics of fluids, and these include reference to the laws of gravitation and motion. In explaining the digestion process, we need to set down the chemical composition of the nutrient substances as well as those of the reagents that act upon them in the alimentary canal of the animal. We then set up the chemical equations that show the reactions and products of those reactions. Scientists call this form of explanation "scientific reduction;" it is a topic we will cover in Chapter 4.

Simpson then turns to a second kind of explanation (p. 34), one that is appropriate to the question: How did it come about? In short: How come? A geologist views a great mountain range with highly complex internal rock structure. How did this great mass we see here today acquire its size and shape, its internal composition and structure? The explanation is primarily historical and must be directed at reconstructing the chain of events that led to the final product. But the explanation must also deal with the natural processes that were involved in each step. For example, crustal compression occurred at one stage; rise of molten rock (magma) at another stage, and so forth. Thus the answer to "How come?" includes answers to repeated questioning as to "How do things work?"

Simpson refers to a third kind of explanation (p. 34). The appropriate question is "For what purpose?" or "What is the use of it?" The first question has no meaning in physical science and in the purely physical areas of natural science. We would never think of asking "For what purpose does a volcano erupt?" We can, however, ask that kind of question about living organisms. For what purpose did warm-bloodedness develop in mammals? Because "purpose" can easily suggest design with an ultimate goal, which in turn can imply a designer with supernatural powers, we should rephrase this question to replace "purpose" with "function." Thus: What useful function does warm-bloodedness perform in mammals? The role of this kind of purpose in the evolution of life forms is something we will examine in Chapter 4.

What do the science philosphers have to say about explanation? In general, they are in agreement with what Professor Simpson has written, but they probe deeper into the logical structure of explanation. Ernest Nagel recognizes four types of explanations, but we will boil them down to three that seem to match Simpson's three questions.

First and most important to all of science is a *deductive model* that relates to the question "How do things work?" (Nagel 1961, pp. 21-22). Let us use one of his examples to illustrate. Why do droplets of water (always) form on the surface of a glass of ice water in warm humid weather? Obviously the popular phrase that the "glass sweats" is completely off target. Notice particularly that we inserted in parentheses the word "always," showing that we intend a universal statement. Isn't that the same as a law? Yes, of course it is, but in this example the law includes some very specific provisions about the container and the state of the surrounding air. These we can call the *antecedent conditions*. Before going further, we need to establish four terms and give them code letters to simplify the text:

EP The empirical proposition or statement that is to be explained. (Formally, it is called the *explanandum.*)

EX The explanation. (Formally, it is the *explanans.*)

C An antecedent condition.

L A law.

In our example, the antecedent conditions (C_1, C_2, C_3, . . . C_n) are contained in EP. The explanation (EX) must therefore lie in one or more laws (L_1, L_2, L_3, . . . L_n) that are more general in scope than the law already contained in EP. Our model requires that EP be explained by deduction from the Ls. Now, by deduction we mean that from a law we derive a prediction; in this case our prediction corresponds with the content of EP. Let's see how this thought process works.

One first step is to upgrade the law in EP to a broader statement: Water always condenses on any cold solid surface from water vapor in the adjacent air when that air has a sufficiently high vapor content and is sufficiently warmer than the solid surface. Going one step further, we can say: Water in the vapor state always condenses into the liquid (or solid) state when it undergoes sufficient cooling to reach the dew point. This eliminates the precondition (C_1) of a cold solid surface. Most persons who asked the first Why? will be satisfied at this point. Beyond this point, however, the explanation (EX) must be carried to even more general laws that concern the phases (as kinetic energy states) of molecules in the vapor, liquid, and solid states and the energy/entropy flows that are entailed in changes from one phase to another. Only at this level will the explanation be likely to satisfy most scientists.

Nagel identifies a second class of explanations under the name *probabilistic explanation*. We will begin with an example: Why did Hurricane Hugo slam ashore near Charleston, North Carolina, at about midnight on September 21, 1989? Right away, you realize that this is a statement of a singular event—a historical event—fixed in place and time. It is impossible to reword this description as a universal statement ("All . . . are . . ."). Laws are not useful in answering the question of why Hugo was there at that time. Instead, we must look to a unique combination of antecedent conditions (the physical causes) that controlled the time and place. One such cause was the location of the jet stream, a high altitude flow of air that steers hurricanes when they penetrate the middle latitudes. In this case we can perhaps say that the jet stream was in such a configuration and location that it was highly probable that Hugo's curving path would bring it to the North Carolina coast. So, what Nagel is implying in his name for this kind of explanation is that the singular event represents a convergence of one or more highly probable antecedent states of the controlling physical causes. Perhaps we should rename this kind of explanation a *historical explanation*. Historians looking back at a significant point in human history will often say "It was inevitable that so-and-so occurred when it did." They are implying a probability approaching 100 percent. In Chapter 4 we return to historical explanations because they are the dominant type in historical geology and its fossil record of biological evolution.

Nagel also discusses Simpson's third kind of question: For what purpose? We have already restated that question as "What is the function of it?" Thus a third class consists of *functional explanations*. They are questions largely in the realm of biology. The explanations are closely akin to the deductive kind, as they also require that laws be invoked in explaining a particular organic function.

In winding up this review of the meaning of scientific explanation, we consider three very general concepts that are involved. First is that explanation can come from either of two basic sources of information: the antecedent conditions (Cs) or the general laws (Ls). For the former, chance plays a major role and the explanation can be seen as probabilistic. For the latter, the explanation is determined by the laws and can be seen as deterministic. This is a subject for further analysis in Chapter 6.

Second, in deductive explanation, prediction is essential. The law that is invoked is used to deduce the phenomenon that is to be explained. To deduce is to predict what is or will be the specific case. So explanation and prediction are logically indistinguishable.

The third idea is a rather disturbing one. In describing laws, we stated that they are derived by induction from sets of repeated observations. Now, however, we have said in effect that the same or similar observations are explained by those very same laws. The two statements make up a circularity of reasoning, or tautology. What this means is that all deductive explanations are automatically true in the logical sense. But must we not be concerned that by the same token they are worthless in the scientific, or empirical sense?

The Scientific Hypothesis, or Theory

Scientific statements that purport to describe and explain specific observed phenomena, and can stand opposed to one or more alternative statements, are known as *hypotheses*. At the outset, we emphasize that the hypothesis is (or contains) a tentative assumption put up for the sake of argument. Thus the concept of incompleteness or uncertainty is essential to distinguish hypotheses from laws of nature.

A particular hypothesis in its simplest form arises in the human mind, using percepts based initially on observation as well as previously collected laws and scientific statements that possess only a small probability of being in error. In some manner that we do not fully understand, the mind spontaneously organizes the bits and pieces of information it has received into chains, networks, or other configurations that may involve a major cause and its effect(s) or many interacting causes and their effects. In other cases the information may be organized into a static description of a physical structure of matter.

At the outset, let us clear up a question that is perhaps trivial, but that can cause minor irritation if not attended to. What is the distinction, if any, between a hypothesis and a theory? My own conclusion is that the two words may be used interchangeably without causing confusion or consternation. On the other hand, some scientists and philosophers have decided that a *theory* is a major hypothesis, one of great scope and importance, giving explanation to numerous and diverse sets of observations. In that view, a theory is a super-hypothesis that has achieved special distinction and has been given a fancy title, much as knighthood or an honorary doctorate degree is conferred on a person in recognition of meritorious achievement. I would prefer to use "hypothesis" as the more useful term, irrespective of the rank of the statement. A lot depends on context. I use "theory" in many places where that agrees with general usage, as we find in the case of Einstein's theories of relativity, the Big Bang theory of the creation of the universe, and Darwin's theory of evolution. Typically, the scientist's name designates the theory.

Another consideration in preferring "hypothesis" over "theory" is that the latter has at least two meanings in common use. Besides referring to the hypothesis itself as a structured statement about a specific phenomenon, "theory" can mean the general (universal) physical description of a widely repeated phenomenon. For example, the nature of flow of heat through any substance can be expressed as a set of mathematical equations having the status of laws, which we refer to collectively as "the theory of heat flow."

Do Hypotheses (Theories) Explain?

Having disposed of something trivial, we turn to something of serious concern. In Chapter 1 we noted that science aims to provide the best possible description and explanation of the phenomena it observes. In the previous section of this chapter we have tried to analyze the various kinds of scientific explanations. Does it follow, then, that a scientific hypothesis must always include both a description and an explanation? Can a perfectly sound hypothesis describe a phenomenon without offering any explanation? One need only look back into the history of science to show that the answer is a clear "yes."

Take, for example, the Copernican Theory that the planets revolve about the sun. Called the *heliocentric theory*, it revived an ancient view of followers of the Greek philosopher Pythagoras and was placed by the Polish astronomer Nicholas Copernicus in direct confrontation with the prevailing *geocentric theory*, that put the earth at the center. The heliocentric theory was greatly strengthened by Galileo's telescopic observations of the the phases of Venus and the revolution of the moons of Jupiter. Then Johannes Kepler worked out empirically, from direct astronomical observations provided by Tycho Brahe, three fundamental Laws of Planetary Motion. These were in the nature of a more detailed mathematical description of what actually happens.

Yet, during all this time there was no real explanation of the heliocentric model in terms of fundamental or universal laws of the behavior of massive objects in motion. It remained for Sir Isaac Newton to furnish what we would regard as an explanation of the heliocentric model. In that case, a set of universal laws of gravitation and motion was brought forward as a form of explanation.

Certainly, then, a scientific hypothesis (theory) can consist of description without explanation. In such cases, however, acceptance of that hypothesis as good science is implicitly conditional upon the eventual achievement of an explanation.

Are Theories (Hypotheses) the Same as Laws?

Assuming for the moment that "theory" and "hypothesis" are synonymous, and that both words refer to a model of empirical reality generated in the scientist's brain, we turn next to the relationship between theories (hypotheses) and laws of science. This is a far more important and difficult question to consider. [Notice that in this section we have elected to use "theory" rather than "hypothesis" because that is the practice among philosophers of science.]

First off, look at samples of "theories" and "laws" in a physics textbook, searching for some obvious and compelling reason that "theory" is used for one and "law" for another. Assume, as we have concluded earlier, that a law of science is a universal statement describing a regularity of nature. The law describes a relationship that holds under all circumstances, in all places, and at all times. Let Newton's first law of motion (stated earlier in this chapter) be the example. Compare it with Einstein's special theory of relativity, as expressed through the classic equation $E = mc^2$ Although very different in their meanings, the two equations use the same fundamental dimensions. Both are propositions expressing regularities in nature through universal statements. Is there any good reason not to switch titles to read: Newton's "first theory of motion" and Einstein's "special law of relativity"?

Philosophers of science have assembled a bewildering collection of interpretations of the possible kinds and meanings of "scientific theory." They have classified theories into as many as five main varieties (see Paul T. Durbin's *Dictionary of Concepts in Philosophy of Science* 1988, pp. 320-22). We shall focus on two concepts that may be useful in relating laws of science to theories of science. One is the extent to which the stated proposition is substantiated by empirical evidence (observation); a second is the breadth of natural phenomena that the proposition encompasses. Except for these differences, we will consider the structure of both to be identical. The idea here is to suggest that laws and theories share differences in quality in two directions indicated in the previous sentence; i.e., (a) in degree of confirmation, or corroboration; (b) in degree of generality of application.

As to point (a), we will characterize a theory as containing a proposition (or collection of propositions) postulating some phenomenon or relationship for which there is (at the time of its inception) no evidence based on observation. Thus at least a part of the theory is hypothetical, a conjecture. Perhaps it is the idea that "something is missing" that, unlike a law, is the necessary condition for a theory. Being derived from observation of a vast number of real cases, all of which are in agreement, a law carries no hypothetical content. As to point (b), laws are universals in the sense of formal logic, meaning that both are statements containing or implying "All (such-and-such) are . . ." or "Such-and-such . . . is always is the case."

We than add to this concept of two directions of difference the idea that a strong odor of hierarchy pervades the interrelationships of laws and theories. They must form one orderly progression starting with the largest proportion of conjecture and passing into a final phase with a full complement of observational data. At the same time, there will independently be a progression from the most limited or specific proposition to the most general and all-inclusive proposition. Thus two independent kinds or classes of hierarchies can be recognized.

Figure 2.1 puts these ideas in a crude graphic form that may be easier to grasp than sentences. Philosophers rarely use diagrams, but for scientists diagrams are helpful models and seem to form spontaneously in the mind. Scaled on the *Y* axis, the vertical dimension expresses the level of generalization of the proposition, whether it be a theory [T] or a law [L]. The range is from narrowly specific propositions (at the bottom) to broad, universal propositions (at the top). Scaled on the *X* axis, the horizontal dimension expresses the level of support by firm evidence based on observation. Here, the range is from weak (at the left) to strong (at the right).

Using our diagram, we can show the history of a proposition of broad scope [T_1] making its debut at the left as one that is initially highly speculative—based largely on imagination or conjecture—and therefore located in the domain of theories. If successful in obtaining support by observation, this proposition moves to the right into the domain of laws, becoming transformed into a general law [L_1]. Another proposition [T_2], having narrow scope (greater specificity) can also progress on its own horizontal track to become a law [L_2]. A variety of possible horizontal

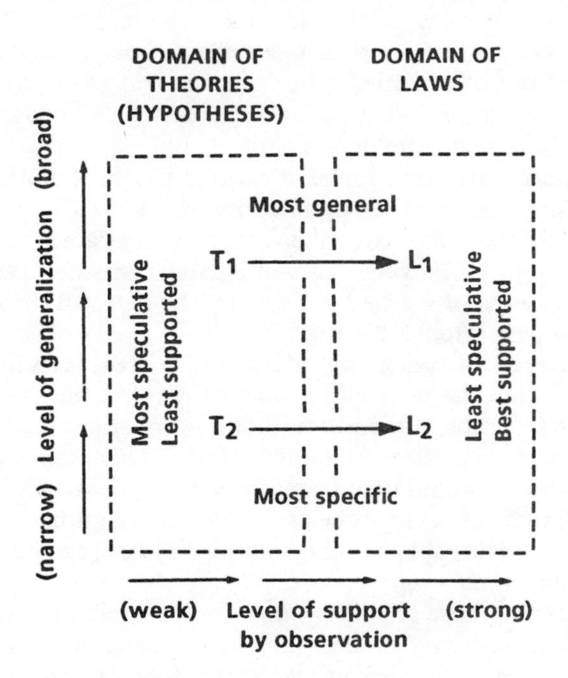

Figure 2.1 How theories (hypotheses) are related to laws.

tracks is possible with a wide range of starting and ending points and of levels from base to top.

The confusion over appropriate naming of the scientific proposition arises because, when T is fulfilled and becomes L, that event is ignored by scientists, philosophers, and historians alike, and they continue to call it a T. Surely, by now, Einstein's special relativity equation should be awarded the status of a law, and so should Darwin's fundamental proposition of descent with modification.

The Hypothesis as a Product of Inspiration

The mental process of hypothesis-making has been described as one of inspiration, as distinct from pure logical reasoning from information. [We now revert to using "hypothesis."] Physicist J. T. Davies observes that a hypothesis "comes from an intuitive leap of the imagination, from inspiration, from induction, or from a conjecture" (1973, p. 12). How remarkable it seems that the most crucial process in all of science is so shrouded in mystery! In this process, the mind is creating a *model* to fit as best it can the *real thing*, which is "what is actually out there." The model is made more nearly complete by inserting imaginary pieces of information to fill gaps in the available supply of input information. This attempted fitting is done to achieve an identity, described as an *isomorphic system*, between the model and the real thing. The degree of isomorphism can range from poor to excellent. Once conceived, the model is transported out of the mind by translation into an organized scientific statement in words that make it available to other minds, and residing where it can be stored more or less permanently.

Astronomers have long found it most difficult, if not impossible to observe directly the birth, or origination, of a star. The event is always obscured by a dark veil of cosmic dust and gas that absorbs all light from the newborn star. Only when the veil is later dissipated can the star be seen. (The existence of a veiled star can now be detected by other forms of radiation that pass through the veil.)

The origination of something, whether of a great scientific hypothesis or of a star, is thus the most elusive part of the scientific quest. Evolution in form, structure, and composition that follows origin is usually open to observation and can often be understood in remarkable detail by the scientific method. I would urge you, however, not to succumb to the belief that formulation of a hypothesis of the origination of a star is a supernatural act, incapable of being analyzed by science and therefore not a part of science itself.

Without doubt, those scientists who have achieved a rating of greatness have had extraordinary talent for constructing new hypotheses of tremendous depth and consequence. Their names adorn the walls of the world's science museums and universities—Copernicus, Newton, Darwin,

Einstein, and Bohr, to name only a few. Those who study the history and philosophy of science in collaboration with psychologists have made many attempts to analyze the creative process in science and to list criteria that set apart the most creative scientists. Despite all their efforts, the nature of that creativity remains obscure.

The views of Sir Karl Popper are interesting in this connection. As one of the most influential contributors to the philosophy of science in the middle decades of our century, his ideas on the origin of scientific hypotheses, as well as their testing and corroboration, have attracted many followers. Popper was born in Vienna in 1902, educated in the University of Vienna, and taught philosophy at Canterbury University in New Zealand during World War II. He then transferred to the faculty of the London School of Economics; he was honored with a knighthood in 1964. Popper was an active member of a famous school of philosophy known as the Vienna Circle, whose program later became known as logical empiricism. Popper argued forcefully against a purely logic-oriented, or rational, method of arriving at general hypotheses of universal importance (1959, pp. 27-32). He offered the following alternative for arriving at what he and his group referred to as *scientific discovery*:

> However, my view of the matter, for what it is worth, is that there is no such thing as a logical method of having new ideas, or as logical reconstruction of this process. My view may be expressed by saying that every discovery contains "an irrational element," or "a creative intuition," in Bergson's sense. In a similar way Einstein speaks of "the search for those highly universal laws from which a picture of the world can be obtained by pure deduction. "There is no logical path," he says, "leading to these . . . laws. They can only be reached by intuition, based upon something like an intellectual love ('Einfuhlung') of the objects of experience." (p. 32)
>
> [Note: Exact source of Einstein's statement is given in a footnote on p. 32. Popper states that the English translation of Einstein's German text gives "sympathetic understanding of experience" as an alternative meaning for "Einfuhlung."]

After reading Popper's statement, including that by Einstein, some readers might infer that hypotheses arrive by Divine revelation, put in the scientist's brain by the Creator for a purpose, and thus God guides hu-man understanding of the cosmos. But science considers any such inference as lying outside its sphere of knowledge.

Because Popper so positively rejects the inductive process as capable of producing an important scientific hypothesis, he might seem to be arguing for creation of ideas *ex nihilo* (out of nothing) or *in vacuo* (in a vacuum). I doubt that Popper would allow such an interpretation. Surely, every scientific idea or hypothesis has empirical content; i.e., some experiential basis drawing on prior knowledge of something in the real world. Note that Einstein recognizes this necessity when he refers to "the objects of experience." Consequently, I would not eliminate inductive input.

Going one step further, I would suggest that filling the missing information gap(s) to produce a complete and coherent hypothesis would be impossible without using information gained initially by observation of nature. Granted that available empirical information may need to be dissected and reduced, reassembled from numerous bits and pieces, or otherwise processed to generate what can be described as a "new idea" that fits perfectly into the knowledge gap, no nonempirical information has been included. Nor does this restriction detract from the remarkable ability of certain scientists to carry out that process. If you wish to call the process "creative intuition," please feel free to do so. Great chefs "create" new dishes in a similar manner, using only edible substances at their disposal.

It is rarely the case that a scientist formulates a hypothesis incorporating new information that no other observer has previously received and recorded. There are, however, occasional occurrences of a totally unpredicted exposure to new information. One example might be the scene that greeted scientists inside the deep-sea submersible vessel *Alvin* when they reached the floor of the Pacific Ocean near the Galapagos Islands to encounter a troughlike depression. From a chimneylike tube of mineral matter attached to the rocky floor of the depression there streamed upward a jet of black water, as if from a smokestack. It proved to be hot water rich in dissolved mineral matter. Surrounding this "smoker" was a community of strange living forms, including giant clams and enormous wormlike things attached to the bottom and waving to and fro in response to water motions. What a shock! The hypothesis the *Alvin* scientists immediately sought to formulate was that a community of unique life forms is in some way being sustained by nutrients emitted by the smoker, all in a cold environment with no light whatsoever.

Purely inductive processes for formulation of a hypothesis can also arise in space exploration of planets and telescopic observations of the universe. Happening upon something totally unexpected and wholly unfamiliar forces the inductive process to take place, but even so, the scientist cannot help but refer to knowledge already gained in different situations to guide the design of a new hypothesis. This form of guidance by analogy is commonplace.

Deduction Takes over the Leading Role

Where factual statements are not immediately available in support of the hypothesis, investigators are forced into an inventive role. They must engage in the process of deduction, a predictive process. The meaning of "deduction" is easier to grasp if we think of it as meaning "anticipation," a reasoning process by which an investigator is led to predict the finding of a relationship or event not previously observed or identified. (Recall from Chapter 1 the propositional form "If *a*, then *b*.") Actually, what is anticipated may have already been observed and placed on record, perhaps noted by someone else in a different investigation, but has not yet come to the attention of the inventor of the hypothesis.

Let us also use as an example the formulation of a major geologic hypothesis—a model of the behavior through time of the earth's outermost rock layer, or crust. By the late 1950s, there had accumulated a body of scientific statements relating to the possibility that the crust is actively spreading apart along a great rift (a crack) that lies at the crest of the mid-oceanic ridge (Figure 2.2). The remarkable symmetry of ridge topography and of ages of sediments deposited on the basaltic crust, in addition to documentation of high heat flow and an outpouring of basaltic magma (molten rock) along the axial rift, strongly suggested a spreading apart of the crust. This was the inductive phase in modeling.

An early hypothesis accepted the initial statement that seafloor spreading occurs while new oceanic crust is continually formed. Going a step further, its inventors deduced that (a) the ocean basins must be widening and thus increasing in areal extent and that (b) at the same time the earth as a whole must be expanding in volume. In a final step, it was deduced that (c) the earth's total surface area must be increasing at a rate sufficient to match the increase in area of new crust. It should be obvious that while the limited scenario of crustal spreading and growth of new crust has only a rather modest probability of being wrong, the concept of an expanding earth is highly speculative and is initially lacking in support from reliable scientific information. To postulate that the earth is expanding (the "balloon hypothesis") requires a great deal of imagination—something that carries a high risk of being wrong. The hypothesis as a whole thus has a substantial probability of being in error, even though some of its parts, standing by themselves, can pass the requirements of low error probability.

A typical complex hypothesis is directed at providing an explanation and an interrelationship of a specific collection of statements of facts and simple scientific statements. Thus, a *scientific deduction* drawn from that hypothesis starts with the specific terms of the hypothesis—the "givens," you might say—then turns to a general statement or law, or a group of such laws, as providing the underpinnings for a rather specific conclusion (Nagel 1961, p. 23). The deductive process results in a prediction of missing parts of the explanation contained in the hypothesis. Prediction of the specific is thus guided and supported by laws of science that have an extremely small probability of being in error.

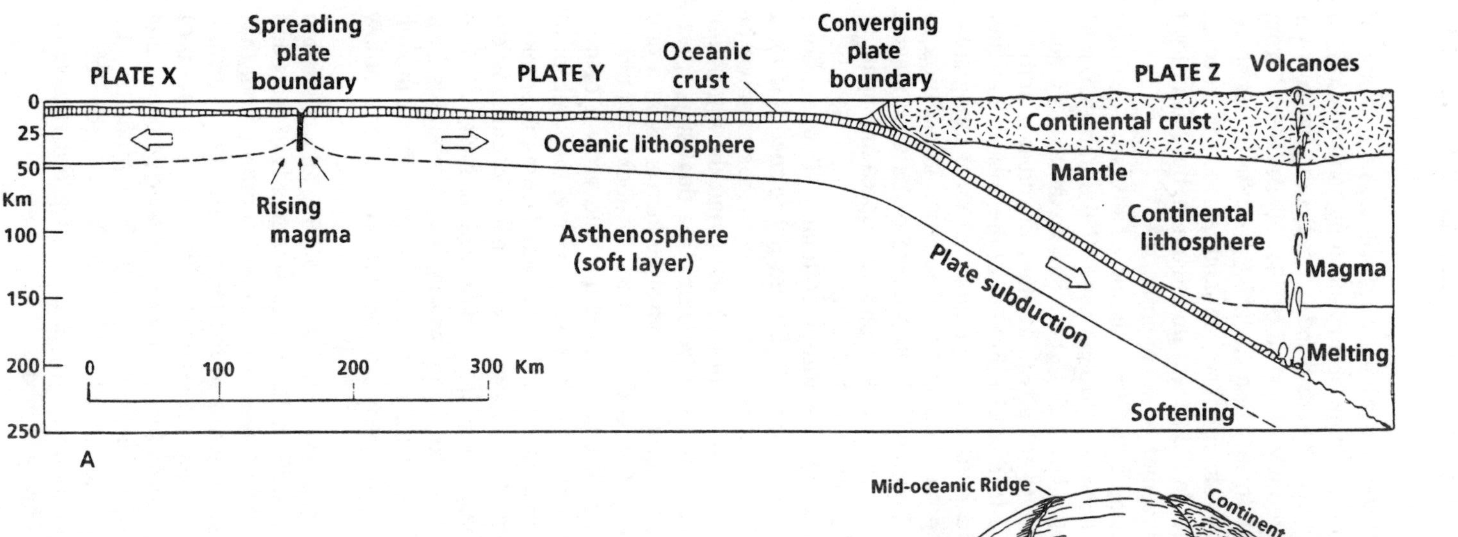

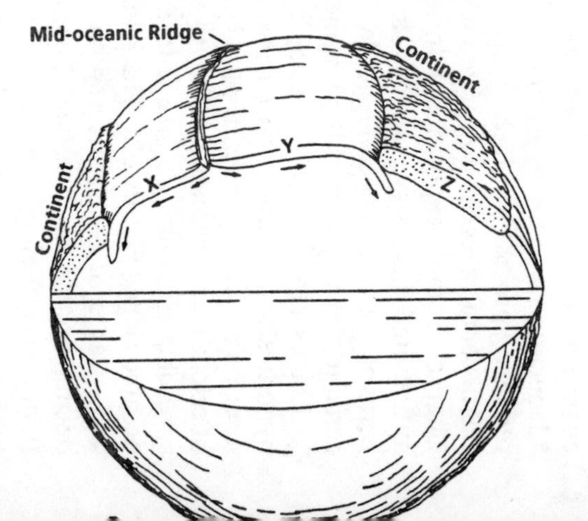

Figure 2.2 Basic components of plate tectonics. (A) Schematic cross-section drawn to same horizontal and vertical scales, with earth's curvature removed. (B) Pictorial diagram of plates on the globe, their thickness greatly exaggerated. (A. N.

Take this example. A geologist is trying to deduce (visualize) the changes that take place in a body of magma (molten rock) as it cools and solidifies several kilometers below the surface. Let us first bring forward the general or universal laws that will apply:

> Law: A fluid exerts upon a body immersed in it a buoyant force equal to the weight of the fluid displaced by the body.
>
> Law: An immersed body of solid matter of higher average density than the fluid will sink because the weight of the body (downward force of gravity) is greater than the buoyant (upward) force exerted by the fluid.

The antecedent conditions are next stated. These may take the form of scientific statements based on observation and experiment; they are inductively obtained:

> From laboratory experiments with igneous rocks heated above the melting point at various pressures, it can be stated that the density of the magma in question lies in the range of 2.8 to 3.0 grams per cubic centimeter (g/cc). Experiments also reveal that the mineral olivine, with a density ranging between 3.3 and 4.4 g/cc, is the first common mineral to crystallize into solid grains as the temperature of the melt falls during experimental cooling.

The deduction we can now make is that in the magma body, crystals of olivine will be formed in abundance at an early stage in the cooling and will sink through the magma until they reach the bottom of the magma chamber; therefore, we can expect to find a basal layer of igneous rock rich in olivine. The geologist now goes into the field to attempt to confirm this deduction. If the base of the igneous rock body can be found exposed or a drill core can be taken, the deduction can perhaps be confirmed.

In looking back over the steps in this example, we note that the strength or value of the final deduction depends upon the general laws that back it up. In practice, a scientist does not actually repeat the laws that apply to a deductive chain of reasoning. Those laws are learned early in the scien-tist's career, are accepted by other scientists, and remain unspoken unless a scientific statement violates one of them. It is assumed that the hy-pothesis first formulated by inductive processes also conforms to the laws of science. Always, those laws are in control of what the scientist says and does.

Getting back now to the expanding-earth ("balloon") hypothesis, if you wish to raise it to a level of high quality, you must begin a search for some consequences that are attached to it. In this case, you may reason that, if the mass of the entire earth has not changed, volume expansion to a larger sphere must be associated with a decrease in density of at least a substantial portion or zone of the earth. Perhaps your first thought is that if some of the earth's mass is displaced outward, farther away from the axis of rotation, the rate of rotation will correspondingly decrease. Now you have a lead to follow. What independent evidence has already been gathered that bears on the possibility of change in rotational velocity over the past 100 to 200 million years? You are forced to search for some structure or property of rock that carries within it the record of length of a day.

You come across a published study in which the lines of banding on a single coral structure are interpreted as records of growth. Each line represents a day, or one rotation of the earth. Duration of one year can also be read from a rhythmic change in the thickness of growth lines in a much longer cycle, representing the year. Thus, the number of days in the year is established, and the duration of the day becomes known. The age of the rock in which the fossil coral is embedded can be determined independently. The investigation shows that, indeed, length of day has been increasing rather steadily for more than 400 million years. This is the confirmation of your deduction! But wait, you also learn that a much simpler answer is available to explain the evidence. Tidal friction that the moon exerts upon the rotating earth can be expected to slow the earth's rotation. Calculations can be made of the rate of slowing from this cause. The calculations seem to fit well with the evidence from the corals. There is no need to associate the lengthening day with slowing due to earth expansion. Your deduction, though confirmed, can be explained easily by a natural process that does not require an expanding earth. Your entire hypothesis remains weak because independent evidence is lacking for an expanding earth.

The emergence of a newer hypothesis of the earth's crust, based in part on the older one, was not long in coming. It retained the requirement of seafloor spreading and the formation of new oceanic crust as two platelike masses pulled apart along the axial rift. Meantime, evidence favoring seafloor spreading was getting even stronger. In the mid-1960s, new statements of low-error probability could be made in support of a scenario of seafloor spreading continuously for as far back as 200 million years. This was evidence of numerous reversals of magnetic polarity recorded in stripelike zones parallel with the mid-oceanic ridge axis and arranged in mirror-image patterns with respect to the axis. Ages of the polarity reversals were determined by independent means and spreading rates were estimated. A short time later, another line of excellent supportive evidence emerged from the observation that the first crustal motion of earthquakes generated along transverse (transform) faults offsetting the ridge axis is always in agreement with a spreading motion. The probability that this statement is in error had been reduced to a very low value by application of established principles of seismology.

The newer hypothesis accepted the high-quality evidence of seafloor spreading, but turned to a mechanism by means of which the oceanic crust could be disposed of as fast as it was formed, and without requiring the earth's volume and surface area to expand. The suggestion had been made many years earlier that a line of deep oceanic trenches close to the continental margins indicated that the crust was being deeply downbent. This configuration indicated that perhaps the oceanic crust could move sharply downward into the underlying soft mantle. For decades seismologists had been recording the presence of many earthquakes originating at great depth in a slanting zone extending from the trenches far under the continents. This finding led to the suggestion that perhaps the oceanic crust, down-tilted as a single slab, is being pushed down far beneath the continents and that internal friction is setting off earthquakes. The pattern of deep earthquakes was discovered as early as 1930 and continued to be reinforced in the ensuing decades, but the information remained largely unused until the mid-1960s, when it became a key element in the newer hypothesis (Figure 2.2).

The newer hypothesis proposes that the *lithosphere*, a thick, strong rock layer of which the oceanic crust is the uppermost layer, moves horizontally as a unit away from the mid-oceanic spreading axis. At the margin of an adjacent continent the lithosphere bends down sharply and descends into the soft, highly heated zone below, called the *asthenosphere*. The descent of the lithosphere is called *subduction*. The plunging slab becomes heated once in contact with the surrounding hot rock, gradually softens or melts, and is thus absorbed to become a part of the asthenosphere. In this manner, lithosphere is recycled into the earth at the same rate that it is formed at the spreading axis (Figure 2.2).

At this point in formulating the newer hypothesis, some key deductions were possible and could be tested. Because the down-plunging slab must be comparatively cold and brittle, its descent should be accompanied by a great deal of minor fracturing (cracking and slipping of the rock), giving off many earthquakes. Existing plots of deep earthquakes were already available to confirm this deduction. Further analysis of earthquake records showed that the orientation of stresses that produce the earthquakes is just what would be expected under the kinds and directions of stresses that the slab would experience. Thus, highly reliable statements became available to support the mechanism of subduction (in itself a single hypothesis) and to support the entire hypothesis as well. The self-reinforcing feedback cycle of induction/deduction we have been describing here was outlined in Chapter 1 (see Figure 1.1).

The total scheme, or model, became known as the hypothesis of *plate tectonics*. It has found broad-based support among geologists as many new deductions have subsequently been made and found to be in agreement with observations. Any scientific hypothesis carrying the strength that the entire plate-tectonics model now enjoys is not likely to be totally discarded and replaced in the future by an entirely different model. It will more likely be continually revised and restructured to decrease the overall probability that it is incorrect. In this way, science gains ground in understanding and explaining the structure and behavior of all parts of the universe. Despite revisions and restructuring, large areas of science enjoy a certain level of permanence because the odds are extremely small that a cataclysmic overturning of scientific statements will occur on a grand scale.

The Principle of Parsimony

A pervasive principle of science, one that we as students exercised in our graduate seminars and debates, applies in the formulation of hypotheses and in setting up scientific explanations of any kind. Commonly called the *principle (law) of parsimony*, it instructs us not to bring in unneeded explanations or causes if those we already have on hand are adequate to do the job. The law has been around a long time, for it is dubbed "Occam's razor," after William of Occam (circa 1285-1349), a Franciscan philosopher who taught the concept at Oxford for a number of years. (His name is also spelled Ockham or Ockam.) In the language of his time, the admonition goes: "What can be done with fewer assumptions is done in vain with more." Most college students of my time memorized the equivalent in Latin: *Entia non sunt multiplicanda praeter necessitatem.* Now that Latin has gone out of style, we can try to remember the English translation: "Entities must not be multiplied beyond necessity." The principle of parsimony is also known as the "principle of simplicity" and the "principle of economy" (Angeles 1981, p. 195).

A good example of the principle of parsimony occurs in the development of our balloon hypothesis of seafloor spreading and an expanding earth. As first formulated, it contained two explanations: first, the production of new oceanic crust by upwelling of magma along a mid-oceanic spreading rift; second, an expanding earth to accommodate the increase in global surface area. The second is in a sense contrived to explain the first (i.e., to make the first possible). If we could apply Occam's razor to get rid of the second explanation, the hypothesis would be more parsimonious and, in that way, would become much stronger. This action was taken when we showed that plate subduction can dispose of the oceanic crust as fast as it is being formed, thus eliminating the need for an expanding earth. A single, unified mechanism suffices, and the hypothesis is internally coherent and consistent.

The principle of parsimony must be used with caution. Logicians will point out that oversimplification can be an equally grievous error. Philosophers Morris Cohen and Ernest Nagel caution us about this error, which they refer to as "the fallacy of simplism or pseudo-simplicity":

> We must guard against identifying the true with the apparently simple. And in fact hasty monism, the uncritical attempt to bring everything under one principle or category, is one of the most frequent perversions of the scientific method. (Morris and Cohen 1934, p. 384)

Quality of the Hypothesis

To summarize the concept of quality of a scientific hypothesis and to give it graphic expression that may make it easier to grasp, I have drawn a rising straight line with a set of dots (Figure 2.3). Think of it as a "ladder of excellence" on which each "rung" denotes a tenfold increase in quality. We must take into account two complementary statements of probability:

P_T = Probability that the hypothesis is true.
P_F = Probability that the hypothesis is false.

The sum of these two probabilities is unity; therefore, as P_T increases, P_F decreases. However, we must keep in mind that neither value can actually reach either unity or zero—those limits can only be approached, so our ladder has no upper or lower end. Gambling odds on the hypothesis actually being a true statement of nature are given as the ratio of P_T to P_F. I have suggested adjectives to describe the quality of the hypothesis, but these are quite subjective because, as with people in general, scientists differ among themselves as to the risks they are willing to take in a given situation.

I gather from reading Karl Popper that there may be problems with the concept of a hypothesis being evaluated on the basis of the probability that it is false or true. One problem that seems genuinely serious was raised by Popper in the 1960s. It concerns the comparison of hypotheses that are of quite different levels of importance in terms of their power to relate diverse phenomena and to yield important deduced consequences. Popper's statement of the problem is as follows:

> I pointed out that the probability of a statement (or set of statements) is always greater the less the statement says: it is inverse to the content or the deductive power of the statement, and thus to its explanatory power. Accordingly every interesting and powerful statement must have a low probability; and vice versa: a statement with a high probability will be scientifically uninteresting, because it says little and has no explanatory power. (1963, p. 58)

If my interpretation of Popper's statement is correct, it says that what I have described as a simple statement, such as "the streak of hematite is always red," would rate highest on the scale of quality; whereas a powerful hypothesis such as plate tectonics, with its enormous explanatory power and capability of producing testable deductions, would automatically get a much lower rating. Popper says that "we do not seek highly probable theories but explanations; that is to say powerful and improbable theories"(p. 58).

How can this difficulty be circumvented? I suggest that we can salvage the probability idea by evaluating the hypothesis on a different basis. We can think, instead, of a certain probability that the hypothesis will survive through time under continued testing and growth in complexity. To this we must add "in competition with alternative hypotheses of comparable complexity and explanatory power." Perhaps we can devise some scale of value that expresses accurately the internal strength of the hypothesis. Consider that the ability of a hypothesis to survive will increase as does

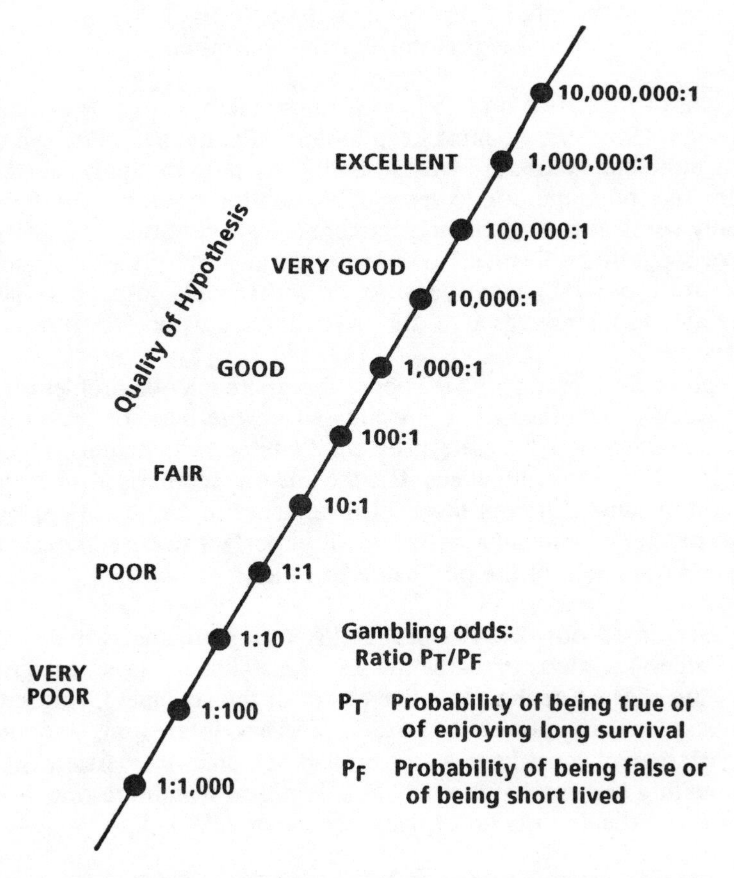

Figure 2.3 An imagined ladder of excellence applied to hypotheses.

its ability to unify diverse phenomena under a single set of basic statements and its flexibility to adapt to the results of testing. Survival is then viewed in the context of actuarial statistics. What are the odds of the hypothesis enjoying strong support twenty years from now, or fifty years from now, or in a hundred years? This actuarial viewpoint takes into account not only the present strength of the hypothesis, but also its potential for internal change to meet stresses imposed by massive inputs of new information and the impact of rival hypotheses. In any case, the concept remains one of probabilities rather than one of absolute acceptance or rejection. We seem to have latched on to the evolutionary concept of survival of the fittest—the fittest hypothesis, that is.

Our next chapter carries us further into the evaluation of hypotheses. We will be particularly interested in how a hypothesis can be strengthened by processes of prediction and corroboration, and how it can be discovered to be in severe difficulty and perhaps even to require total rejection.

Credit

1. From John M. Ziman, *Public Knowledge,* Cambridge University Press, New York, pp. 5-27. Copyright © and reproduced by permission of The Cambridge University Press.

CHAPTER 3

Prediction, Testing, Corroboration, Falsification

From our example of a working hypothesis and its deduced consequences, leading to its partial rejection and the substitution of a new working hypothesis, it is evident that the scientific method is made complex by feedback mechanisms of both empirical observations and deductions. Let us now take another and closer look at the process by which a scientist generates a working hypothesis.

The Hypothetico-Deductive Process

In almost every field of science, explanatory hypotheses have evolved through numerous stages of deduction, testing, and revision. The scientist becomes familiar with accumulated hypotheses and empirical knowledge from student days—the content of professors' lectures, textbooks, journal articles and treatises, laboratory exercises, and, finally, the master's essay and the doctoral dissertation. Research early in one's career is commonly an extension of research carried on for years by the student's supervising professor. It is almost impossible to say which is arrived at first first in scientific research: empirical data gained by induction from experiment and observation, or an imaginative and creative hypothesis. Some modern interpreters of philosophy of science follow Karl Popper in placing emphasis on the latter phase; they characterize the scientific method as being a *hypothetico-deductive* process. Professor Francisco J. Ayala, a distinguished geneticist who has done extensive evolution research using the fruit fly, *Drosophila*, describes the hypothetico-deductive process of science as follows:

> Science is a complex enterprise that essentially consists of two interdependent episodes, one imaginative or creative, the other critical. To have an idea, advance a hypothesis, or suggest what might be true is a creative exercise. But scientific conjectures or hypotheses must also be subject to critical examination and empirical testing. Scientific thinking may be characterized as a process of invention or discovery followed by validation or con-

> firmation. One process concerns the acquisition of knowledge, the other concerns the justification of knowledge. (1977, p. 478)[1]

Ayala clearly gives the initial position to the creative act of formulating the hypothesis; validation through the deduction of consequences and their testing by empirical methods follows. But surely the hypothesis could not have been formulated in total absence of a rather large supply of knowledge, a supply that could only have flowed into the scientist's brain from outside sources, prior to the imaginative act.

Ayala pursues his point at greater length but then, finally, in his last sentence he gives inductive activity its just credit:

> Hypotheses and other imaginative conjectures are the initial stage of scientific inquiry. It is the imaginative conjecture of what might be true that provides the incentive to seek the truth and a clue as to where we might find it (Medawar 1967). Hypotheses guide observation and experiment because they suggest what to observe. The empirical work of scientists is guided by hypotheses, whether explicitly formulated or simply in the form of vague conjectures or hunches about what the truth might be. But imaginative conjecture and empirical observation are mutually interdependent processes. Observations made to test a hypothesis are often the inspiring source of new conjectures or hypotheses. (P. 478)[1]

Hypotheses and Prediction

Professor Ayala has set down a list of four different activities that are involved in evaluating the quality of a scientific hypothesis (p. 479). First, the hypothesis must display internal consistency. It must be logically well formed and must not contain self-contradictory statements. Second, it must have explanatory value. The hypothesis must call upon causes or forces that are at work to produce what is observed to happen. Merely rephrasing the problem by use of synonyms to create a tautology is not acceptable as an explanation. Third, the hypothesis must be evaluated to see if it really makes an advance in the state of knowledge as compared with existing hypotheses. For example, it must explain some observations that other hypotheses have not been able to explain, while at the same time being always consistent with basic laws of science that pertain. Fourth, the hypothesis must be vulnerable to being rejected through the use of empirical tests. The hypothesis must have the power to make "predictions" that can be put to such tests. Let us look into this fourth requirement in more detail, for its meaning must be understood.

Philosophers of science are prone to repeat that a hypothesis must carry the power "to make predictions." At first, I found it difficult to understand just what they meant by a "prediction." Most of us think of a prediction as a statement of some event that has yet to happen (i.e., a prognostication). Suppose that a scientist has devised this hypothesis: An earthquake is caused by a sudden slippage along a surface between two large masses of rock, releasing a large quantity of energy that has accumulated as a buildup of elastic strain, storing energy in increasing amounts over a long period of time. Let us suppose that testing of the hypothesis has produced scientific statements in agreement with the deduced consequences of the hypothesis, based on observations taken before and after several earthquakes. Using the hypothesis a scientist is then able to predict an earthquake in the future by pointing out those changes that can be observed as precursors to the event: for example, observed strain (bending) accumulating in rock on both sides of a fault.

Philosophers of science have another meaning for the word "prediction"; it is the ability of a hypothesis to lead to deductions of scientific statements that were not anticipated when the hypothesis was formuated. This meaning of prediction we have already dealt with in some detail in our account of the deduction of consequences by means of which a hypothesis can be strengthened or weakened. Those deduced consequences do not refer to future events; they deal with phenomena that are presently observable or have already occurred. Although this is what the science philosophers mean by "prediction," the word can mislead the general public. Most persons don't associate "prediction" with a statement of what has already happened. The dictionary defines "predict" as "to declare in advance, to foretell." What the philosophers of science really mean by *prediction* is to foretell what the scientist will find if he or she goes out and looks for what is deduced as a consequence of the hypothesis. When the prediction is of a past event or state of things that has not yet been discovered or verified, it is sometimes called a *retrodiction.*

The question is, must one or more deduced consequences always be associated with a scientific hypothesis? If the answer is yes, we must say: A statement that cannot produce a deduced consequence is not a scientific hypothesis. Karl Popper has made this pronouncement and it has been repeated by his students and followers for many years. In published arguments over whether organic evolution is a scientific hypothesis, we find the word "predictability" kicked around like a ball between opposing soccer teams. It is certainly fair to conclude that a scientific statement sufficiently complex to be called a hypothesis is almost always capable of yielding a number of deduced consequences that should be investigated by empirical methods.

Corroboration and Falsification

In reading various articles and published letters about the nature of science and how it differs from religion, I frequently came across the statement that a scientific statement or hypothesis must always be subject to the possibility of being *falsified* (whereas a statement of religious belief cannot be falsified). I immediately took the word "falsified" to mean that a scientist could put forward a scientific paper containing false statements—falsehoods, that is—in support of a hypothesis. The scientist, I assumed, could invent and record fictitious data from a laboratory experiment or a field study. I was puzzled because I felt sure that a religious leader or prophet could just as easily invent a false religious experience to gain support of followers and make new converts.

My puzzlement quickly came to an end when I realized that the verb "to falsify" can be misleading as used in the papers I was reading. Although *Webster's Ninth New Collegiate Dictionary* does define "falsify" as "to prove or declare false," it also gives alternative definitions: "to represent falsely (misrepresent)" and "to tell lies." It was the latter meaning that I was reading into the text.

A typical statement on the requirement of *falsifiability* and the process of *falsification* is this one given by Professor Ayala:

> The critical element that distinguishes the empirical sciences from other forms of knowledge is the requirement that scientific hypotheses be empirically falsifiable. . . . If the results of an empirical test agree with the predictions derived from a hypothesis, the hypothesis is said to be provisionally corroborated, otherwise it is falsified. . . . A hypothesis that is not subject, at least in principle, to the possibility of empirical falsification does not belong in the realm of science. (1977, p. 479)[1]

Ayala is evidently a faithful disciple of Popper, the philosopher largely responsible for publicizing these concepts of prediction and falsifiability. Ayala goes on to state:

> The requirement that scientific hypotheses be falsifiable rather than simply verifiable may seem surprising at first. It might seem that the goal of science is to establish the "truth" of hypotheses rather than attempt to falsify them. (1977, p. 479)[1]

My complaint with the Popper terminology arises from the observation that the words "verify" and "falsify" are absolutes. This quality of absolute certainty about a statement being correct or incorrect is exactly what I tried to throw out in my earlier discussion of the words "fact," "actuality," and "truth." Instead of such absolutes, I proposed a probability concept—the odds of being in error when a statement is declared to be true or false. I stick to that concept here, for it would be most brash and nonobjective for a scientist to declare a scientific hypothesis to be false or true, no matter how decisive the outcome of the empirical tests might appear.

Popper's requirement of possibility for falsification of a scientific hypothesis raises questions much more serious than merely one of definitions. I argue that, just as a scientific statement cannot be proven to be true, neither can it be proven to be false. The most that can be done as science is to evaluate or estimate the probability that a statement is false or that it is true. The two values of probability are complementary. If the probability that the statement is false is valued at 1 percent, the probability that the same statement is true is 99 percent.

We have here two complementary values: in Popper's terminology, they are falsification and verification. In my probability terminology, they are, respectively, a large probability of being in error versus a small probability of being in error. My system is open-ended. Popper's is closed. (More about this distinction follows.) I have no doubt his terminology will prevail; it is already deeply embedded in the literature of philosophy of science.

Questions of terminology aside, there is an important point to be made about the alternatives with respect to a given hypothesis: shall it be accepted or rejected? These two alternatives are not symmetrical in power. If a deduced consequence should, on empirical examination, be identified as predicted, the hypothesis itself would not stand as proven correct. The reason is that some alternative hypothesis might also stand to be verified by the same test data. But, if the deduced consequence should be shown to differ from observation, the hypothesis from which it derives would automatically be shown also to be incorrect. Note that I have used absolutes—"correct," "incorrect"—for the sake of the logic of the thing. We continue, using the language of logic.

For example, Hypothesis A leads to the formulation of four deduced consequences, C_1, C_2, C_3, and C_4.The consequences are investigated and found to exist as deduced. That the consequences are all true does not prove that the hypothesis is true, because there may be a second but quite different hypothesis, Hypothesis B, for which the same deduced consequences apply. Thus, tests that are passed by Hypothesis A simply permit it to stay alive. If a fifth deduced consequence, C_5, would prove not to exist when tested empirically, Hypothesis A would be destroyed. We conclude that a hypothesis can be shown false (can be falsified), but can never be shown true.

Converting back from "absolutes" to "probables," we can say that a deduced consequence affirmatively shared by two or more conflicting hypotheses can do little or nothing to improve the quality of those hypotheses. If, however, a particular deduced consequence required of a favored hypothesis is incongruous to other rival hypotheses, that particular consequence is crucial; it sets up the test whose result can drastically increase or decrease the probability that the favored hypothesis is correct. Crucial tests that are successfully passed will tend to downgrade or to eliminate rival hypotheses for which a different test outcome is predicted. Thus, one hypothesis can improve in quality as compared with rival hypotheses.

Scientists are sometimes tempted to push a strong hypothesis beyond the limits of relative and probable validity to an absolute statement, such as "corroborated (or established) beyond reasonable doubt" or "shown to be a fact." See how Professor Ayala pushes toward this limit in evaluating the hypothesis of organic evolution; note, however, that he inserts the word "reasonable" before "doubt," implying that the judgment is not absolute:

> The larger the variety of severe tests withstood by a hypothesis, the greater its degree of corroboration. Hypotheses or theories may thus become established beyond reasonable doubt. The hypothesis of evolution, that new organisms come about by descent with modification from dissimilar ancestors, is an example of a hypothesis corroborated beyond reasonable doubt. This is what is claimed by biologists who state that evolution is a fact rather than a theory or hypothesis. In ordinary usage, the terms "hypothesis" and "theory" sometimes imply a lack of sufficient corroboration. The evolutionary origin of organisms is compatible with virtually all known facts of biology, and has passed a wide variety of severe tests. (1977, p. 481)[1]

Evidence that evolution took place through geologic time rests largely on fossil remains of plants and animals in rock strata whose ages can be determined independently. The evolutionary "tree" establishes the order of appearance of each class of life form as a "branch" of the tree. To date, all fossils that have been found can be represented by a single-trunked tree with numerous branches. Biologists will point out, however, that the possibility always exists that fossils of a given life form will be discovered in the same strata with other forms that have heretofore always been limited to a different relative position and age. For example, it is conceivable that a scientist will some day discover human bones among dinosaur bones in such a relationship that it is judged highly likely that humans and dinosaurs lived at the same time. Such a finding would deal a crushing blow to the widely favored hypothesis of a unique evolutionary sequence. In Popperian language, the hypothesis of evolution would be falsified. So far, no incongruous relationship of this type has been found, but there remains the remote possibility that such a find will be made and, if so, its impact would be to greatly increase the probability that the hypothesis of evolution is incorrect.

Geologists and biologists don't seem very worried that a downgrading of the evolutionary model will occur, and some have gone so far as to state bluntly to the public and press that "evolution is a fact." When arrogance like that begins to take over from skepticism among scientists, they are inviting trouble. As one scientist has remarked, "Science is based upon skepticism" (Root-Bernstein 1981, p. 1446). The best way to practice skepticism is to look at every scientific hypothesis as having, from the moment it is formulated, a built-in probability of being in error. With such an outlook, the scientist is much less likely to fall in love with his or her hypothesis and become blinded to whatever defects may appear in it from time to time.

Naive Falsification

In recent years, the principle of potential falsification as the absolute criterion that every scientific hypothesis must satisfy has come under fire from philosophers of science. This topic is reviewed by science philosopher Philip Kitcher of the University of Vermont in his book *Abusing Science* (1982). His discussion is prompted by the fundamentalist creationists' claim that the theory of evolution is incapable of being falsified, and thus cannot be accepted as empirical science. Here we need to reexamine Popper's rather firm and inflexible declaration that for a statement to be a scientific hypothesis it must always be falsifiable. When couched in the strict language of formal logic, Popper's principle may seem obviously sound. Kitcher points to pitfalls in application of the now-traditional falsifiability criterion in empirical science (1982, pp. 42-44). He says:

> The time has come to tell a dreadful secret. While the picture of scientific testing sketched above continues to be influential among scientists, it has been shown to be seriously incorrect. (To give my profession its due, historians and philosophers of science have been trying to let this particular cat out of the bag for at least thirty years. See, for example, Hempel 1941; Quine 1952.) Important work in the history of science has made it increasingly clear that no major scientific theory has ever exemplified the relation between theory and evidence that the traditional model presents. (P. 42)[2]
>
> [Note: References for Hempel and Quine are given by Kitcher.]

The subject can be extremely difficult to grasp, unless one is highly skilled in the language of logic, but Kitcher provides a clear example of a naive application of the concept of falsifiability that could have done irreparable damage to Newtonian laws of gravitation and motion. Newtonian celestial mechanics proved remarkably successful in predicting the orbits of the planets—successful, that is, until observations were improved in precision to make it obvious that the orbit of Uranus consistently failed to follow the calculated orbit. Uranus was at that time the outermost known planet of the solar system. Using the *naive falsificationist criterion*, it would have been necessary to declare the Newtonian laws false, this in spite of their excellent predictive performance for all the other known planets. A less drastic alternative was, however, available. Another as yet undiscovered planet could be postulated to exist in orbit beyond Uranus; its orbit could be calculated in such a way as to account for the seeming discrepancy in the observed orbit of Uranus. Here, then, was a prediction independent of the major hypothesis itself, but embodying the same laws and principles. Not long after, the missing planet was spotted by telescope. Named Neptune, its observed orbit agreed with the calculated orbit. The main hypothesis was saved.

Perhaps the game of chess can serve as a device to show how science can stave off the fate of falsification. Your opponent has put your king in check (meaning that your king is vulnerable to being taken off the board in opponent's next move). Your opponent, seeing no way for you to get out of the situation, calls "Checkmate." Momentarily, you see no escape and seriously consider accepting defeat. Suddenly you see a way out: moving your knight between your king and opponent's queen can take you out of check. A few moves later, your opponent maneuvers you into another check; again you move a piece and save the situation. That move may result in loss of one of your pieces. This form of evasion can go on more or less indefinitely. In a similar manner, a complex scientific hypothesis involves many "chess pieces," each of which can be manipulated independently. Falsification can be avoided, perhaps indefinitely, by successive manipulations of only one piece at a time, or simply allowing one piece at a time to be taken off the board. As in the game of chess, science allows the hypothesis to be restructured, within certain rules, without limit. Thus naive falsification is circumvented and it may be futile to attempt to achieve it. But there is usually a final situation of checkmate in the game of chess. Is not an absolute and final falsification also possible during testing of a scientific hypothesis?

Are Laws of Science Falsifiable?

Kitcher asks us to consider the possibility that individual laws of physical science, treated apart from other laws, are neither testable nor falsifiable. He argues: "On their own, individual scientific laws, or the small groups of laws that are often identified as theories, do not have observational consequences" (1982, p. 44).[2] However, he does not explain to his readers just why this should be so. Take, for example, the essential assertion of the law of gravitation—that masses attract each other—and consider how you might wish to test it on the tentative assumption that it is a hypothesis.

You can, of course, repeatedly raise a paperweight to shoulder height and release it, noting that it always drops to the floor. This act would not, however, qualify as a test, because the predictive statement "it will drop every time" is not an independent hypothesis; it is merely a special case of the statement of the hypothesis itself. Why is the experiment not an independent test (i.e., an observational consequence) of the hypothesis? Perhaps, in terms of logic, the answer is that the law itself was derived as a generalization of repeated observations that unsupported small objects always fall through the air toward the earth's center (more or less) unless other directed forces of greater intensity (such as strong updrafts of air) exceed the downward pull of the earth's mass upon the small object. Granting that, as Popper and Einstein seem to be telling us, Newton's brilliant generalization "all masses attract each other" may have come as sheer free-floating inspiration, it is more realistic to suppose that observation of nature drove Newton to his conclusion. In that case, to repeat the observations that initially forced the universal statement creates a circularity of reasoning. As long as one cannot get out of the tight little circle—observation, induction, deduction, and return to the same observation—it is perhaps impossible to falsify the universal law.

More on Naive Falsifiability

This may be a good point at which to attempt further distinctions between universal laws, on the one hand, and hypotheses, on the other. "Laws," it is sometimes said, "are discovered, not made" (Hospers 1980, p. 106). The first point is, therefore, that a particular law "tolerates no alternative," so to speak. Hypotheses are synthesized within the human mind. In contrast to a law, a given hypothesis is an artifact that can be matched off to an alternative hypothesis (or several such alternatives) referring to the same set of observations.

Second and more important, perhaps, is that hypotheses most commonly are propositions designed to describe specific classes of events or states of matter. If relating to an event—a "happening"—the hypothesis tells us details of the specific succession of events that occurred. If relating to a specific state or structure of matter—the atomic structure of the diamond crystal, for example—the hypothesis describes a unique crystal lattice of carbon atoms. It may also tell us how the crystal lattice came into existence; that, too, was a event or happening. A third difference, yet to be presented, is that universal laws explain nothing about themselves, whereas hypotheses usually include explanation, and the explanation ultimately reaches down to the universal laws. Now that we have further elucidated (or obfuscated?) the status of laws as distinct from hypotheses, we are ready to return to Kitcher's thesis that complex and powerful theories (hypotheses) are not easily, if ever, completely falsified. He says:

> This crucial point about theories was first understood by the great historian and philosopher of science Pierre Duhem. Duhem saw clearly that individual scientific claims do not, and cannot, confront the evidence one by one. Rather, in his picturesque phrase, "Hypotheses are tested in bundles." (1982, p. 44)[2]
>
> [Note: No reference to Duhem is cited. Pierre Duhem (1861-1916) ranks as one of the founders of physical chemistry.]

Most major scientific hypotheses, Kitcher says, are "large bundles" of statements. Certainly this is true for organic evolution and plate tectonics. They include many smaller or lesser statements that are the prime targets for testing as independent hypotheses. If a lesser or secondary statement appears to be falsified by its disagreement with a test result, only that secondary statement need be eliminated or revised. We need not throw out the entire "bundle," for to do so would be to lose most of the knowledge it has gained for science. Successful testing of the larger portion of the bundle can preserve this knowledge.

Kitcher and those who share his views about falsifiability seem to be telling us to play down the Popper principle of falsification, which usually turns out in practice to be an unsupportable naive falsification, and to concentrate instead on something positive. Although logic can indicate that a hypothesis can never be proved true as a result of tests that turn out to be positive, surely the testing does more than just keep the hypothesis alive. Successful tests, involving as they do independent hypotheses, draw in new information. Added to the main hypothesis, the successes of test hypotheses increase the total scope and explanatory power of the main hypothesis. Popper grasped this idea in his "positive theory of corroboration" (1959, pp. 265-69). When a hypothesis stands up to numerous rigorous tests, it acquires a degree of *corroboration*, which is a quality of strength independent of considerations of whether it is falsifiable. Popper ties in the degree of positive corroboration with the level of logical probability, for he says: "we can also say that an appraisal of corroboration takes into account the logical probability of the statement in question" (p. 269).

As the knowledge content of the entire hypothesis is increased, the hypothesis will tend to unify that knowledge under a single set of laws and principles. Thus both organic evolution and plate tectonics have great internal unity and are remarkably successful in explaining highly diverse phenomena by a single set of basic statements. Kitcher points to another positive feature of successful major hypotheses—their fecundity, shown in the capacity to reveal new areas of research suggested by the independent tests.

Kitcher, in reviewing the newer views of the philosophy of science, has emphasized three positive characteristics of "successful science":

> *Independent testability* is achieved when it is possible to test auxiliary hypotheses independently of the particular cases for which they are introduced. *Unification* is the result of applying a small family of problem-solving strategies to a broad class of cases. *Fecundity* grows out of incompleteness when a theory opens up new and profitable lines of investigation. Given these marks of successful science, it is easy to see how sciences can fall short, and how some doctrines can do so badly that they fail to count as science at all. A scientific theory begins to wither if some of its auxiliary assumptions can be saved from refutation only by rendering them untestable; or if its problem-solving strategies become a hodgepodge, a collection of unrelated methods, each designed for a separate recalcitrant case; or if the promise of the theory just fizzles, the few questions it raises leading only to dead ends. (1982, p. 48)[2]

The Method of Multiple Hypotheses

Pursuing a single hypothesis and standing up for it under attacks from fellow scientists can have some unwanted and unpleasant side effects. As the inventor of the hypothesis, you are not very receptive to alternative hypotheses; you may develop some blind spots, which is to say that you tend to lose objectivity in carrying on your investigation. This is something seen all too often in the past as scientists, pitted against each other in defense of a favorite hypothesis, debated publicly. Often one debater threw in a personal attack on the character of the opponent, just as candidates for political office are prone to do.

To curb the tendency to lose objectivity, a scientist will do well to formulate as many different reasonable hypotheses as possible to apply to a given input of scientific data. Consequences are deduced for each hypothesis and tested accordingly. This approach has been referred to as the *method of multiple hypotheses.* It can be successful only if the scientist is genuinely devoted to problem solving, as compared with developing a public image as a warrior in defense of a cause. Scientists working as a team can practice this method by sharing and testing the hypotheses its members contribute.

The method of multiple working hypotheses was powerfully presented around 1900 by Thomas Crowder Chamberlin, a professor of geology at the University of Chicago. He is most widely known today for having collaborated with astronomer F. R. Moulton to produce a rather unique hypothesis of the origin of the solar system. That hypothesis enjoyed a good deal of popularity, perhaps because it was the first strong contender to replace the then-ruling hypothesis formulated by the French astronomer Laplace over a century earlier.

In 1904, Chamberlin reviewed his ideas about the method of multiple hypotheses in an address given before the International Congress of Arts and Sciences in Saint Louis (Chamberlin 1904). We can imagine him as a formidable figure, wearing cutaway coat and striped trousers as befitted a university professor in the European tradition. His language was appropriate to the formalities and excrescences of the Victorian era he had outlived.

Chamberlin first denigrated what he called "the method of the ruling theory," in which a scientist uses an array of what are deemed to be "facts" to arrive by an inductive process at a theory to explain those facts. Quoting from his address: "As soon as a phenomenon is presented, a theory of elucidation is framed. Laudable enough in itself, the theory is liable to be framed before the phenomena are fully and accurately observed" (1904, p. 68). He went on to show how such a theory begins to dominate the thoughts and actions of its inventor, who tends to ignore new lines of evidence that should be used to test the theory:

> Soon also affection (for the theory) is awakened with its blinding influence. The authorship of an original explanation that seems successful easily begets fondness for one's intellectual child. This affection adds its alluring influence to the previous tendency toward an unconscious selection. The mind lingers with pleasure upon the facts that fall happily into the embrace of the theory, and feels a natural indifference toward those that assume a refractory or meaningless attitude. Instinctively, there is a special searching-out of phenomena that support the theory; unwittingly also there is a pressing of the theory to make it fit the facts and a pressing of the facts to make them fit the theory. When these biasing tendencies set in, the mind soon glides into the partiality of paternalism, and the theory rapidly rises to a position of control. Unless it happens to be the true one, all hope of the best results is gone. The defects of this method are obvious and grave. (p. 68)

Chamberlin mentioned no examples, but perhaps he had in mind a ruling theory prevalent in his time. Lord Kelvin (Baron William Thompson), a leading English physicist, strongly endorsed the theory put forward by astronomers and philosophers of the eighteenth century that the earth as a planet came into existence from a mass of high-temperature gas that condensed from a gaseous state into a molten state, then cooled to reach a solid state. Kelvin calculated the age of the earth on the basis of this ruling theory and concluded that an age of 20 to 40 million years was a reasonable figure. This pronouncement was a devastating blow to Charles Darwin, whose ruling theory of evolution required much more time than Kelvin's could allow. The story of how Kelvin's theory was demolished by the discovery in the early 1900s of the phenomenon of radioactivity is particularly fascinating because that discovery not only revolutionized all thinking about the age of the earth and the forces that power internal geologic processes, it also gave new life to Darwin's scenario of the evo-ution of species.

The method of the ruling theory has not gone out of style, despite Chamberlin's oration against it. We have a good example of it in the field of modern anthropology. An American school of cultural anthropology, led by Franz Boas, emerged during the 1920s and flourished for decades thereafter. Boas formulated a theory to the effect that each primitive group of humans develops its particular form of social behavior, including customs, mores, and religious practices, entirely independent of heredity; that culture is not under biological control of genes. Of those who sought evidence to support this "nurture theory," the person best known to Americans was Margaret Mead. Through her personal observations of a primitive society in Samoa, begun in the 1920s, she attempted to document Boas's theory of culture. You should be aware that the nurture theory had arisen to replace an extreme theory of genetic control over human behavior that flourished in the early 1900s. That biological theory turned sour when some of its proponents advocated the improvement of the human species by elimination of persons supposed to be genetically inferior.

The ruling nurture theory of Boas was strongly challenged in 1975 by Harvard biologist Edward O. Wilson, a specialist in insect societies. In a book titled *Sociobiology; The New Synthesis* (Wilson 1975) he revived the long dormant "nature theory" that biological forces acting through genetics control (to some extent at least) the culture patterns of any given primitive group of humans (Rensberger 1983). Human behavior thus comes under biological guidance by an adaptive process of evolution forced by changes in culture. Adding fuel to the fire of controversy, field research methods used by Margaret Mead were strongly attacked in 1983 by anthropologist Derek Freeman and the debate was in full swing (Marshall 1983). Will a new ruling theory displace the current one? Will "nature" replace "nurture" as the paradigm of modern cultural anthropology?

As Professor Chamberlin continued his address on methods in science, he turned to the method of the multiple working hypotheses. Here his eloquence surged to new extravagances of expression. We quote selectively from his text:

> The effort is to bring up into distinct view every rational explanation of the phenomenon in hand and to develop into working form every tenable hypothesis of its nature, cause or origin, and to give to each of these a due place in the inquiry. The investigator thus becomes the parent of a family of hypotheses; and by his paternal relations to all is morally forbidden to fasten his affections unduly upon any one. In the very nature of the case the chief danger that springs from affection is counteracted.

> The investigator thus at the outset puts himself in cordial sympathy and in the parental relations of adoption, if not of authorship, with every hypothesis that is at all applicable to the case under investigation. Having thus neutralized, so far as may be, the partialities of his emotional nature, he proceeds with a certain natural and enforced erectness of mental attitude to the inquiry, knowing well that some of the family of hypotheses must needs perish in the ordeal of crucial research, but with a reasonable expectation that more than one of them may survive. . . . Honors must often be divided between hypotheses. (1904, pp. 69-70)

Is it conceivable that if the culture school to which Margaret Mead belonged had used the method of multiple hypotheses, we would have seen the simultaneous testing by that school of two or more working hypotheses, each with its own set of deduced consequences to guide programs of field research? Must science continue to be, as in the past, a succession of title bouts in which one boxer after the other takes to the ring to dethrone a champion? Perhaps it is this very combative system that brings new blood into the study of science. If the champion of a ruling theory had, by chance, not become a scientist, would he or she have instead become a champion in the combat ring of politics or industry? Would the evenhanded application of the method of multiple hypotheses favor as its recruits those with less drive for success and less ability to achieve it? Interesting questions!

Like most touted prescriptions for doing science "the right way," the method of multiple hypotheses puts forward the "ought" with little attention as to whether anyone is actually making it the "is." One critic of Chamberlin's oratory, J. G. Johnson, a geologist, considers that it is unrealistic to expect scientists to use the method, because of the wasted time and effort in carrying out laboratory testing of all the predictions they would yield (Johnson 1990, p. 44). He describes the method as an outmoded "chimera," a position perhaps as far out in left field as Chamberlin's is in right field. Formulating at least one reasonable alternative hypothesis, or even two, requires only a few minutes of mental effort and most investigators do this routinely. Any number of highly improbable explanations can be quickly generated and as quickly dismissed for good reasons; they require no expensive, time consuming laboratory or field tests. For each reasonable hypothesis and for most of the key deductions they generate, empirical data for at least preliminary testing are already available in print. Considering viable alternatives, even if they are never mentioned in print, is simply part of assuming a self-protective stance.

Keep also in mind that multiple hypotheses include those put forth by others in competition with yours. If you can devise a single test that shows promise of falsifying your opponent's hypothesis but not yours, you would be a fool not to carry out that test. The method of multiple hypotheses can be carried out by multiple investigators, each paired to one hypothesis. Chamberlin does not mention this arrangement, but it could only intensify the beneficial effects he ascribes to multiplicity of hypotheses under the private consideration of only one person.

Scientific Revolutions

We have seen that, as explained in Philip Kitcher's description of the fate of scientific theories, a widely held theory may begin to wither in the glare of new information that it lacks the power to explain. When this happens, it becomes vulnerable to dramatic collapse in the face of a new and more powerful theory. The king is dead; long live the king! The community of scientists then rallies around the new theory, giving it their almost undivided support and research effort.

Such is the scenario envisioned by science philosopher Thomas S. Kuhn, whose seminal work, *The Structure of Scientific Revolutions*, was published in 1962 and quickly caught the attention of the scientific community. Kuhn brought into widespread use a word that had little previous exposure among scientists and, for that matter, throughout the public in general, whereas today it is a buzzword freely tossed about in erudite discussions on almost any subject.

Kuhn chose *paradigm* to designate a scientific theory of great breadth and explanatory power within a particular area of science. A paradigm can be equated with what we have referred to earlier as a ruling theory. The word carries strong implications of "pattern" in the sense that it is the general plan to be followed in more specific applications. In *Webster's Third New International Dictionary*, the word is illustrated by anthropologist Margaret Mead's phrase "mistaken the paradigm for the theory." This suggests that paradigm and theory are on two levels, the former more general than the latter; yet this distinction disappears when Kuhn cites as examples of paradigms such great scientific theories of the past as geocentricism (Earth at the center of the universe), replaced by heliocentricism (the Sun at center of planetary orbits). Other examples of paradigms given by Kuhn are the oxygen theory and the electro-magnetic theory; these are highly specific theories (1962, p. 150). I suggest that you not worry too much about fine distinctions between "paradigm" and such more easily understood terms as "dominant theory" and "ruling hypothesis."

As Kuhn sees the history of science, it is punctuated by crises that are resolved by *scientific revolutions.* A crisis is preceded by a period of decline in the then-current paradigm, as more and more new observations appear that are not accommodated by that paradigm. Conflicting observations of this type are described by Kuhn as *anomalies.* Despite these indications of approaching moribundity, the scientists remain loyal to their paradigm. In Kuhn's words:

> Though they may begin to lose faith and then to consider alternatives, they do not renounce the paradigm that has led them into crisis. They do not, that is, treat anomalies as counterinstances, though in the vocabulary of philosophy of science that is what they are. (p. 77)

Kuhn then goes on to say that "once it has achieved the status of a paradigm, a scientific theory is declared invalid only if an alternate candidate is available to take its place." Kuhn's use of the words "lose faith" and "renounce" lead to a new insight into his intended meaning of "paradigm"; it is a theory that has gradually come to involve belief (i.e., the theory carries the truth). Emotion has become involved in holding on to the dying paradigm. Ultimately, however, it is relinquished. Kuhn explains:

> The decision to reject one paradigm is always simultaneously the decision to accept another, and the judgment leading to that decision involves the comparison of both paradigms with nature and with each other. (p. 77)

This sounds like a tough and objective approach to the decision problem, the way we like to think scientists act. But elsewhere Kuhn seems to be emphasizing that belief and emotion are heavily involved in solving a crisis:

> The transfer of allegiance from paradigm to paradigm is a conversion experience that cannot be forced. Lifelong resistance, particularly from those whose productive careers have committed them to an older tradition of normal science, is not a violation of scientific standards but an index to the nature of scientific research itself. . . . And it is only through normal science that the professional community of scientists succeeds, first, in exploiting the potential scope and precision of the older paradigm and, then, in isolating the difficulty through the study of which a new paradigm may emerge. (P. 150)

I see in the last sentence the idea that the new paradigm always arises from the corpus of the previous one; it could not arise independently from some outside field of knowledge. Previous knowledge is not simply thrown out as garbage. Instead, it is largely recycled and restructured.

What Kuhn says about crises and revolutions in science certainly applies to my own field, geology. Plate tectonics was the new paradigm, replacing one holding that crustal movements through geologic time have been predominantly vertical (up and down). We thought of mountains as moving more or less straight up along with upward movements of molten magma. Crustal blocks could also move down in fault basins. Erosion constantly tended to lower all exposed surfaces. The paradigm of the time had no name, but today we call it "vertical tectonics" to distinguish it from the "horizontal tectonics" of plate motions. True, there were some very nasty problems to fit into vertical tectonics. Crumpling of once-horizontal strata into tightly folded structures in the European Alps and elsewhere clearly required great horizontal motions. These structures were the anomalies Kuhn describes. They were, however, treated as difficulties to be solved without abandoning the paradigm of vertical tectonics, whereas we see them now as counterinstances that could easily have served to falsify the theory.

The old paradigm endured for over a century, but it never worked very well. On reflection, it perhaps lacked the substance needed even to identify it as a paradigm. When Alfred Wegener proposed his theory of continental drift, the precursor of a revolution began to appear, but in it there was not really a new paradigm to contest the old one. As new information poured in about the crust of the ocean floors, the pressure for a new paradigm began to intensify sharply. The information logjam was set free in remarkably short order by a coherent new paradigm, which we described on earlier pages. Rather than encountering what Kuhn considers the typical slow and reluctant acceptance of new paradigms, plate tectonics swept the fields of geology and geophysics with mass conversions of its scientists. With incredible dispatch, these scientists fitted almost every category of secured knowledge of the old paradigm into the new one, and amazingly enough, nearly all of that knowledge was retained in its restructured setting.

Think about Darwin's theory of evolution through natural selection in the context of crises and revolutions. Surely the theory of evolution was a new paradigm, but did it displace an earlier one? Two quite distinct bodies of theory were then prevalent, either of which might be taken as a prior paradigm. One, which philosopher/biologist David L. Hull has called *idealism*, was an obscure secular theory that explained "natural phenomena in terms of timeless general patterns or 'archetypes' " (Hull 1988, p. 41). It was a concept that had gained a considerable following in Germany and France, but Darwin seems to have steered clear of direct confrontation with its supporters. We will focus, as did Darwin, on the long-prevailing religious belief that species are divine creations through miracles.

Evolution was the naturalistic (secular) alternative to catastrophism, whether as a series of divine creations alternating with catastrophes, or a single creation followed by the catastrophic Flood of Noah. The latter view was a program revealed through the Scripture. Thus the creation/

catastrophe view of life on earth was not science in the sense that we define it today (i.e., empirical science); instead, it was religious dogma, depending on faith in the supernatural. In this sense, it is not eligible for consideration as a scientific paradigm.

Recognizing the religious dogma of creation/catastrophe as simply a historically prior explanation stated in a nonempirical form of knowledge, we can perhaps conclude that Darwin's theory was the one and original paradigm of what Kuhn calls "normal science." For every field of what is now normal science there was a first paradigm; before it was the "pre-paradigm" period, as Kuhn calls it (1962, p. 162). Kuhn says that whereas individuals practice science in the pre-paradigm era, "the results of their enterprise do not add up to science as we know it." The idealism theory of archetypes can thus qualify as being pre-paradigmatic. Individuals were practicing good descriptive science in the field of biology long before Darwin's theory hit the world in 1859. Plants and animals, both living and fossil, had been described and classified in great detail in the previous hundred years. What was lacking, Kuhn suggests, was evidence of progress. By that, I presume he means absence of the coherent and universal explanation that the first paradigm was to provide. Lacking a paradigm—which is to say, lacking a powerful and fecund hypothesis, or theory—the mechanism for deducing consequences and performing tests was lacking. Without such stimulus the science could not easily grow in new directions.

While Kuhn's view of crises, revolutions, and paradigms had enormous impact among science philosophers and upon the practitioners of science, there were some sticky side effects, arising from some of Kuhn's characterizations of the way the science community carries on its business. Kuhn's picture of the way the system works is not exactly flattering to scientists (p. 163). Working in isolation, directing efforts solely to an audience of colleagues, and selecting only problems that appear readily solvable are among the features cited that make science look more self-serving than perhaps it should be to maintain a pure and lofty public image.

What seems to have happened is that critics of science have used Kuhn's description of the scientific establishment itself (as distinct from the theory of crises, revolutions, and new paradigms) to attack the scientists in areas of morality and ethics, the implication being that scientists as a group are generally behaving in a socially undesirable manner. Philip Kitcher sensed this reaction and sought to correct its effects:

> Thomas Kuhn's book *The Structure of Scientific Revolutions* has probably been more widely read—and more widely misinterpreted—than any other book in the recent philosophy of science. The broad circulation of its views has generated a popular caric-

> ature of Kuhn's position. According to the popular caricature, scientists working in a field belong to a club. All club members are required to agree on main points of doctrine. Indeed, the price of admission is several years of graduate education, during which the chief dogmas are inculcated. The views of outsiders are ignored. Now I want to emphasize that this is a hopeless caricature, both of the practice of scientists and of Kuhn's analysis of the practice. Nevertheless, the caricature has become commonly accepted as a faithful representation, thereby lending support to the Creationists' claims that their views are arrogantly disregarded. (1982, p. 168)[2]

We seem to have strayed a bit from the historical matter of paradigms and scientific revolutions into the sociology of science, a fascinating subject dealt with at some length in Chapter 6. In our next chapter we will be right back on course, investigating science.

Credits

1. From Francisco J. Ayala, Chapter 16, pp. 474-516 in T. Dobzhansky, F. J. Ayala, G. L. Stebbins, and J. W. Valentine, *Evolution*, W. H. Freeman and Company, San Francisco. Copyright © 1977 by W. H. Freeman and Company. Used by permission.

2. From Philip Kitcher, *Abusing Science: The Case Against Creationism*, The MIT Press, Cambridge, Mass. Copyright © 1982 by The Massachusetts Institute of Technology. Used by permission.

Chapter 4

The Complex/Historical Sciences

Science, like an oriental carpet, has great strength because of the common threads that form the warp and woof. What we see, however, is a great variety in the patterns that result from the use of various kinds and lengths of fibers held in place by those common threads. Each branch or area of science has its special character, leading to different patterns of investigation and different modes of statement.

The Fields of Empirical Science

Classifying the fields of science and arranging them in some sort of hierarchy of generality and specificity is an exercise in frustration, but it can lead to productive introspection about the nature of science. Figure 4.1 is my attempt to make some order and perhaps to provide some understanding of the intrinsic nature of the categories and classes of knowledge that science contains. The schematic includes all members of the family of empirical science. Perhaps you notice right away that mathematics and logic are missing. A full explanation of their absence will come in Part Two. They are knowledge fields not based in observation of external reality; instead they are based in concepts originating in the brain and are inhabitants of what we will refer to later as the transempirical, or ideational realm.

Our schematic contains a hierarchy of four classes (Roman numerals). Class I is all *empirical science*; it consists of Class II, containing *pure science* and *applied science* as mutually exclusive subclasses. Pure science is conducted without the necessity of having any normative or prescriptive function. Applied science is overtly and blatantly normative and prescriptive, meaning that it assigns an external or ulterior purpose to investigation and funnels it either toward a moralistic goal—to do good or evil—or just to make money (which may entail good or evil). Applied science accomplishes the "oughts" that humans believe should follow upon the "is"es of pure science. Applied science is thus a collection of weapons forged by humans to achieve human goals.

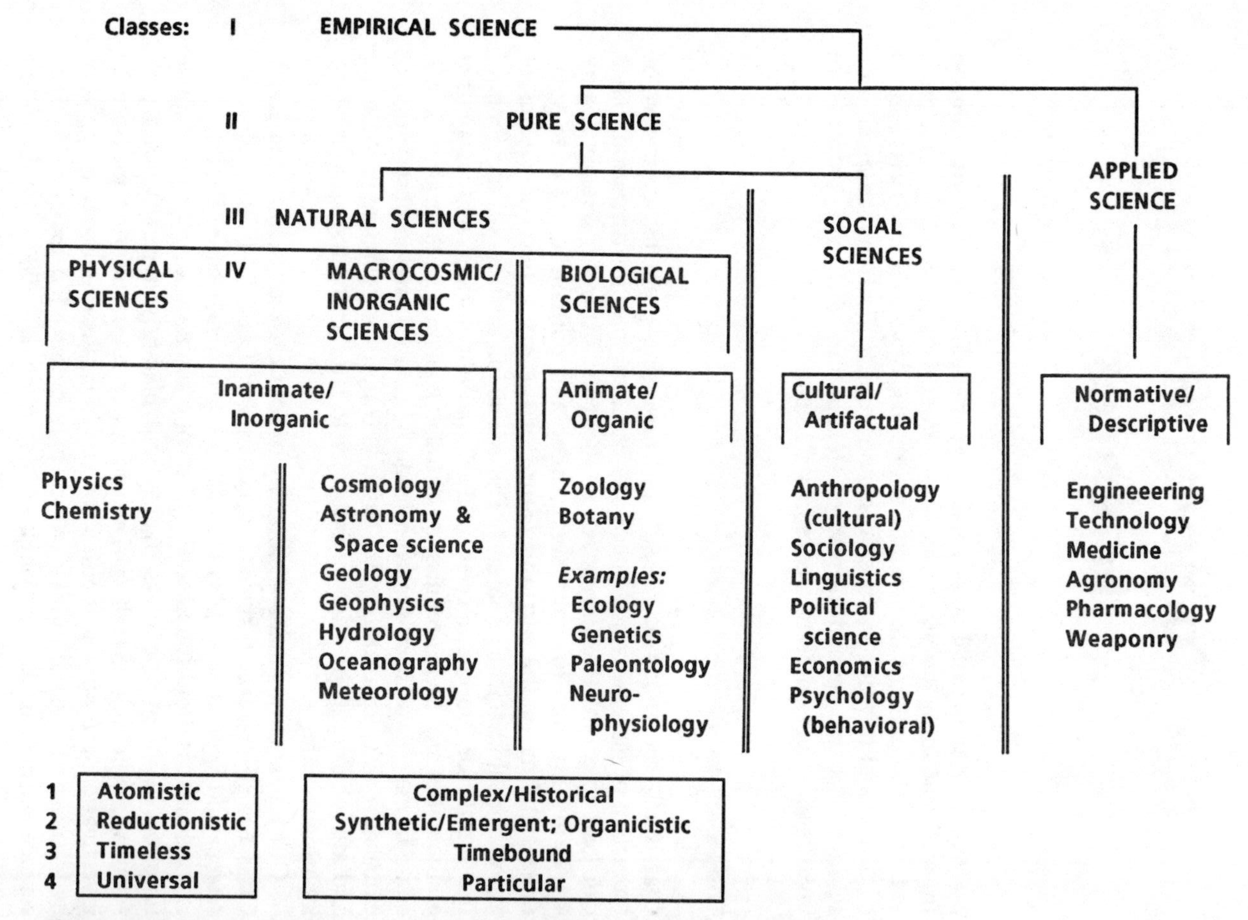

Figure 4.1 A flawed attempt to organize and classify the numerous branches and twigs of empirical science.

The aura of moral imperative envelopes much of applied science, and while many persons find this aura positive, exciting, and stimulating, a few others find it neutral, negative, or disturbing. Funding of pure science has much or most of its source in the shops of applied science, so that the word "pure" may be just a laughable euphemism.

Class III has two members: the *natural sciences* and the *social sciences.* The latter is separated from the former by a fairly satisfactory boundary, or criterion. (Vertical double rules indicate epistemic/ontic barriers.) The social sciences are human-related, being concerned with that which is cultural, and hence artifactual. These artifacts include the empirical content and history of economic, political, and social structures. Psychology is traditionally placed in the social sciences but, like anthropology, it contains biological content (neuropsychology) clearly belonging in the biological sciences. These components need to be carefully sorted out.

At level IV lie the subclasses of natural science. The *physical sciences* are usually set apart from the other natural sciences on grounds that they seek to understand the basic nature of matter and energy and their interactions; whereas the others study complex and unique systems of matter and energy that occupy specified and fixed positions in time and space. The physical sciences, consisting of traditional and modern (quantum) physics and chemistry (physical/inorganic), are concerned with general (universal) laws that can be expressed mathematically or by other symbols. The method of the physical sciences is almost entirely experimental, including not only laboratory experimentation but also repeated instrumental observation in a natural environment. The physical sciences are strongly quantitative, meaning that they collect numerical data arising from closely controlled experiments and express their laws in the form of mathematical equations. Astronomy (including cosmology) has traditionally been named as a physical science, but I challenge that assignment.

One of the other two branches of natural science, the *biological sciences* (biology, or life sciences), poses no serious difficulties of classification and content. Naming the second, however, is a tough assignment. It was traditionally named the *earth sciences* and a half-century ago consisted of geology, oceanography, and meteorology. But Earth is only one of several planets, and there are other kinds of sun-orbiting objects. Add to this that there are no sufficient generic grounds to exclude stars, or even galaxies and all other cosmic structures from this class. So both astronomy and cosmology can be allowed to come into the tent, and the spatial scale becomes cosmic. No single, simple name will encompass this menagerie, and I have suggested the class title *macrocosmic/inorganic sciences.* "Macro" distinguishes the scale of its relatively large objects from those of atomic and quantum physics; "inorganic," from the organic (animate) contents of biological science.

Actually, the natural sciences at the level IV comprise a three-ring circus, with shared properties relating the first and second rings but excluding the third, and others relating the second and third but excluding the first. These relationships are shown on the diagram by two sets of horizontal brackets and two vertical barrier symbols. Two criteria are expressed by the brackets: The inanimate (inorganic) character of both the physical sciences (PS) and macrocosmic inorganic sciences (MIS) distinguishes them as a pair from the biological sciences (BS), which are animate (organic) in character. (Numerous laboratory synthesized "organic" molecules are not found in organisms; hence the preference for "animate.") As shown below, PS has fundamental differences from both MIS and BS. These are shown in tabular form below:

PS	MIS and BS
(1) Atomistic	Complex/historical
(2) Reductionistic	Synthetic/emergent (organicistic)
(3) Timeless	Timebound
(4) Universal	Particular

(1) Borrowing from the Greek atomists (*atomos*, indivisible), the concept contained in the term *atomistic* as descriptive of PS is emphasis on a hierarchy of particles, one set contained within the other. Starting with the atom as composed of sets of electrons and nuclear particles (neutrons and protons), the hierarchy descends to the quantum levels of forces and masses, all being particulate, i.e., leptons and hadrons, weak and strong forces; quarks and messenger particles. For MIS and BS, *complex/historical* combines two attributes. The complexity of molecular structure and of larger structures made of molecules is obvious. The historical quality is covered under (3).

The physical structures with which MIS and BS deal are enormously complex aggregations of atoms and molecules. A single organism may contain millions or billions of molecules, while many kinds of organic molecules contain thousands of atoms. Even a single living cell is many orders of magnitude more complex than one of the atoms or molecules it contains. A chunk of rock contains millions or billions of atoms, often of many different elements, locked into geometrically complex systems of chains, layers, and lattices. The same statement about complexity applies to huge objects that make up the universe: stars, galaxies, gas clouds, and black holes.

(2) *Reductionistic* in PS refers to the analytical process of taking a whole structure apart and laying out all of its discrete components. *Synthetic/emergent* for MIS and BS carries the message of a reverse process of building of larger structures from smaller ones, along with a

quality of uniqueness that has its origins in history of the universe. These qualities are expressed in the noun *organicism*, defined in *Webster's Ninth New Collegiate Dictionary* as "the explanation of life and living processes in terms of the levels of organization of living systems rather than in terms of their smallest components." The adjectival form, antithetical to reductionistic, would be *organicistic.* (Reductionism is discussed later in this chapter.)

(3) *Timeless* in PS emphasizes the repetition of changes (events) free of any absolute framework in time. *Timebound* in MIS and BS emphasizes the requirement of specified time (absolute) in specified space in every description of events that produce or change complex structures (also discussed later in chapter).

(4) *Universal* in PS describes the universal nature of many or most of its propositions in logical terms; i.e., the universal affirmative form "All . . . are . . . ." *Particular* in MIS and BS identifies most of its propositions as being of the class of particular affirmatives, "Some . . . are . . ." Here, the underlying idea is that PS yields a collection of universal laws, whereas MIS and BS generally does not yield any laws, because the probability of exact repetition of any complex and timebound event is negligibly small (discussed below).

The Social Sciences—The "Soft/Hard" Debate

Placing the social sciences (SS) in comparative isolation from the natural sciences within the class of pure sciences may require some explanation, or at least discussion, of that position. Perhaps the first problem that comes to mind is whether segregation of the social sciences (SS) from the natural sciences (NS) solely or largely on grounds of the artifactual nature of the content of the SS is warranted. In Chapter 9, under the discussion of human history, we will try to show that it is futile to attempt to separate products of human making from natural products in any meaningful way. One example we give is that of chemically synthesized organic compounds, created in the chemical laboratory and otherwise unknown in nature. Surely, we wouldn't move organic chemistry over into the SS column for that reason. As to psychology, we have qualified it in the SS column as being "behavioral," meaning that there is school of psychology that explains all human behavior as a cultural acquisition, entirely free of genetic controls. We have inserted "neuropsychology" as its counterpart in the natural sciences to express the neurological functioning of the brain and nervous sytem in connection with human (and other animal) behavior.

We couldn't get very far with moving the list of social sciences into the biological sciences (BS) column, even though it is understood that humans are biological things—animate and organic. Perhaps there is some historical reason for a protest to such a proposal, and that is precisely what we find in a long drawnout reluctance by the natural scientists even to accept the SS into the fold of empirical science.

My dictionary (*Webster's Ninth New Collegiate*) has for its first definition of SS: "a branch of science that deals with the institutions and functioning of human society and with the interpersonal relationships of individuals as members of society." Perhaps the words "interpersonal relationships" give a clue as to the deepseated reluctance of members of the self-designated "hard sciences" to accept their colleagues in what by contrast are referred to with intended denigration as the "soft sciences." Thus taunted, the social scientists fight back fiercely. An excellent and highly readable article on this feud appeared in 1987 in *Discover*. Its author, Jared Diamond, is a physiologist on the faculty of the UCLA Medical School. His field of membrane physiology lies at the "hard" end of biology. His title bears his theses: "Soft Sciences Are Often Harder Than Hard Sciences." What makes Diamond's analysis particularly credible is that his wife, Marie Cohen, practices clinical psychology, a field considered as "very soft" science. The adjective "hard" refers to the use of mathematics; i.e., to a highly quantitative science, as in the physical sciences, and (for many) includes mathematics itself. Diamond convincingly explains how mathematics, including statistical testing of quantitative sample data, is today widely used in many areas of the social sciences. He scores a direct hit by saying: "Ecology and psychology and the social sciences are much more difficult and, to some of us, intellectually more challenging than mathematics and chemistry" (p. 39). I can go him one better by naming as more challenging all of the complex/historical sciences in addition to ecology and psychology. Virtually every one of these complex/historical fields is today strongly infused by mathematics and mathematical statistics.

There remains for many observers, however, the lingering worry that social scientists may unconsciously or unintentionally infuse their hypotheses of interpersonal relationships with value judgments ("oughts," that is) that belong in the class of ideational knowledge. The temptation to make this infusion would perhaps be strongest in sociology and political science, for there the sociopolitical ideologies under examination are expressions of human ethical and moral values. Perhaps the strongest pressure to avoid such mixing of perceptional and ideational forms of knowledge comes from the social scientists themselves. Self-policing by peers within each category of science is one of the great strengths of science.

The Role of History in Natural Science

Besides dealing with highly complex structures, the macrocosmic/ inorganic and biological sciences (MIS/BS) investigate complex time sequences of events, something we do not find in basic physics and chemistry. Examples are the life cycle of an individual of a plant or animal species, the evolution of life on earth over hundreds of millions of years, or the birth and life history of a star or a galaxy. Thus the quality of history pervades MIS/BS.

History in MIS/BS commonly takes the form of a sudden initiation (origination) of a system, followed by a much longer period of progressive change of the system (evolution). We are, of course, using the word "initiation" in a purely naturalistic sense, entirely devoid of any suggestion that a supernatural agent is involved. We can think of initiation as a relatively rapid reorganization of matter into a new system, derived from the substance and energy of an existing system but nevertheless uniquely different. An example would be the formation of a star by the rapid gravitational collapse of a diffuse mass of dust and gas. There follows evolutionary change in the star over vastly longer spans of time than its initiation required. As used here, "evolution" carries no implication that the change is directed to some ultimate purpose or goal; the change simply occurs through the action of physical laws. A planet, such as our Earth, seems to have experienced a rapid initiation by gravitational collapse of dust and gas to form a solid sphere. This event is estimated to have taken about 200 million years, whereas the physical evolution of its internal structure has lasted over more than 4 billion years. An old star may end its long life by a cataclysmic event—an explosion, or supernova—that initiates a new kind of structure, a neutron star, and a spreading cloud of gas and dust.

In biological evolution, an important hypothesis states that a new class of life—an order, family, or species—appears rather rapidly or even suddenly. This brief initiation phase is followed by a much longer period of smaller or minor changes, as organisms respond more or less gradually to environmental influences and other factors that control changes in gene composition. Among geneticists, this style of evolutionary tempo is called *punctuated equilibrium*. The term was introduced by paleontologists Niles Eldredge and Stephen Jay Gould, but expresses an idea that has been around since the 1940s. I like the wording used by geneticist G. Ledyard Stebbins to describe this tempo. Referring to the evolution of the human brain, he says that increase in brain size took place in "quantum jumps followed by periods of relative stability" (Stebbins 1982, p. 357).[1] He also uses the term "quantum bursts" for the events of rapid evolutionary change (p. 358).

Although laws of the physical sciences (PS) describe the kinds of activity associated with change in pure forms of matter and energy, the universal statements that represent those laws are not uniquely positioned in time. Professor Walter Bucher, an American geologist, referred to this form of knowledge as *timeless knowledge* (Bucher 1941, p. 1). In contrast, knowledge of the origin and evolution of systems studied in MIS/BS he would have described as *timebound knowledge*.

Looked at in a somewhat different light, Bucher's timeless knowledge consists of statements that describe events or relationships with an extremely large probability of being repeated. The radioactive disintegration of the nucleus of the uranium atom occurs somewhere in the universe in exactly the same way countless times each hour, minute, and second. Countless streams of photons are emitted in exactly the same way throughout the universe; they travel through space constantly, all with the same speed. The probability P of exact repetition of the phenomenon thus approaches unity, or 100 percent ($P \rightarrow 1.0$). When, however, we look at the evolution of life on earth, as displayed in the fossil record, we are viewing a long, complex chain of events that took place in a fixed span of time. As an entirety, the chain is unique in its structure; it is almost inconceivable that an identical chain of structures and events will ever occur on any other planet in the universe, or has ever occurred. What we are really saying is that the probability of the particular evolutionary chain being repeated in exactly the same way approaches zero ($P \rightarrow 0.0$). I find satisfaction in this application of the theory of statistical probability to the distinction between PS and MIS/BS. It focuses attention on the uncertainty that attaches to every scientific statement, whether it be simple or complex, general or specific.

In presenting the above analysis in 1954 to a session of the National Academy of Sciences, I added the following terminology: Historical investigation can be defined as referring to the analysis of complex states of matter of *low recoverability*; investigations of general laws of physics (dynamic investigations) can be defined as referring to states of *high recoverability* (Strahler 1954). These terms fit well with Boltzmann thermodynamic principles of uniform/nonuniform states of distributions of gas molecules in an isolated system.

Do the Complex/Historical Sciences Generate Laws?

Do the macrocosmic inorganic and biological sciences (MIS/BS) have laws, such as those of the physical sciences (PS)? For some scientists an answer of no would be sufficient to banish these branches entirely from the fold of natural science. In biology certain natural organic processes can be reproduced rather closely by laboratory experiments under controlled conditions. Examples would be numerous in the reproductive cycling of plants and animals. Mendel's experiments with hereditary characteristics of pea plants would perhaps be an example. One might consider this activity as an illustration of timeless knowledge. Both in biology and geology statements called "laws" have been made about the typical outcomes of specific processes. That such statements should be considered laws comparable to those in physics is a debatable question.

Evolutionary biologist Ernst Mayr has evaluated the status of so-called laws in biology and, particularly, in evolution (1982, pp. 37-38). He refers to them as generalizations that describe regularities or trends but stops short of referring to them as "laws." The reason, he notes, is that these statements tell only what is most commonly the case or most likely to occur: "Generalizations in biology are almost invariably of probabilistic nature. As one wit has formulated it, there is only one universal law in biology: All biological laws have exceptions" (pp. 38-39). Similar statements would apply to generalizations made about repetitious processes and events in geology and other branches of inorganic earth science. The element of chance variation is always present in the real world of nature and usually it is a very important element. We will discuss this subject on later pages. What we need is a probabilistic model of nature that accommodates the range of chance variations we expect to find.

As to timebound knowledge, referring to complex historical sequences of events and changes that actually occurred at a particular place during a specified segment of time, the possibility of formulating laws is rather more formidable than for timeless knowledge. If something happened only once in all of cosmic time can it be described by a law? Not likely, unless you can give good reasons to predict that there will be repetitions of the same thing in the future. George Gaylord Simpson, as a paleontologist, had one foot in biology and the other in geology; he took a dim view of historical laws:

> The search for historical laws is, I maintain, mistaken in principle. Laws apply, in the dictionary definition "under the same conditions," or in my amendment "to the extent that factors affecting the relationship are explicit in the law," or in common parlance "other things being equal." But in history, which is a sequence of real, individual events, other things never are equal. Historical events, whether in the history of life, or recorded human history, are determined by the immanent characteristics of the universe acting on and within particular configurations, and never by either the immanent or the configurational alone. . . . It is further true that historical events are unique, usually to a high degree, and hence cannot embody laws defined as recurrent, repeatable relationships. (1963, p. 29)
>
> [Note: For Simpson, *immanent* is equivalent to "timeless"; *configurational* to "timebound" or "historical."]

I'm not altogether happy with Simpson's rejection of historical laws, even though I accept the logic of his position. In all of complex/historical science, the search must be made for generalizations that have a reasonably good probability of being repeatedly applicable. If we fail to identify such generalizations in the study of natural history—if our published accounts are no more than descriptions of the unique—we are vulnerable to the charge that we are not doing science at all. Even if that charge be withdrawn, we can be charged with doing a lower-grade or less-meaningful brand of science than science that produces laws or probabilistic generalizations.

Let me return to the geologic hypothesis used as an example in earlier pages. In modern geology, the basic mechanical activities of the earth's outer layer (the lithosphere) are encompassed by the general hypothesis of plate tectonics. Evidence is strong that the continents of North and South America, Eurasia, and Africa were at one time joined together in a single, great continent called *Pangaea.* About 200 million years ago, this single continent began to split apart along great fractures, known as "rifts." As the rift valleys widened, they became seaways, which expanded into deep ocean basins. New crust was continually being formed by rising molten rock (magma) along a central oceanic rift. Today a great Atlantic Ocean basin separates the American continents from those of Africa and Eurasia. Consider, now, that 200 million years is not a very long span of time in comparison with the age of an earth formed over 4 billion years ago. The opening of the Atlantic basin occupied only the last one-eighth fraction of the total earth history. Have continents split apart before in that earth history? Evidence is strong that it has happened several times before. There is also evidence that ocean basins close up and are eliminated by the underthrusting mechanism of subduction, which we used earlier in our exemplary hypothesis. Subduction can cause an entire ocean subfloor to disappear into a deep, hot mantle region below the lithosphere. When this happens continents slam together like halves of a double sliding door—continental collision, it's called. Evidence is good that the closing of an ocean basin is a repeated event in geologic history; it alternates with the process of opening of a basin. A distinguished Canadian geologist, J. Tuzo Wilson, outlined a synthesis of alternate opening and closing of ocean basins; it is widely recognized today as the Wilson cycle.

In general terms, the sequence of events in one Wilson cycle has probably been very much like that in other cycles throughout at least three billion years. If only one cycle has been in operation at a given time on the globe, there has been time for perhaps six complete cycles. Does this give enough repetitions to let us assign to the idealized Wilson cycle the status of a scientific law? No, you say? Then are we permitted to call it a "principle" of geology? Perhaps so. In any case, we have enough geologic evidence to permit us to sketch a *model cycle* that embodies the most likely features of a natural cycle (i.e., the important features most instances share in common). The model is a hypothesis of a particular kind: it is a historical hypothesis that asserts what is likely to happen repeatedly in nature, including time yet to come. This historical hypothesis differs from a unique hypothesis that is a description and perhaps also an explanation of a single (singular) event.

Prediction—Logical or Temporal?

The possibility of a fundamental distinction between the physical sciences and the complex/historical sciences, over and above the distinctions I have already pointed out, has been put forward by philosophers of science, among them Karl Popper. Molecular geneticist Francisco Ayala explains:

> Some philosophers of science have claimed that evolutionary biology is a historical science that does not need to satisfy the requirements of the hypothetico-deductive method. The evolution of organisms, it is argued, is a historical process that depends on unique and unpredictable events, and thus is not subject to the formulation of testable hypotheses and theories. (1977, p. 486)[2]

Ayala strongly disputes the claim that the hypothesis of evolution cannot be tested; he says that claim is based on "a monumental misunderstanding." He then points out that biological evolution has two scientific phases. One deals with causal questions; in other words, how biological processes or mechanisms work. This phase corresponds with the "timeless knowledge" of Bucher, and with my category of events or relationships with a very high probability of being repeated. The other phase of biological evolution is historical—the tree of life as it grew through geologic time with its various branches. This historical aspect corresponds with Bucher's "timebound knowledge," and with my category of evolutionary sequences having an infinitely small probability of being repeated elsewhere in space or time. It is with the historical phase that the dispute is concerned. Philosophers of the logical-empiricist school at one time went so far as to claim that the history of evolution as it has been reconstructed by biologists and paleontologists is not a scientific hypothesis because it has no power of prediction.

Recall my account on earlier pages of how I was confused by the intent of the meaning of the word "prediction." It finally dawned on me that two meanings are in common use by scientists, but that the authors usually do not specify which one they intend. Confirmation and clarification came when I happened on the following statement by Professor Ernst Mayr of Harvard University, an eminent scientist in the field of evolutionary biology:

> The word *prediction* is being used in two entirely different senses. When the philosopher of science speaks of prediction, he means *logical prediction*, that is, conformance of individual observations

> with a theory or a scientific law. . . . Theories are tested by the predictions which they permit. . . . Prediction, in daily usage, is an inference from the present to the future, it deals with a sequence of events, it is *temporal prediction*. (1982, p. 57)

George Gaylord Simpson uses the word "prediction" in the temporal sense (1963, p. 36). In discussing the physicists' narrowly circumscribed view of science, he notes that "some philosophers and logicians of science have concluded that scientific explanation and prediction are inseparable." He elaborates:

> Explanation (in this sense) is a correlation of past and present; prediction is a correlation of present and future. The tense does not matter, and it is maintained that the logical characteristics of the two are the same. They are merely two statements of the same relationship. This conclusion is probably valid as applied to scientific laws, strictly defined, in nonhistorical aspects of science. . . . But we have seen that there are other kinds of scientific explanations and that some of them are more directly pertinent to historical science. It cannot be assumed and indeed will be found untrue that parity of explanation and prediction is valid in historical science. (p. 36)

The temporal sense of prediction has been used more recently by Ernst Mayr. During an interview with science writer Roger Lewin, Mayr observed that evolution, as a historical process, cannot predict future evolutionary changes. He is quoted as saying:

> If you had stood on the earth at the beginning of the Cretaceous (135 million years ago) and seen dinosaurs all over the place, you could not have predicted that the miserable little things* that came out only at night would eventually take over when Cretaceous came to an end. You can predict the next appearance of Halley's comet, but you can't predict changes in biological diversity. Such uncertainty is typical of evolution. (Lewin 1982, p. 719)
>
> [*Shrewlike ancestors of the mammals.]

Fundamentalist creationists, who seek to discredit science, search the scientific literature incessantly for statements made by prominent scientists that appear to imply weakness or inadequacy of the theory of evolution. These believers in the Genesis scenario of six-day divine creation would latch onto Mayr's statement in an instant, for it supports a theme they have been trying to put across to the public for years. Richard K. Turner, an attorney for the creationists in their 1981 suit before the

California Superior Court to force teaching of the biblical creation version in state schools, gave us a perfect example of this tactic. As reported by science writer William J. Broad (1981, p. 1332), Turner misinterprets Karl Popper's use of "prediction" (as deduction of consequences) to mean that a scientific hypothesis must have the ability to predict what will happen in the future. Switching from logical to temporal meaning of "prediction" allows the creationists to claim that because evolutionary theory cannot predict what specific evolutionary changes will occur in the future, the theory of biological evolution is not scientific. From there, the creationists go on to argue that the hypothesis of evolution is actually a belief system—an expression of scientism—and hence forms the basis of a kind of religion of its own.

Professor Ayala does not buy the creationists' version. Instead, he maintains, "even the study of evolutionary history is based on the formulation of empirically testable hypotheses" (1977, p. 486). He gives an example from the evolutionary tree that includes hominids (humans) and the great apes (chimpanzees, gorillas, orangutans). The ways in which the branches of that tree can be drawn differ according to two current hypotheses. Using the logical sense of "prediction," he says: "A wealth of empirical predictions can be derived logically from these competing hypotheses" (p. 486). "These alternative predictions provide a critical empirical test of the hypotheses" (p. 487).

Philosophers of science have disagreed as to whether logical prediction (the deduction of some past event in support of a historical hypothesis) can be accepted in the absence of any means of making temporal predictions (what will happen in the future) from a given historical hypothesis (Hempel and Oppenheim 1953). They have argued that both past and future must be predictable from the same hypothesis. This is a principle of logical symmetry that certainly must apply to laws of physical science that are in the timeless category.

But it has also been argued that, because of the extreme complexity of most historical hypotheses (timebound knowledge), prediction of what will happen in the future is neither required nor feasible (Scriven 1959). That argument is based in formal logic, which I will not attempt to go into here. Scientific temporal prediction depends for its success on recurrence or repeatability, which is the nature of timeless knowledge (Simpson 1963, p. 38). The kinds of events that characterize timeless knowledge can "move about freely in time," so to speak, from past through present to future. While it is true that a complex historical hypothesis contains both timeless and timebound classes of information or knowledge, the timeless con-stituents—common events, that is—are not what make the hypothesis unique. It is the uniqueness of the succession of different kinds of events in a given history that counts—events that could not have been predicted if an observer had been there at the time, and can therefore not be predicted by today's observer looking into what is, to us, the future.

It is often pointed out by science philosophers that different branches of science use radically different means to formulate and test hypotheses. I would like to think that such statements are generally unwarranted. Special ground rules for special groups of scientists is not something with which I can be comfortable. Excluding formal logic and mathematics, all of empirical science ("factual" science) must play by the same rules. From what I have reviewed of biologists' assessments of the scientific nature of the study of evolutionary history, I am satisfied that a single set of rules will work for all. Ernst Mayr has described the common aims of science in these words: "All sciences, in spite of manifold differences, have in common that they are devoted to the endeavor to understand the world. Science wants to explain, it wants to generalize, and it wants to determine the causation of things, events, and processes. To that extent, at least, there is a unity of science" (1982, p. 32). Quantum physics does, however, pose some serious questions about such matters as reality in the empirical sense being applicable to the behavior of particles—quarks, for example. Perhaps there is a discontinuity, or cutoff point, beyond which the fundamental frame of reference has to be radically restructured for quantum mechanics. Einstein's general and special theories of relativity seem to conform satisfactorily with traditional models of matter and spacetime, and that may perhaps mark the boundary of traditional science.

The Uniqueness of Living Matter

In the interview with science writer Roger Lewin (1982), referred to in an earlier paragraph, Professor Mayr discussed some ideas he had covered in his book, *The Growth of Biological Thought* (Mayr 1982). Mayr had been disturbed by the misunderstandings that many physical scientists have about the field of evolutionary biology. He felt that physical scientists frequently do not appreciate important differences between the living world and the inanimate world. It is not that the same physical laws are not followed in both physics and biology; the basic laws are common to both fields. As Mayr put it, "There isn't a process in a living organism that isn't completely consistent with any physical theory. Living organisms, however, differ from inanimate matter by the degree of complexity of their systems and by the possession of a genetic program" (Lewin, p. 719). Of these two differences, the greatest level of uniqueness attaches to the second point—the possession of a genetic program.

One distinctive quality of the living organism, not found in the inorganic environment in which it lives, is the ability of the life form to achieve extraordinary molecular complexity, accompanied by a capacity to store energy in chemical forms. Living matter is organized into *systems*, which are usually rather clearly defined by a tangible boundary. The single cell of a plant or animal is perhaps the simplest example; another would be a single body organ, such as the kidney or liver. The total individual is also a clearly defined single system. We can go on to consider a community of plants or animals (or a mix of both) in their environment as a single system—an *ecosystem*. These organic systems import and export both energy and materials through their boundaries; they are *open systems*. Open systems are identifiable in all other areas of natural science, but these are purely inorganic systems (Strahler 1980).

Consider a single plant cell as an open system of energy and matter. Nutrients enter through the cell wall as ions or molecules, while waste products of metabolism leave the cell through the same wall. Energy needed to power the biochemical reactions enters the cell as radiant light energy. The dominant process of chemical synthesis in the plant cell is photosynthesis, in which the light energy is used to combine carbon dioxide (a gas) with water molecules to form carbohydrate (compounds made up of carbon, hydrogen, and oxygen). In this process, solar energy is stored in the carbohydrate molecules as a form of chemical energy. Some of this, in turn, is used in the synthesis of more complex organic molecules—proteins, for example. Again, energy goes into storage in those larger molecules. Among the most complex organic molecules synthesized in the cell are those of DNA and RNA; their information content is enormous. At the same time that complex molecules are being constructed, others are being broken down into simpler components that are released to the outside environment. The reverse process to photosynthesis is respiration; it releases heat energy to the environment.

Life Systems and Energy

Analysis of the organic cell as an open system was presented to scientists of North America by Ludwig von Bertalanffy, an Austrian biochemist, in a now-classic paper titled "The Theory of Open Systems in Physics and Biology" (1950). Upon first seeing his material soon after publication, I was struck by how important his systems analysis could be for all areas of natural science, and I incorporated it into one of my papers on natural systems of water erosion, published a year later. The basic energy equation given by von Bertalanffy is a bit on the difficult side for most readers since it uses the differential calculus. At the risk of submerging many readers, I can state the equation in words as follows: The time-rate of change of energy concentration within a very small element of an open system is equal to the sum of the rate of change of production of energy (produced in biochemical reactions) and the net rate of outflow of energy from the element. By "element" is meant any very small unit volume in the cell; you can think of it as a tiny cube or sphere. The same statement could apply equally well to the entire cell as an open system. For a cell, three energy states are possible with this equation. First, the cell system can be experiencing a net gain in stored chemical energy. Second, the system can be experiencing a net loss in stored energy. Third, the quantity of stored energy within the system can be held constant with time; this represents a *steady state* of the open system. Biologists call the condition of steady state *homeostasis.*

An important point to note in connection with the theory of open systems and steady states in biology is that it seems—on the surface, at least—to violate one of the most sacred laws of physics: the second law of thermodynamics. This law includes the dictum that in isolated systems the tendency is for the matter inside the system always to go in the direction of greater disorder. A concept of ordered states and disordered states is involved here, something not previously mentioned in our description of the cell system. In thermodynamics, the complex organic molecules in the cell are considered to represent a high level of order (orderliness) of the atoms. In contrast, the collection of rather simple inorganic molecules and ions that the cell uses to construct the complex molecules is viewed as representing a low level of orderliness. The property of lack of order is disorder. Therefore, it can be said that within the growing cell, matter is being transferred from a state of disorder to a state of order, but this direction of change is forbidden by the second law. Fundamentalist creationists use this conclusion as the basis for claiming that the theory of organic evolution violates the second law; therefore, that evolution is false.

For the moment, it is enough to be aware that the laws of thermodynamics forbid creation of new energy supplies from nothing (*ex nihilo*). Moreover, in biological processes, energy cannot be derived from mass by particle collisions. We must, instead, interpret the storage of chemical energy in cells by means of an energy cycle, or energy balance, in which all forms of energy entering and leaving the system are accounted for, much as you account for the flow of money in and out of your bank account.

Synthesis of large organic molecules in a living plant cell uses energy from outside the cell, transforming the sensible heat from solar sources into chemical energy that can be temporarily stored. During a stage of increase in stored energy, external energy is consumed by the cell, but a part of that energy is expended in the storage process itself. The important point is that for energy to be stored in a form capable of doing useful work, a great deal of energy must also be "burned" (i.e., lost as heat dissipated into the environment). A steam locomotive, in pulling a heavy train of cars up a long, steep grade, burns a great deal of coal or wood and dissipates a great deal of heat into the atmosphere; yet, at the top of the grade, the train has gained a large supply of stored energy (as potential energy). A cell, its energy input diminished, undergoes a degradation (chemical breakdown) of its complex, energy-storing molecules, releasing heat energy and resulting in a collection of raw materials of lower chemical energy.

The ratio of energy stored to energy consumed in open systems, such as a living cell, has been intensively studied. (See Strahler and Strahler 1974, pp. 19-23, Appendix 2.) Charles A. Coulomb, the French scientist (1736-1806), was interested in evaluating the mechanical power capability of the human male. This was a subject of general interest at a time when men provided a great deal of the power of industry, using devices such as treadmills that would turn wheels to grind grain, lift water, or operate a crane. Coulomb observed that a porter who brought firewood up to his apartment, a vertical ascent of about 40 ft (12 m), had a maximum work capacity of about six wagonloads per day. He made 66 trips per day up the stairs, carrying an average load of 150 lbs (68 kg). Suppose that the porter carried only one or two sticks of wood per trip. He could have made many more trips per day, but since much of his work consisted of lifting his own body weight, the ratio of stored potential energy (stored work) to energy expended in moving his body (processing work) would have been quite small. On the other hand, if he attempted to lift a much heavier load, his progress up the stairs would have been painfully slow, and the total energy stored would be small by the end of the day. Somewhere in between was the optimum ratio of load to body weight, such that the maximum stored work was done in ratio to processing work. Evidently, the porter had learned from experience the optimum load of firewood that would get the largest total weight upstairs in one day.

Without going into an explanation, we can simply state here that, as a rule of thumb, the maximum rate at which energy can be placed in storage is achieved when 50 percent of the energy is stored and 50 percent is expended in processing work needed to accomplish the storage function. This 50/50 division of power into the two functions for optimum storage is known as the *Darwin-Lotka law*. The law is named for Sir Charles Darwin and A. J. Lotka; the latter in the early 1920s analyzed quantitatively the role of energy expenditure and storage in organic evolution (Lotka 1922; Odum and Pinkerton 1955). From an evolutionary viewpoint we simply comment that an organism capable of storing energy at the fastest possible rate would have advantages over other individuals of the same species with lesser abilities, since the food supply might be severely limited and it would be advantageous to be able to convert it into stored energy in the least possible time.

Genetic Programs and Feedback Systems

Professor Mayr's second point of uniqueness of living matter is the genetic program that directs the formation of each individual: "The genetic instructions packaged in an embryo direct the formation of an adult, whether it be a tree, a fish, or a human. The process is goal-directed, but from the instructions in the genetic program, not from the outside. Nothing like it exists in the inanimate world" (Lewin 1982, p. 719). But Mayr goes on to explain, the genetic package is much more than just a set of instructions; it is the product of a descent through evolution (see also Mayr 1982, pp. 55-56). In this sense, the genetic program is history—millions upon millions of years of evolutionary history following a continuous chain of small steps. As I have emphasized earlier, the probability of such a chain being duplicated at any place in the universe, at any point in past or future time, is almost infinitely improbable. It is in this sense that the genetic program is unique.

Essentially the same point has been made by George Gaylord Simpson, who stresses that no two organisms are exactly alike, not even identical twins. He explains:

> Each is the product of a history both individual and racial, and each history is different from any other, both unique and inherently unrepeatable. These aspects of biology deal not with the immanent, the inherent and changeless characteristics of the universe, but with contingency, its states, fleeting and in ceaseless change, each derived from everything that went before and conditioning everything that will follow. (1969, p. 10)[3]

A unique feature of vertebrate animals, setting them apart as a class from other organisms and all inorganic natural systems and structures, is that they have built-in *feedback systems*, specifically developed to transmit information from sensory organs through neurons to a central clearing-house—the brain—where a response is triggered and sent through neurons to the appropriate body organs for action (Simpson 1963, p. 26).

The term "feedback" needs to be understood in each of two meanings intended in science. In purely mechanical systems in nature we can often recognize a self-adjusting or self-correcting mechanism that serves to keep the system on an even keel—a steady state, that is. An example would be an ocean beach of sand and pebbles, shaped by the action of breaking waves that generate the alternating uprush of a sheet of turbulent water and its downslope return by gravity. As many of you know from experience, the beach changes seasonally in form and position. In summer an embankment (a berm) is built seaward by relatively weak waves and swells; in winter the berm is eroded away by waves of high energy and the beach assumes a different profile. Geologists who study this cycle recognize that seasonal changes in beach form are responses to seasonal changes in the energy input; they refer to the self-adjustment process as "feedback," although it has nothing whatsoever to do with information transmission. In the higher plants (the vascular plants) feedback can be recognized as carried out by physical and chemical processes, but not involving information as such.

The type of feedback we are talking about in animals is designated *cybernetic feedback*; it is found in biological, psychological, and social systems, as well as in a host of mechanical and electronic devices invented by humans (Strahler 1980, pp. 25-26). *Webster's Third New International Dictionary* gives as one definition of feedback: "a return to the input of a part of the output of a machine, system, or process and leads to a self-correcting action." The word "cybernetic" refers to the control system itself. A familiar example in mechanical systems is the thermo-static control mechanism of a home heating unit. The neurological feedback system in vertebrate animals, evolved over hundreds of millions of years, is vastly more complex and efficient than any cybernetic feedback system invented by humans, electronic computers included. The human brain is also the control center for elaborate social feedback systems in which appropriate human responses are conceptualized to regulate society and keep it functioning on an even keel.

We can add a final unique quality or property of living organisms; they exhibit numerous behavioral properties and physical forms, functions of body, and organs that serve specific purposes in remarkably successful ways. Eyes are wonderfully adapted to the purpose of seeing, ears to hearing, hands to grasping. The process of organic evolution makes possible adaptations of form and function of the organism; these are genetically controlled features and they have undergone changes through time. Changes that increase the chances of the individual to survive and to reproduce tend to be preserved in the species. This is the natural selection process.

When we, as humans, observe these remarkable adaptations in our species and in other species, we think immediately of purpose being fulfilled. We think of a particular body organ as being designed for a useful purpose. Such thoughts are described by the term *teleology*, which simply means "to explain something in terms of fulfillment of purpose." A natural teleological explanation applies correctly only to features of living organ-isms. In the case of humans, it may be extended to human artifacts—a knife or fork, for example—which are deliberately designed to serve a purpose. This is artificial teleology. We are not particularly interested here in artifacts (including culture in general), but rather in attributes of organisms that are genetically determined.

Before Charles Darwin, natural teleology was, in the Hebrew/Christian tradition, closely tied to theistic religion. God created man and all other organisms exactly as they are today, along with their physical environments. All we humans needed to do was to sit back and admire the handiwork of the omniscient and omnipotent Creator. True, there was that bothersome organ, the appendix, that seemed to serve no good purpose and often dealt a lethal blow to its owner. According to these creationists, nothing so marvelous as the human species could have reached such perfection through the blind and unthinking process of organic evolution. This "argument from design" has for centuries been a favorite with all theists of the Judeo-Christian faiths and has served them as a proof of the existence of God (see Chapter 11).

The Question of Reductionism in Biology

A philosophical issue currently undergoing vigorous debate concerns the relationship between physical science and the other natural sciences. The assertion is made that the underpinnings of the complex/historical sciences are those laws and relationships established in physical science—physics and chemistry. Therefore, it is argued, each branch of natural science can be reduced to statements of physics and chemistry. The philosophical concept involved is called *reductionism*, defined as "a procedure or theory that reduces complex data or phenomena to simple terms" (*Webster's Ninth New Collegiate Dictionary*). Conversely, it is argued that all elaborate theories and laws of the complex/historical sciences can be derived from simpler, more general theories and laws of physics and chemistry.

Offhand, reductionism may sound like a great idea. To make things simpler makes them more easily understood by more people. Those natural sciences dealing with inorganic (inanimate) objects and structures easily lend themselves to being reduced to descriptions of physics and chemistry. The form of a mineral crystal reduces very nicely to an arrangement of atoms in a three-dimensional lattice. A flowing river easily becomes an exercise in the mechanics of flow of a viscous fluid. The Big Bang in cosmology is nothing but an intense flux of elementary particles, such as photons and neutrons.

Reductionism (or reductivism) also has broader meaning in the philosophy of science: "the belief that all fields of knowledge can be reduced to one type of methodology, or to one science, which encompasses principles applicable to all phenomena" (Peter Angeles 1981, p. 242). The ontological model of mechanistic monism, described in Chapter 11 and illustrated in Figure 11.1, might serve as an illustration of this broad meaning of reductionism by the manner in which it eliminates the supernatural realm from an explanation of the universe.

Reductionism also enters into the mind/body problem, an age-old philosophical problem that we briefly reviewed in Chapter 1. Recall that modern science has disposed of the dualistic ontology that recognized both a mind (or soul) and a body, each in its separate realm of reality, replacing it with a monistic-materialistic model in which the mind is physiologically one with the body. In this case, reductionism takes the form of simplification by elimination of one of two entities by combining their functions.

In science, reductionism usually refers to the process of reducing all observable (empirical) phenomena to the simplest (or smallest) components of mass and energy. Thus, under modern quantum mechanics, everything in the universe is reduced to the four fundamental forces (weak and strong, carried by messenger particles) and subatomic particles (leptons and hadrons, composed of quarks). Included in this reduction program are both the physical descriptions of the particles and the formulation of laws associated with their behavior. Thus laws of science are arranged in a hierarchy in which each layer of laws elucidates the layer above it and is in turn elucidated by laws in the layer below it.

One needs to be extremely cautious in inferring that this kind of reduction performed by physicists constitutes a scientific explanation of the complex forms or phenomena that are reduced. For example, if we disassemble an automobile into all its separate parts, we cannot claim that we have "explained" the automobile as a functional whole. Nor would the reduction of a single part of that automobile to its component atoms "explain" that part as a whole functional unit. Looking back at our discussion of scientific explanation in Chapter 2, it is obvious that knowledge gained in the reduction process is effectively applied in Professor Nagel's deductive model of explanation of physical laws, but for the probabilistic and historical kinds of explanation, reduction may be meaningless or irrelevant. In one sense all the particles that inhabit the world of quantum mechanics are products of cosmic history. The elements that make up the universe came into existence in a historical sequence that involved stellar evolutionary processes, and these are to some extent now understood and documented. So perhaps physics, like biology, is dependent on a unique cosmic history for whatever level of explanation may be possible.

When it comes to the study of live organisms, the process of reduction becomes a bit tacky. Is a living cell simply an aggregation of atoms and molecules carrying out processes of pure physics and chemistry? Is that all there is to life? Or is there some special force, quality, or substance in a living cell that stands apart from the domain of physics and chemistry? A "Yes" answer to the last question has been voiced for centuries, starting with Aristotle, by a class of philosophers called *vitalists*; their doctrine was *vitalism*. Vitalism was opposed by *mechanism*, which rejected the idea that there is "something more" in living cells than in nonliving matter. Vitalists postulated the existence of a mysterious, unique component of life described in such terms as "entelechy" (a perfecting principle), *élan vital* (vital force), or "radial energy." The vitalist/mechanist controversy raged from the time that René Descartes (about 1640) opted for the mechanistic view, through to the early 1900s, when vitalism as a scientific view was put to rest. The reason for its demise is fairly obvious: vitalism invoked a supernatural concept, or at least a non-empirical concept, that cannot be tested scientifically. Vitalism therefore rested on faith alone.

The place of vitalism as an answer to the question "Is that all there is to life?" was taken by *ontological reductionism*. Recall that ontology is the branch of philosophy that asks "What is reality?" When a living cell is broken down into its material components, they prove to be nothing but atoms. Energy that the cell cycles through its metabolic processes proves to be no different from forms of energy known to physics. This is a purely naturalistic or mechanistic picture. As molecular-biologist Francisco Ayala says: "Ontological reductionism also implies that the laws of physics and chemistry fully apply to all biological processes at the level of atoms and molecules" (1977, p. 488). He goes on to say:

> Ontological reductionism does not necessarily claim, however, that organisms are nothing but atoms and molecules. The idea that because something consists of "something else" it is nothing but this "something else" is an erroneous inference, called by philosophers the "nothing but" fallacy. Organisms consist exhaustively of atoms and molecules, but it does not follow that they are nothing but heaps of atoms and molecules. (Pp. 488-89)[2]

What Ayala seem to be getting at is that ontological reductionism does not concern itself with the question: How did the organism get to become the complex system it is? If we took all the individual atoms of which the cell is composed and dropped them into a beaker, what would we do next to "persuade" those parts to form a complete living cell? Relying on natural processes, aided or unaided by humans, the probability that all the components would spontaneously come together to form a cell is infinitely remote. Even if we assembled all the atoms into complete molecules and provided exactly the correct kinds and numbers of them, they would not spontaneously form a living cell. Ayala sums it up: "Living processes are highly complex, highly special, and highly improbable patterns of physical and chemical processes" (p. 489). Emphasis should be given the word "improbable"; better to say "infinitely improbable by chance alone." We mean, of course, "improbable that it would happen now." The cell has a long and complex history of evolution from simpler states of matter.

If we turn to consider the properties of a living cell as compared with the properties of the component atoms, the two sets of properties are not alike. The cell has its own unique set of properties, referred to as *emergent properties*. Now the philosophical question takes on a new twist: Can the laws and theories that account for the behavior of the cell (a complex system) be logically derived solely from the laws and theories that govern the behavior (functioning) of the atoms or the molecules as separate component parts? One particular brand of reductionism, called *epistemological reductionism*, considers the question as to whether "the laws and theories of biology can be shown to be special cases of the laws and theories of the physical sciences" (Ayala 1977, p. 491). Ayala gives a full explanation as follows:

> The connection among theories has sometimes been established by showing that the tenets of a theory or branch of science can be explained by the tenets of another theory or branch of science of greater generality. The less general theory (or branch of science), called the secondary theory, is then said to have been reduced to the more general or primary theory. Epistemological reduction of one branch of science to another takes place when the theories or experimental laws of a branch of science are shown to be special cases of the theories and laws formulated in some other branch of science. The integration of diverse scientific theories and laws into more comprehensive ones simplifies science and extends the explanatory power of scientific principles, and thus conforms to the goals of science. (P. 492)[2]

Several major successes have been achieved in epistemological reduction throughout the history of science. For example, much of what was the science of chemistry (especially about how compounds were formed from elements) prior to about 1900 yielded to a new and revitalized chemistry making use of knowledge the physicists furnished about the way in which electrons move in orbitals about the nucleus of the atom. More recently, a large part of what was the theory of genetics prior to the 1960s succumbed to explanation by organic chemistry—the structure and function of the DNA molecule. Dramatic as these reductions were, they have not been judged as completely successful, for there is always some unexplained content of the more complex system that fails to yield to the more general system.

Whether the processes of complex/historical sciences will ever be fully explained in any meaningful way by reduction remains a debatable question. Successes will doubtless continue in that phase of empirical science that concerns processes and mechanisms (timeless knowledge). I cannot see reduction having any success whatsoever in the historical branches of natural science, e.g., historical organic evolution, evolution of the earth's crust, and stellar-galactic-cosmic evolution. These systems developed by a long, unique succession or chain of physical/chemical reorganizations. Once broken, such a chain could almost never be reproduced. Thus, while we can explain the general behavior of the complex natural system by basic laws of physics and chemistry, we cannot actually derive the more complex system from the simpler one.

Biologists are particularly skeptical of the possibility that epistemological reduction can succeed in their area of science. Can the natural teleological phenomena seen in life forms be completely reduced to processes of physics and chemistry? Professor Simpson does not think so: "In physical sciences it is not legitimate, indeed it is downright silly, to ask what things are for or what good they are" (1969, pp. 10-11). Physics and chemistry do not ask what possible purpose is served by, say, oxygen combining with hydrogen to produce water. We would not think of saying that this chemical union occurs in order to produce a liquid compound that slakes the thirst of vertebrate animals. "But in biology it is not only legitimate but also necessary to ask and answer questions teleological in aspect, concerning the function or usefulness to living organisms of everything that exists and occurs in them." He goes on:

> The structures and processes of organisms are useful, they perform functions, and they would not exist or occur if that were not true. They can never be understood or explained by the most complete and exact specification of the chemical reactions involved. Those reactions themselves are meaningless except as they relate to the organisms and populations and ecosystems in which they occur. (P. 11)[3]

Professor Mayr, in his interview with science writer Lewin, had some cogent remarks to make about reductionism in science; he is skeptical of the dominant role assigned to it by physicists. He said to his interviewer: "This reductionism has led to what David Hull (no reference given) calls the arrogance of physicists. . . . They say, yes, you biologists deal with complex things, but the ultimate explanation will be supplied by the level at which we study" (Lewin 1982, p. 719; see also Mayr 1982, p. 33). Mayr doesn't think that an understanding of particle physics will provide an understanding of everything else in the real world. Instead, Mayr thinks "complex systems have to be studied at high levels of complexity. New properties turn up in systems that could not have been predicted from the components, which means you have to study things hierarchically."

This chapter has stressed fundamental differences between physical science and the other natural sciences. As theoretical physics delves deeper and deeper into the nature of the fundamental forces, setting as its goal the achievement of a grand unified theory, physical science seems to drift farther and farther apart from the complex/historical sciences. Chapter 5 seeks an area of common ground that serves to unify rather than fragment science as a whole: it is the pervasive role of blind chance in the world of nature, encountered whether we are looking at the motions of individual atoms of a gas or at differences between individual organisms within an animal species.

Credits

2. From Francisco J. Ayala, Chapter 16, pp. 474-516 in T. Dobzhansky, F. J. Ayala, G. L. Stebbins, and J. W. Valentine, *Evolution*, W. H. Freeman and Company, San Francisco. Copyright © 1977 by W. H. Freeman and Company. Used by permission.

3. From George Gaylord Simpson 1969, *Biology and Man*, Harcourt, Brace & World, New York, pp. 10-11. Used by permission of Harcourt Brace Jovanovich, Inc.

CHAPTER 5

The New Philosophy of Science

Judging from this chapter title, you might expect to find here an entirely new philosophy of science, above and beyond what has been described in the foregoing chapters. Not so; both old and new versions have been introduced already, and perhaps in places clearly identified as such. For example, Professor Kitcher's explanation of "naive falsification" is within the new philosophy of science. Much more remains to be said about the new philosophy.

First, let's clarify the distinction between science itself and the philosophy of science. Think of an ant hill, or better still, an ant farm in which a colony of ants is housed in a box and its tunnels exposed to view through a plate glass wall. Science is represented by the ant farm; the process of science, by the activities the ants are carrying out. Philosophy of science is represented by humans observing from outside the ants at work and the structures they make and maintain, including where they get their food, how they reproduce, and how they fight off intruders. The ants, of course, are oblivious to the presence of the watchers.

The ants carry out remarkably complex activities, dictated by genetic inheritance from past evolution—they are programmed to do what they do. As for the observers, they have the difficult task of figuring out exactly what the ants are doing, and that's where the problems begin to enter. No sooner does one team of observers publish their account of what goes on in the ant colony than a new team of observers arrives on the scene and comes up with a radically different account. Which account is the more realistic? Does one version automatically invalidate the other? Or does the second account build and expand upon the first?

Logical Empiricism—The Old Philosophy of Science

If a new philosophy of science is clearly recognizable, what preceded it? For our purposes, the predecessor will be *logical empiricism*, a school of philosophical thought mentioned in Chapter 2. Logical empiricism was an outgrowth of *logical positivism* that originated in the 1920s. At that time a professor of philosophy in the University of Vienna, Moritz Schlick, attracted a group of kindred spirits—the Vienna Circle. They based their philosophical view of science on the writings of an earlier generation of scholars, particularly Ernst Mach, Gottlob Frege, and Bertrand Russell, who had developed a new mathematical logic as a successor to the time-honored classical, or Aristotelian, logic. Two more immediate predecessors were G. E. Moore and Ludwig Wittgenstein. The latter, especially, through his work *Tractatus Logico-Philosophicus* (1922), provided a sort of "bible" for the logical positivists. Wittgenstein had studied under Russell, then worked in solitude to develop his philosophical concepts. Later, he lived in Austria, where his work came to the attention of the Circle members.

Logical positivism (LP) can be described as "hardnosed" through its claim that only strictly empirical information deserved to be called knowledge; all other propositions (transempirical, or "nonempirical," concepts) were meaningless utterances and scarcely better than intellectual "trash." This meant that any concepts generated by your mind through imagination were worthless, at least to science, unless they dwelt upon things perceivable to the senses. Strangely, the fields of pure mathematics and formal logic, which are nonempirical, were not only accepted by the LPs, but considered essential to a science. And, of course, by the same token belief fields such as religion and ethics (as emotive utterances) were not to be equated to true knowledge. The Vienna Circle was broken in the 1930s and its members dispersed geographically by the arrival of German Nazism. By about 1940 this school of thought had been replaced by its derivative system, logical empiricism (LE).

Before attempting to compare or contrast LE with the new philosophy of science (NPS), consider the following dialogue between a philosopher and a reporter who asks the questions:

Q: Basically, what was "wrong" with LE? Why wasn't it working?

A: LE was telling us what scientists are supposed to be doing and thinking in their pursuit of science. Philosophers call this the "prescriptive" or "normative" view, meaning that LE was telling the scientists what they ought to be doing.

Q: How, then, is NPS different, and why is it the "right" way?

A: NPS tells how scientists really, truly, actually do science. Philosophers regard this information as purely descriptive, i.e., it tells what *is* the case (not what *ought to be* the case). Philosophers today are claiming that LE had painted an unrealistic, or even false, picture of what scientists actually do.

Q: If LE and NPS make different, independent statements about two different aspects of science, can't they both be "right"?

A: Good question! Two sightless persons feeling opposite ends of the same elephant will come up with unlike descriptions of the beast, each of which may be quite accurate within its tactile area. But there's a real question as to whether LE is actually "right" (sound) in its own territory. If LE can be shown to be faulty, it shouldn't be prescribed for scientists as the way they should do their science. You might say that the *ought* should be tailored to agree with the *is*, and not the other way around. After all, scientists are not about to exchange their successful, highly productive ways for some idealized program that might prove to be far less effective.

The Program of Logical Empiricism

Logical empiricism (LE) arose as a transformation within logical positivism (LP), a change carried out largely by former members of the Vienna Circle. Although we often associate LE with Karl R. Popper, it would be more accurate to give credit to Rudolph Carnap, whose 1936-37 journal article, *Testability and Meaning*, "can reasonably be viewed as the founding document of logical empiricism" (Harold I. Brown 1977, p. 23). Whereas LP held that the truth of a hypothesis could be established through its successful agreement with repeated observations (i.e., by corroboration), Carnap rejected that claim, saying that it is impossible conclusively to verify any scientific proposition (Brown 1977, p. 23). To this is added the provision usually attributed to Popper, that a hypothesis or theory can be falsified once and for all by the disclosure that a single one of its predictions disagrees with observation of nature. In short, logic dictates that a hypothesis can be falsified, but never verified. That a hypothesis can be falsified requires that it be testable, and thus testability is always cited alongside of falsifiability as comprising the essence of LE. Popper's concepts were set forth in great detail in his work *The Logic of Scientific Discovery*. Originally written in German (*Logik der Forschung*); it was published in Vienna in 1934. The English edition appeared in 1959, at which time Popper was living in England. His second major work, *Conjectures and Refutations*, a collection of many of his published papers and lectures, appeared in 1963. A third major work, *Objective Knowledge*, was published in 1972 and revised in 1979.

Logical empiricism (LE) has been rather extensively described in our Chapters 2 and 3. The key features we covered are the formulation of hypotheses (theories), the hypothetico-deductive process, prediction and testing, corroboration and falsification. In Chapter 1 we covered other essential requirements of LE, namely, perception and conceptual thought, science and language, facts of science, scientific observations, and induction versus deduction. These topics clarified what can be considered the *presuppositions* of LP, which are that (a) *formal logic* (especially mathematical logic) is in the saddle at all times—like a sheriff of the Old West, enforcing the laws at gunpoint—and (b) *empiricism* rules supreme: all science rests in observation of "what's out there." ("What you see is exactly what you get.") These presuppositions, or axioms, are embedded in the name "logical empiricism."

In Chapter 3, in the section on Naive Falsification, we used Philip Kitcher's exposition of the real problems of accomplishing falsification of a complex hypothesis, *à la* Popper. Kitcher's account showed that scientists have ways of saving their hypotheses from fatal falsification. Here, we will zero in on the falsification principle, using logic as our weapon to destroy its credibility. Refer back to Chapter 1 under the heading of Induction and Deduction in Science; reexamine Figure 1.1, showing the induction/deduction feedback cycle. A similar diagram, Figure 5.1, adapts parts of this cycle to our needs here. From the hypothesis a test statement is deduced that can lead to falsification, i.e., a potentially falsifiable statement.

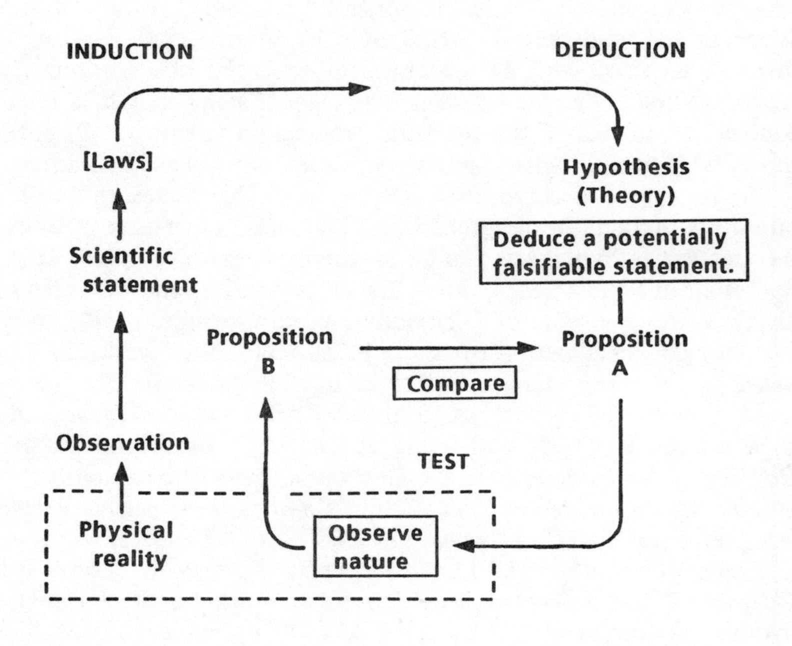

Figure 5.1 Testing a hypothesis within the induction/deduction feedback cycle.

In carrying out the test, the content of this test statement is sought in the realm of physical reality through the process of observation (induction process). Let us say that the statement is confirmed by independent observation and stands alongside the deduced version, where the two are judged to be identical. The hypothesis has escaped being falsified.

If, on the other hand, the observation process leads to the clear conclusion that the test statement is not what is observed to be the case, the deduced statement is denied and the hypothesis stands falsified. What is essential here is that the observation statement—a proposition—be in the relation of a counter-proposition to the deduced proposition. Let A be the deduced proposition, and B be the observed proposition. The relation of A to B must be (1) "If A, then *not* B" and vice versa, (2) "If B, then *not* A." In empirical science decision rests with B, so we use the second form of the proposition. If, on the other hand, observation produces a proposition, B, identical with A, the relation is "If A, then B" or vice versa, and this relation would constitute corroboration.

Now, there is a serious problem in accepting the above falsification. It is, in fact, fatal (see Brown 1977, p. 73). The test that leads to falsification makes use of evidence derived by observation, but we have already shown that it is logically impossible for statements derived by induction from nature to be either proved or disproved. Thus, the evidentiary statement itself lacks the power to falsify (or verify) any deduced statement. You might try to get out of this trap by deducing a second falsifying proposition from the failed first one. That could only lead to an infinite regression of such failed subordinate propositions. There is no way out.

Harold I. Brown, a philosopher of science, explains how the logical empiricists try to get around the difficulties of falsification: "How, then, can falsification be accomplished? For Popper falsification takes place only after scientists agree to accept a basic statement as adequately corroborated" (Brown 1977, p. 73)[1]. A "basic statement" is a scientific statement based on observation, in this case, used as the falsifying statement. Now, corroboration cannot logically lead to the conclusion that a statement is true, but there is a case for the contention that after a sufficient number of instances in which the statement agrees with observation, the reliability (probability) of the statement rises to some level at which we are willing to gamble that it will continue to agree with observation. That leaves the falsification mechanism of LE in a very weak and ambigious position.

Popper's solution seems no different than an overt admission that LE has failed miserably in practice, for he writes:

> From the logical point of view, the testing of a theory depends upon basic statements whose acceptance or rejection, in its turn, depends upon our *decisions*. Thus it is *decisions* which settle the fate of theories. (Popper 1959, p. 108)

Well, there you have it. Popperian falsification is not only logically vulnerable to being judged impotent, but fails to describe what scientists actually do in the practice of science. The new philosophy of science (NPS) recognizes that decision making is a necessary part of conducting science. Judgment as to what is a sound decision is not, however, entirely arbitrary. Simple scientific statements of the existential form can and do enjoy a high probability of holding true in application. Statements judged to be highly reliable are used to judge as false relatively simple hypothetical statements. If it were required that decisions to accept or reject hypothetical statements be determined by the toss of a coin—50 percent true versus 50 percent false—the pursuit of science would simply need to be abandoned. In judging the reliability of a scientific proposition, prior knowledge based on experience carries considerable weight, just as it does in events of our lives. Prior knowledge or experience can change the odds drastically, so that instead of their being 50-50 at the outset, they may actually be 99 to 1 that the decision is a correct one (or 1 in 99 that it is incorrect). The role of prior knowledge in modifying the probability of an observation statement being true or false is a topic we will examine in greater depth in Chapter 6.

So here we have an introduction to at least one example of what goes into the NPS. Brown tells us why a new view of science philosophy was necessary:

> The new approach to the philosophy of science that we now turn to has a different history [than that of LE]. It emerged largely in response to the growing sterility of logical empiricism, its failure to achieve adequate solutions to its own problems and to further clarify the nature of science, as well as from the many anomalies raised by new work in the history of science. (1977, p. 77)[1]

Perception—The New Model

One very distinctive feature of the new philosophy of science relates to perception, the process by which empirical knowledge is received by the brain and is incorporated into percepts. This is a topic we presented in Chapter 1 under the heading "Perception and Conceptual Thought." Now is the time to ask: Did we present the older view of the logical positivists (LPs) and the logical empiricists (LEs), or is it the new philosophy of science (NPS)? Perhaps now you will be savvy enough to look at those publication dates of the cited works of Marx Wartofsky (1968), Bernard Campbell (1985), and Theodosius Dobzhansky (1977). They speak the new philosophy of science. So, what then would be the old philosophy of the LPs and LEs?

Many of the later LE philosophers viewed perception as a very simple process of seeing (sensing) precisely what is external to the observer. This was also the view of most research scientists of that period and probably prevails widely today. You receive the sensory input and simply enter it directly into the data storage of your brain. What this means is that any number of observers can sense the same real object and all will store an identical percept of it. Those percepts, when put in the form of propositions, are identified as "empirical facts," and all would read alike. You might say: "They all see it in the same way, as the same thing, with the same meaning." Because that is what is held to be the case, the philosophers and scientists could claim: "Empirical facts guarantee the objectivity of science." Moreover, these empirical facts are entirely independent in content and origin from whatever theories or hypotheses they may be required to confront as evidence for corroboration or falsification.

Psychologists engaged in research on the process of perception quickly found fault with the older model. Experiments easily show that each of two different observers will present a different version of what each perceives when viewing the same object—a drawing, for example. At this point, we can repeat Professor Wartofsky's statement that sensory perception "is largely a function of intent and context, and depends to a great extent on frame of mind, attention, and what we know to look for" (1968, p. 332). Bernard Campbell divided perception into two information sources: (a) input from the senses and (b) memory of previous experience (1985, p. 330).

"Memory of previous experience" can be contained in the general term *prior knowledge*. The sensory input often but not always provides new knowledge. The new knowledge is not just added to the store of prior knowledge, it is also shaped, formed, or modified to relate to the prior knowledge. Thus no two humans will manufacture identical percepts from the same sensory input. In scientific investigation, this uniqueness in an individual's store of prior knowledge would seem to guarantee that there cannot be any true objectivity in science. So, the new philosophers of science would tell us: Down the drain goes another myth about the scientific method.

Prior knowledge can consist not only of empirical, inductively derived, knowledge, but also of theoretical or hypothetical concepts not immediately based in observation. Recall our conclusion that the theory (hypothesis) always contains one or more hypothetical propositions that are in part original conceptions of the mind. It is from these invented restructured concepts that deductions are made, setting in motion the search for empirical supporting evidence. The new philosophy of science (NPS) claims that when a scientist devises a hypothesis and wants fervently to have it substantiated (and thus reap lots of applause and a few medals), he or she is heavily biased with that desire and it cannot help but enter into the shaping of percepts. Philosophers of science refer to this bias or attitude as being *theory-laden*. The percepts themselves thus become theory-laden. Actually, Professor Thomas C. Chamberlin said much the same thing in his address of 1904 on the method of multiple hypotheses, recounted in Chapter 4. Recall his elegant phraseology: "Soon affection (for the theory) is awakened with its blinding influence." "The authorship . . . easily begets fondness for one's intellectual child." Today we just use the buzzword "theory-laden." Chamberlin also said: "The defects of this method are obvious and grave." So, what's new?

Philosophers are interested in a consequence that arises from the new view of perception and its relation to opposed theories. It is the charge that scientists indulge in *relativism*, a practice that is bad science, and we have been discussing it in the foregoing paragraphs.

If, as we have already stated, the new or modified percepts we fashion are a synthesis of both our existing theories and our new sensory perceptions, different scientists will interpret new empirical evidence in various different terms or meanings or frames of reference; i.e., the content of their arguments and evidentiary statements (pro or con) stands *relative* to (in relation to) how they perceive the subject or phenomenon they are investigating. The charge of practicing relativism need not carry the demeaning conclusion that science cannot generate reliable new knowledge and plausible theories. Again, the practical solution lies in making a decision in favor of what seems to be the most probable theory. The decision will tend to be slanted to reflect the opinions of those individuals having the greatest prior knowledge of the subject involved. Several or many opinions need to be weighed within the community of scientists. Peer reviews are sought; scientists debate freely in open forums or in published contrary opinions and rebuttals. The other scientist listeners or readers who have no vested interest in any of the alternative hypotheses will also play a part to the degree that one alternative gains group favor, while another loses it. But here we are getting ahead of ourselves and venturing into the sociology of scientific research.

Relativism—What Does It Really Mean?

At this point we need to insert an explanation of relativism, because the term has at least two distinct meanings in philosophy, and they can be and are easily confused, both unintentionally and intentionally. The first definition refers to sensory perception, as in the above paragraphs: Percepts differ among individuals observing the same object. This idea we have already made clear. Traditionally, in what is called "Protagorean relativism" (after Protagoras the Sophist, ca. 490-429 B.C.) this form of relativism goes further and implies that each person's percept of some object is just as sound (correct, valid) in the sense of being "true," as that of the next person. Professor Peter Angeles gives the following example:

> X says "The wind is cold." Y says "The wind is warm." Neither statement is incorrect. Neither X nor Y is uttering false statements. Both statements are true *relative* to how X and Y perceive (feel) the wind. No method or standard exists which transcends those perceptions and which can be used to determine which statement is true and which is false. (1981, p. 245)

(The italics on "relative" are mine.) Relativity of sensory perception—*perceptual relativism*—is unfortunately used as a weapon for science-bashing by those holding antiscience bias—the New Age people who promote pseudoscience, for example—and the Christian creationists who hope to show that "after all, both religion and science are belief systems."

A second definition of relativism applies to ethics and morals: "Ethical truths depend on the individuals and groups holding them" (*Webster's Ninth New Collegiate Dictionary*). Here, in *ethical relativism*, value theory is involved, and it (like religion) falls into the belief field of knowledge. Ethical values "differ from society to society, person to person; are conditioned by the pecularities of the society in which they arise; are not universally applicable at all times or in all places; are correct or incorrect, desirable or undesirable only relative to whether or not they conform to a common norm or to common acceptance" (Angeles 1981, p. 244). We will refer to ethical (moral) relativism in Part Two (Chapter 12). Be on the lookout for a grevious error of logic, in which an author will deliberately switch the two meanings of "relativism" in order to deceive you, the reader.

There is an important response to perceptional relativism that has to be dealt with in science, but that is not present in ethical relativism. In science, when two observers come up with different percepts of the same object, that difference cannot be simply ignored and left standing. The difference must be examined, analyzed, and explained, and this process may lead to the rejection of either or both percepts. It may prove that each observer's brain is responding to a different attribute of the same object, and in that case the information content of those percepts may be additive without being mutually exclusive. We can think of two differing or seemingly contradictory percepts as two scientific hypotheses in competition. Perhaps it would be fair to say that, far from being a defect in the scientific method, the resolution of conflicting percepts is the main business of science. When such conflicts arise, science draws in more scientists to bring to bear their independently acquired knowledge and judgment.

Misused, "relativism" leads to such absurd and demeaning statements as: "Science is really no more reliable than witchcraft." When we understand its importance and learn to cope with it in whatever context it appears, "relativism" can become a useful concept, not just a dirty word.

The Role of Presuppositions

The new philosophy of science (NPS) is full of references to *presuppositions*, which are things we take for granted—things we assume to be so. They need to be cast in the form of propositions, required for the application of formal logic. They are the first or initial propositions to be set down, and logic does not require that they must be derived from experience or any kind of empirical evidence. Presuppositions are commonly described by the adjective *a priori* (L. "from the former"), one meaning of which is "formed or conceived beforehand" (*Webster's Ninth New Collegiate Dictionary*).

Some of the presuppositions of science consist of certain existential propositions necessary to do the business of science, for example: matter exists; space-time exists; the human brain exists; matter and space-time exist externally in relation to the human observer. The term "presupposition" is used quite loosely to refer to any proposition or set of propositions that is assumed to be "true." Thus, any scientific hypothesis (theory) can serve in the role of a presupposition if only we assume (postulate) that it is "true" and go about our business of testing it against observation statements.

Before going further, it is important that you have a good grasp of the subject of scientific revolutions, covered in Chapter 3. Thomas Kuhn's concepts of normal science, revolutions, and paradigms fall into the dawn period of the new philosophy of science. His ideas were set up in contrast to the view of the logical empiricists that the finding of a single counter-instance during the testing of a hypothesis would immediately result in its falsification and summary rejection. What Kuhn seems to have been telling us is that although scientists are fully aware of counter-instances (i.e., cases in which evidence clearly runs contrary to what the ruling hypothesis requires to be the case), they may simply ignore them, or they may consider them minor problems to be investigated and solved, or they may modify the hypothesis to accommodate them. The last move amounts to changing the presuppositions that reside in the hypothesis. The scientists are extremely reluctant to give up their ruling hypothesis, or paradigm, because it has become a presupposition carrying more weight in their minds than contrary evidence from observation. The logical empiricists would scold the scientists for their reluctance to discard the defective hypothesis; they would say: "It is very naughty of you to let your presuppositions govern your actions. You know very well that a single counter-instance requires you to dump the hypothesis immediately."

Scientists generally pay little attention to what the philosophers of science say, and perhaps there has been in the past good reason for them to take that attitude. Something in their collective experience seems to tell them that the hasty abandonment of a hypothesis is unwise; that it is well worth the effort to work around the problem in order to save the hypothesis. Harold I. Brown gives examples of how this reluctance to dump the ruling hypothesis eventually paid off in a satisfactory solution, which turned out to be the later discovery of new evidence that changed the counter-instance into a conforming instance (1977, pp. 95-101).

One example has already been described in Chapter 3. It is the case of the initial failure of Newtonian mechanics to account for the observed orbit of the planet Uranus. Rather than dump the Newtonian theory, astronomers reasoned that the existence of another planet, not as yet observed, could account for the anomalous orbit of Uranus. Two astronomers, Leverrier and Adams, working independently, calculated the orbit of this missing planet, and it was eventually observed as predicted—the planet Neptune. Holding off on rejection of the Newtonian theory paid off; adherence to it as a powerful presupposition had proved the wiser response to the presence of the anomaly.

Notice in the above case how the presupposition of the theory (hypothesis) served to guide the investigators in their research. Harold Brown considers the role of a presupposition to be a positive force in guiding scientific research (1977, p. 97). Guidance by presuppositions is a characteristic of normal science, but this is a phenomenon of science denied importance by the logical empiricists, who would regard that guidance as an act of faith unworthy of empirical science. So there you have a clear point of distinction between the NPS and the LE.

Brown goes further to point out that having on hand the presuppositions of a hypothesis enables the investigator to decide which phenomena to study and which are irrelevant (1977, p. 100). I think of it this way: Would you go to a lumber yard to buy materials for a new house without having any prior plan of that house? Having a set of plans with you enables you to concentrate on the materials you need, disregarding many other items of no significance. Brown summarizes the points we have made as follows:

> In order to carry out meaningful research we require a research problem and some criteria of what evidence is relevant to its solution. More fundamentally, we require some basis for deciding what research problems are worth pursuing. It is our accepted theories, the systems of presuppositions to which we are already committed, which provide this basis. And because we always do research within a system of presuppositions, both our problems and our data are thoroughly theory-laden. (1977, p. 101)[1]

Presuppositions seem to come in levels of generalization or abstraction. At the uppermost level are the axiomatic propositions (a priori propositions), such as the existence and nature of space and time, or the causality presupposition reading "Every event must have a cause." These can be thought of as "necessary, eternal truths; i.e., there is no process by which they can be changed" (Brown 1977, p. 102). Below that level are presuppositions that are parts of limited hypotheses (theories) about specified subject areas, such as the universe, a star, or an electromagnetic field. The higher level can be said to govern the level below it. Thus, the causality presupposition sets the overall pattern for scientific research, which is to look for causes.

At the level of major hypotheses (theories), consider these examples of a presupposition contained in a hypothesis. For Copernicus, the prevailing presupposition was that the orbits of all celestial objects are circular. Before him, the prevailing geocentric (Ptolemaic) view was that the earth lies at the common center of those circular orbits. In suggesting that it is the sun, rather than the earth, that lies at that center, Copernicus continued to adhere to the presupposition that all celestial motions are circular. It remained for Kepler to reject that presupposition and to put in its place the new presupposition that celestial orbits are elliptical. Turning to Einstein's relativity theories, the previous Newtonian theory of mechanics had taken as its presupposition the Euclidian view of physical space. Einstein found it necessary to replace that presupposition with a new one, namely, one based in Riemannian geometry (Brown 1977, p. 105). The lesson seems to be that when a counter-instance is discovered that threatens falsification of a hypothesis, disaster can perhaps be avoided by changing one of the most important and fundamental of its presuppositions.

Has all of the above discussion of presuppositions added anything new or important to what we previously presented in Chapter 3 about the evaluation and modification or replacement of hypotheses? Perhaps we have only added a buzzword that we could have done without. There is, however, something of value in knowing about the philosophers of science and their writings on presuppositions. Science has come under increasingly severe attack over the last two decades or more by the pseudoscientists and others—religious fundamentalists, for example—who regard science as an evil institution to be feared. They look for arguments they think will weaken science and perhaps reduce it to impotence. The motives seem to be emotionally based, and in that sense not empirically rational. These science-bashers have latched on to the buzzword "presuppositions" and claim that science rests on faith in its presuppositions, just as, say, religion rests on faith in the presuppositions that God exists and that God is omniscient and has other supernatural attributes. They say, in effect: "Your science is a belief system just as ours is. You give us a lot of fancy talk about explaining the real world, but your explanations are rooted in belief in presuppositions that have no empirical basis." They go on to say that scientists change their presuppositions as easily as a person might buy a new set of clothes and trash the old ones just to keep in style. They can quite legitimately ask: "Are we to believe what science says today, when tomorrow a completely new set of theories and explanations will be forthcoming?" This criticism is augmented, not diminished, by the claims of the new philosophy of science that all science is fallible, admitting that what we say is the case today may, indeed, be set aside tomorrow. It's all very well to pass off this criticism by saying that, "after all, scientists are humans like anyone else and can make mistakes like anyone else," but that disclaimer does little to soften the claims against the integrity of science. In earlier paragraphs we reached the conclusion that since neither corroboration nor falsification works, science boils down to making group decisions in a manner that seems suspiciously subjective. Is that any way to instill confidence in a public that knows precious little of the content of science? These are perhaps gratuitous questions, but you should be thinking about them if you are or plan to become a scientist.

Changing Concepts in Science

We have seen that the new philosophy of science places great emphasis on a historical rhythm of change that we might describe as punctuated equilibrium (taking our cue from evolutionary biology). Scientific revolutions—episodes of great turmoil and rapid change—punctuate long periods of stasis (little or no change) when normal science prevails. We have thus far emphasized the change in presuppositions that occurs in a scientific revolution. To this we now add that revolutions typically require also that there be conceptual change—a change in relevant concepts that results in key terms getting new meanings or being placed in new classes. Harold I. Brown states the changes accompanying a revolution:

> On the deepest level two kinds of changes take place: both the presuppositions of a science and the concepts used in it are transformed and, as a result of these transformations, the world, or meaning structure, within which the scientist works and his research problems are also altered. (1977, p. 111)[1]

To understand what is meant by changes in concepts we need examples. Brown gives us several. One is the revolution in which Copernicus replaced the geocentric theory (Ptolemaic theory) with the heliocentric theory. Previously, we noted that the presupposition of circular orbits remained intact. So, what then was the significant change that occurred other than the obvious main proposition of the theory? Under the older theory two physically different systems were recognized. One was the terrestrial sphere (our planet) made up of fire, air, water, and earth (the four "elements"). The other system was the realm of the heavens, which consisted of spherical concentric shells, each containing a planet or or moon, and an outermost one on which the stars were positioned. The shells rotated on a common axis, so that the attached objects (stars, planets, moon) necessarily followed circular paths. More important, perhaps, was the concept that this heavenly system consisted of an entirely different "element" than the four in the terrestrial sphere. This strange element was an etherial substance ("ether"), thought to make up all heavenly space.

So far, we understand that the presupposition of circular motion of planets and stars is common to both theories. (We will for the moment disregard other preconceptions that required a change.) What, then, was the conceptual change that occurred? Before Copernicus, the earth was not a member of the class of revolving objects (the planets, loosely speaking), but instead was a member of the class of nonrotating objects fixed in space; whereas with Copernicus it became a member of the planetary class. Before Copernicus, the sun had been a member of the class of planets; afterwards, it was a member of the class of nonrotating objects fixed in space. (Disregard the sun' s rotation on an axis.)

The above changes are considered by Brown to be conceptual changes. A problem here is that the words "concept" and "conceptual" are very broadly defined. (Refer to Chapter 1 for definitions of "concept" and "percept.") Presuppositions are also concepts, as distinct from percepts. Changing over to the idea (concept!) of *classes* and definitions of *terms* within classes will be useful here. To change a thing from one class to another amounts to redefining it. According to Brown, the changes in class of the earth and the sun, described above, amount to changes in the meanings of the terms "earth" and "sun."

We are now ready to compare logical empiricism (LE) with the new philosophy of science (NPS) with respect to the meanings of terms. Brown writes:

> The thesis that the meaning of scientific concepts [terms] changes as a result of a scientific revolution has been regarded by many logical empiricists as one of the most outrageous claims of the new philosophy of science. It has long been a central doctrine of empiricist thought that the meanings of terms are completely independent of the propositions in which they occur and that we can accept or reject propositions without this having any effect on what we mean by the terms which occur in them. (1977, p. 116)[1]

As I see it, this supposed major difference between LP and NPS may be dissolved, or at least mitigated, if only we agree that a scientist who decides to transfer an object or thing (a term) from one class to another, and thereby require a change in its definition, be very careful to include the new definition and call attention to what is being done. Perhaps a better solution would be to coin a new word for the restructured term.

Many and perhaps most words in our dictionaries are shown to have two or more meanings, and these are carefully numbered and spelled out. Language is an artifactual product of human history and it would be completely unrealistic to think, as the LPs did, that science can by fiat free itself of semantic problems.

Let us follow up this discussion by another example—one from twentieth-century science. The subject is the meaning of "mass" in each of two paradigms, one having succeeded the other through a scientific revolution. Mass, under Newtonian mechanics, is an independent variable in all equations—force, energy, momentum, work, et al.—in which it appears. Moreover, mass of a given combination of protons and neutrons never changes; i.e., the atomic mass number of a given isotope of a given element is always the same, regardless of the velocity or acceleration by which this mass term is multiplied. Thus, in the equation for kinetic energy, (1) $E_k = \frac{1}{2}mv^2$ (energy equals one-half of the product of mass and the square of the velocity), if we preselect a unit of mass, such as 1 kg, the mass term is effectively a constant in the equation. Under Einstein's special theory of relativity, the situation is quite different, because mass itself varies with the velocity. The classic equation is (2) $E = mc^2$ (energy is equal to the product of mass and the square of the velocity of light, c). The two equations are identical in structure and in the dimensions of the terms they contain, but they have two quite different meanings; i.e., each is telling us a different story. Equation 1 tells how a change one variable relates to a change in either or both of the other two variables. Equation 2 tells us that energy and mass are interchangeable entities, i.e., it defines energy in terms of mass and mass in terms of energy. In the newer paradigm, we can say that mass always increases as its speed of motion increases (approaching infinity as the speed of light is approached). Believe it or not, energy can be converted into mass, and mass into energy—conversions that are observed both in nature and in laboratory experimentation. Clearly, "mass" has different meanings in the two equations. Consider also that under relativity both time and distance (length) shrink as speed increases, so those terms must also be redefined. We can say that in this scientific revolution, not only did presuppositions change (from Euclidian space to Reimannian space), but terms going by the same words (mass, time, length) underwent drastic changes in meaning. Perhaps Brown's term "conceptual changes" is apt at this point.

This is a good place at which to emphasize the parallel relationship between philosophy of science and science itself. Both of these subjects can experience a revolution that alters both their presuppositions and their concepts. Harold Brown compares these two subjects:

> The notion of a scientific revolution is a philosophical, not a scientific, notion. It is a notion used in constructing a theory of science. But just as scientists make use of data supplied by observation and

> experiment in constructing their theories, so the philosopher of science makes use of the data supplied by the history of science in attempting to construct a philosophical theory of science. (1977, p. 127)[1]

Brown goes on to say that the NPS was precipitated by increasing realization by philosophers that LE contained a number of severe anomalies that it could not resolve within its presuppositions. LP simply did not agree with observations of what scientists were actually doing in the pursuit of science. Obviously, the scientists were not about to change their successful methods just to make them agree with the tenets of LE. Something had to give, and of course it was the philosophy that yielded. The paradigm shift from logical empricism (LE) to the new philosophy of science (NPS) is thus a revolution in the philosophy of science.

The New Epistemology of Science

Epistemology, the study of knowledge, confronts at least three basically different problems. One raises the question of what it means "to know" or "to have knowledge," a question related to the formation of percepts and the formulation of concepts. A second question asks whether knowledge must be true rather than false (infallible versus fallible) in absolute terms. A third asks us to classify and describe the various fields of knowledge. The first question was dealt with in Chapter 1 under the heading of perception and conceptual thought. The second question we will tackle in Part Two. We discussed fields of knowledge early in Chapter 1, noting two great classes of cognitive fields—research fields (empirical fields) and belief fields—and giving some examples. Of course, we could not have set up such a system without having some presuppositions in answer to the first two questions.

In recognizing a wide range of cognitive fields, we were using a presupposition as to what is included in the meaning of the term "knowledge." We used a very broad base of meaning, so broad indeed that ideas generated by our imagination and totally divorced from any external reality (to the extent that this is possible) must be included within the definition of knowledge. Thus, what is commonly known as "pure fiction" (say, a novel about imagined humanlike beings on an unknown planet a million years ago) is recognized as a form of knowledge. It follows, then, that such fiction is quite as legitimate as knowledge for which a substantial base of observational evidence exists. The only criterion of rejection of any proposition or set of propositions as knowledge would then be that the statements made are self-contradictory and/or violate the established meanings of words. We could, even so, simply classify such nonsensical statements as a special form of knowledge, to be known as "pseudo-knowledge."

Professor Mario Bunge's broad-based recognition of classes of knowledge, described briefly in Chapter 1, represents the new view of epistemology. As such, it can be incorporated into the new philosophy of science (NPS). It does not, however, deal with the question of infallibility/fallibility of scientific knowledge, all of which must be of the empirical class, based on observation of nature.

The old view of knowledge goes back to Plato in claiming that if a proposition is false, it is not knowledge at all. By the same token, if a proposition can be shown to be true, it is genuine knowledge. By this view, infallibility is "a defining characteristic of knowledge" (Brown 1977, p. 145). The logical positivists went along with this ancient criterion of genuine knowledge. They were obsessed with the model of mathematical logic set up by Russell and Whitehead and others, not much earlier. Wittgenstein, using this model, thought he had found a foolproof way to arrive at the truth of any empirical proposition. The steps were clearly laid out and needed only be followed, as a child learns to do arithmetic by a set of rules (by use of algorithms). Each deduced consequence of a theory could be matched to an observation of nature, and thus absolute truth was possible of attainment by a totally objective process of corroboration. Logical empiricism, Popper-style, changed the rules to admit that only falsification was possible, but that process simply allowed a statement to be booted out of the truth column and relegated to the false column. The concept of attainable truth persisted.

The new epistemology sees truth and falsity only as unattainable goals—to be yearned for but never to be found. They are targets to aim at, but they will forever recede beyond the reach of the bullet fired at them. Only the probability that a proposition may be judged true or false is an entertainable concept, the new epistemology states. The key word in the foregoing sentence is "judged." It means "informed judgment." The key idea is a method of "application of informed judgment," to which we add "by scientists who have the same kind of brain as anyone else."

How is informed judgment to be applied? Brown tells us that this is done through the application of *reason* (1977, p. 148). The judgment must be *rational.* Reasoning does not necessarily require conscious adherence to the rules of formal logic, and in that sense is not rote adherence to an algorithm. Brown does not, however, define or otherwise make clear what it means "to be rational."

Of Brown's two words (reason and rational), "to reason" (reason, reasonable) is the more general. A carefully composed definition of *reason* and *reasoning* (as verb forms) might read: to engage in a conscious mental activity purposefully to achieve a goal, such as to persuade, deceive, or instill doubt in one's own mind or that of another. Typically, the synonym for the adjective "reasonable" is given as "rational." "Rational" (to rationalize; rationality) is typically defined by a reverse tautology: *Rational* (adj.) means "possessed of reason, or conforming with reason." The synonym given for rational is "reasonable"; for *rationality*, "reasonableness."

In the new epistemology, decision based on informed judgment, arrived at by the rational process, is the crucial mental activity of science. It is the application of prior knowledge that sets apart the new philosophy of science from its predecessors. At the same time, application of informed judgment forever sets aside the establishment of truth or falisity. Banished from science and all empirical knowledge, truth and falsity find a safe haven only in the belief fields of knowledge. There, truth and falsity run free and unconstrained in such fields as religion, ethics and morality, and the arts. And, of course, truth and falsity are key concepts in pure mathematics and formal logic, both of which lie outside of empirical knowledge.

To the above, we need only to add that informed judgment gains in strength (in the sense of reliability) in proportion to the number of trained and experienced scientists (in the stated field of specialization) who pool their individual judgments and as a group arrive at tentatively acceptable statements. This is the principle of consensus. It means more than mere additive strengthening, because it involves exchange of information, which may relate to validity of observational data or may consist of challenges aimed at interpretations of data.

Seen in the above light, science has been emancipated from the intolerable burden of establishing ultimate and enduring truths. Emancipation has its price, which is in the demotion of scientists from the class of gods to the class of mortals. Most scientists can adapt to this new image; they are easily made happy by access to unlimited research and travel funds, the latest in hardware and software to play with, enough journal space to publish all the papers they write, a supply of plastic graduate students, and no end of interesting problems to solve.

As pursued by ordinary mortals, scientific research is impacted by many of the same forces that all knowledge fields experience. Scientific research is influenced by and responds to forces exerted by sociopolitical ideologies and ethical/religious systems. Some of these influences are investigated in the remaining chapters of Part One.

Philosopher of science Harold I. Brown, on whose major work on the new philosophy of science I have relied rather heavily for guidance and insight, gives us a good statement with which to close this section:

> Our central theme has been that it is ongoing research, rather than established results, that constitutes the life-blood of science. Science consists of a sequence of research projects structured by accepted presuppositions which determine what observations are to be made, how they are to be interpreted, what phenomena are problematic, and how these problems are to be dealt with. When the

presuppositions of a scientific discipline change, both the structure of that discipline and the scientist's picture of reality are changed. The only permanent aspect of science is research. (1977, p. 166)[1]

The New Sociology of Science

So far in this account of the new philosophy of science I have not presented the view from sociology. There is a sociology of science. Sociologist Norman W. Storer, a professor in Baruch College of the City University of New York, has made a specialty of the sociological study of science and scientists. To give us that perspective, he begins with "a sketch of science as a community, or as a coherent social system" (1977, p. 29). He finds a clear distinction between this community and other sectors of society; its principal product is "organized, certified empirical knowledge." As with the members of any social group, "the distinctive relationships found among a set of people occupying certain social positions are due to their sharing a set of norms—standards of proper behavior—that tell them how to behave with respect to each other" (p. 30). Storer draws upon the work of Robert K. Merton (1973) to identify four norms "that are central to the ethos of science."

First, the value attached to a scientific statement must in no way be connected with the personal characteristics of the scientist who makes that statement. Merton's term for this principle is *universalism*, but it does not seem to me to give any clue to the norm intended, namely, that the strength or weakness of a hypothesis proposed by a scientist must be considered strictly on its scientific content and supporting evidence. It should make no difference whatsoever that the scientist is of a certain race, religion, sex, age, political affiliation, and so forth. Perhaps we should call this norm "depersonalism" or "apersonalism."

Second, findings made by one scientist must be shared freely and openly with the entire scientific community. Publication of such findings is thus a moral obligation. This is the principle of *communality*. Third, scientists must practice *organized skepticism*. Each scientist must scrutinize the publications of others in the same area of specialization and express his or her criticism in print, in journal articles, reviews, and letters, as well as orally from the floor of a meeting room or a seat on the debating stage. This activity is a form of mutual policing needed to sustain a high quality of published scientific information. Perhaps the most important part of the policing action occurs through peer reviews of articles submitted to scientific journals. Reviewers must take their job seriously; they must search closely for errors in observations and weaknesses in arguments. They receive no monetary reward for this service, which draws time from their own research programs, but it is to the mutual benefit of all.

A fourth norm recognized by Merton is what he calls *disinterestedness*, meaning that a scientist's research should not be guided by desire for personal rewards. He refers to such rewards as private economic gain, glory in the eyes of the nonscientific public, and even the honors and medals awarded by scientists to each other. We must be careful here to emphasize that such personal rewards are essentially excrescences or trappings that do not always accurately measure the quality of the scientific work of the individual. We of university experience know that nearly every senior professor has a following of former students who conspire to get the "old prof" a medal or prize. Award committees rely mostly on the number of nominating letters received in support of a candidate for the honor.

Of the four norms presented here, this last one is least likely to be observed within the scientific community, and is so often flagrantly violated that it is perhaps little more than a sham. I can assure readers who are unfamiliar with the academic profession that nearly every scientist seeks to maximize private economic gain in one way or another, and many try to get public exposure through the news media. Many (with thinly veiled understandings of reciprocity) encourage colleagues to come through with an honor.

The third and fourth points, taken together, are embodied in science philosopher John Ziman's proposition "Science Is Public Knowledge"—the title of an essay he first published in 1968. Regarding this assertion, he wrote:

> What I mean is something along the following lines. Science is not merely *published* knowledge or information. Anyone may make an observation, or conceive a hypothesis, and, if he has the financial means, get it printed and distributed for other persons to read. Scientific knowledge is more than this. Its facts and theories must survive a period of critical study and testing by other competent and disinterested individuals, and must have been found so persuasive that they are almost universally accepted. The objective of science is not just to acquire information nor to utter all non-contradictory notions; its goal is a *consensus* of rational opinion over the widest possible field. (1980, p. 40)[2]

Obviously, consensus is arrived at by a complex social activity within the scientific community.

Storer turns next to examine the driving force behind scientific research and publication (1977, p. 31). He follows Merton in identifying it as "the scientist's interest in acquiring professional recognition." Professional recognition has a meaning here quite distinct from the personal rewards listed above. Recognition is judged primarily in terms of acceptance of one's scientific reports for publication in journals operated by peers in one's own field of specialization. Peer reviews serve to let pass only the highest-quality products, while the excess in number of submitted manuscripts over the number capable of being accommodated makes competition severe. Journals that can be the most choosey confer the highest value upon the papers they publish. Thus, faculty committees whose members must evaluate a colleague for tenure appointment tend to place higher value on the candidate's articles that have appeared in the more prestigious journals. Another source of professional recognition comes from the citation of a scientist's published works in the texts and bibliographies of other scientists' works. A high frequency of citation is equated to high value of the product. The common practice of citing papers written by one's friends, students, or those of shared opinions, even when such citations are not essential to the content of one's paper, pays off in the favor being returned by getting more citations of one's own work in the works of other authors.

Professional recognition also comes from the general excellence of the chair that a professor occupies. The prestige of the university attaches to the chair and its occupant, as does the prestige that the particular science department enjoys, both nationally and internationally. The same effect operates in terms of one's position on the staff of a private or public research institution—for example, appointment to the Institute for Advanced Studies at Princeton, New Jersey.

What in the above description of science as a social activity can be taken as the old philosophy of science—particularly logical positivism (LP) and logical empiricism (LE)—as distinct from the new philosophy of science? Much of what Storer has said in his description of four norms of science applies in full force back to the beginning of our century and includes the scientists who were contemporaries of Wittgenstein, Popper, and others of the Vienna Circle. Of course, LP and LE were directed at physicists, almost exclusively, but the account would fairly include scientists of all graduate science faculties of European and North American universities. Nevertheless, changes in the philosophers' conception of sociology of science can be recognized as being at least concurrent with the arrival of the NPS.

John Ziman recognizes changes in the perceived sociology of science associated with the shift from LE to the NPS (Ziman 1980). The logical positivists and empiricists seem to have simply ignored the entire social setting in which scientists were working then, as now. Instead, these philosophers wrote only of a methodology of processing empirical data—a prescription best described as application of a set of algorithms to be followed in total isolation and complete insulation from influences of other scientists working and publishing on similar classes of problems. Referring to the LP/LE program of research as the "logico-inductive metaphysic of Science," Ziman comments: "How can this be correct, when few scientists are interested in or understand it, and none ever uses it explicitly in his work?" (1980, p. 40)[2]. Again, we end up with the conclusion that LP/LE was a prescription, or norm (the *ought*), egregiously at odds with what scientists actually do (the *is*); whereas the NPS has focused on the *is*, abandoning any pretensions of formulating an *ought*. Perhaps this is better stated as a new philosophy that simply says: "Scientists are doing what they ought to be doing."

Is there a New Sociology of Science (NSS)? Indeed there is! What's more, the NSS has a strange, disturbing quality as seen by the working members of the scientific community. Merton's version of the "old" sociology of science was easy to grasp as an ethical system prescribed for research scientists. In the words of philosopher Mario Bunge, Mertonian sociology of science "distinguished the conceptual content of science from its social context" (Bunge 1991b, p. 534). The suggestion that natural science is "ideologically committed" would never occur to those of us who in the 1950s investigated astronomy, geology, or organic evolution. How could the study of the evolution and extinction of the dinosaurs some 65 million years ago possibly be influenced—let alone controlled—by the prevailing sociopolitical systems of our respective native countries? (Yes, there was Lysenkoism, a deviant theory of genetics thoroughly discredited in the West, imposed upon bioscientists of the Soviet Union by its political rulers, but insistence by a police state that its political or racial views be scientifically supported and confirmed is not what we have in mind.)

Bunge tells us that the NSS, and in particular an extreme variety known as the "strong program," holds "that *all* knowledge is shaped by society and moreover is somehow *about* society, that is that it has a *social content*—whence ultimately there would be no content/context distinction" (1991b, p. 534). This program goes by the name of *externalism* (p. 524). The strong program of the NSS has even been applied to mathematics; i.e., the proposal that pure mathematics is *about* society (p. 535). We have, of course, made clear that pure mathematics is in the ideational knowledge field—a construct of the human mind—but the NSS lumps mathematics and science into one category, along with sociopolitical ideologies. Perhaps the strong program of NSS is applicable in some degree to the sociology of social science, by obvious reason of the social content of the latter.

In contrast to the externalism of the NSS, Bunge recognizes the *internalism* of the Mertonian sociology of science. The latter, as we have seen, emphasizes that science "has an ethos of its own," summarized in Merton's four characteristics of science discussed earlier in this section: universalism, communality, disinterestedness, and organized skepticism (Bunge, p. 532). Moreover, Merton's school "assumed the *uniqueness* of science" that derives from those four characteristics. That uniqueness, says Bunge, is denied by the NSS (p. 533).

In the strong program of NSS there is also an unsavory odor of *rela-*

tivism, defined here by Bunge as "the thesis that there are no objective and universal truths" (p. 524). Taking this package as a whole, the aim of the NSS seems to be to degrade and demean science by much the same flawed relativistic argument used by the New Agers to downgrade science to the same level as their pseudoscience (see Chapter 8).

Bunge summarizes very clearly the absence of any meaningful interrelationship of natural science with social structures:

> But, of course, no one has ever discovered anything about social structure by studying, say Maxwell's equations or the way in which electromagnetic field intensities are measured. It so happens that natural science is not built in the image of society and with the aim of reinforcing social order, rather, natural science is supposed to explore and represent nature . . . (1991b, p. 540)

Turning to the question of whether the internal operating mode of science has undergone a change, I suspect that nearly all scientists now in the seventies and eighties of their lives can see many changes in that aspect of sociology, some of which they will rate as good and others as bad. Compare 1940 with 1990. In 1940 many (perhaps most) researchers worked as individuals in exploring specific problems, and this is reflected in the preponderance of single-author papers in the journals of that time. Funding, when there was any, was largely from the university itself or from the scientific societies through small research grants. Government funding through contracts with universities was almost unknown, and private for-profit corporation funding was limited to projects directly connected with industrial development and technology. Following completion of one's formal graduate study program, the doctoral research was pursued as an individual enterprise directed, of course, by one's professor, but often at one's own expense. Two ways to proceed were open. One was to take a teaching job and perpetuate the academic support cycle. Another was to take employment with a government agency involved in research that often included basic problems of science and had an adequate budget for lab or field work. For geologists, the state and federal geological surveys exemplified this pathway. The doctoral dissertation could sometimes identify with a government research need and thus enable it to be completed. One salient feature of this era was the ready availability of prime journal space for one's research papers. Perhaps this total program could be characterized as one of individual effort and limited resources compensated for by considerable freedom of choice of problems to be studied. To this might be added that within one's own chosen specialty there existed only a small circle of colleagues and it was possible to meet with them at the annual conventions and to exchange correspondence freely. All and all, it was a small (and wonderful) world!

In 1990, perhaps the most conspicuous features are these: Science is done in tightly knit groups, each a mix of senior (staff or faculty) persons and their students or junior staff, all of whose names appear as authors on a single journal publication of their results. Enormously escalated costs, forced by the need to use complex machines and often to deploy them in remote places, can only be met by large-scale funding by government agencies, both civilian and military. Whereas in the 1940s it was typical of the university to support the research of its faculty members, in the 1990s it is typical of the faculty members to be required to generate their own funding and that of their students (and even to fund their own salaries). What used to be easy access to journal space at no cost is now a desperate search for acceptance by almost any journal and on top of that to pay exhorbitant page charges for the privilege.

Science and Society

When we think about the social nature of science, it is helpful to realize that two basically different subject areas are up for consideration. In foregoing paragraphs we have been mostly concerned with "within-science" social relationships; i.e., what goes on within science departments and research laboratories and the problems those scientists face in carrying out their objectives. A quite different subject area is the relationship of science to all of the society that contains it. Here we ask such questions as: What, if anything, is science supposed to do for the benefit of society as a whole? Does a science funded largely by taxpayers' dollars have an obligation to direct its research to a shopping list of societal problems to the neglect or exclusion of interesting problems almost totally unrelated to the contemporary scene? (Examples of the latter might include the origin of galaxies, the process of macroevolution, or the existence of a suspected new quark.)

On this problem of obligations there will be two sets of questions and answers. (1) What do the scientists as a group say on this point? (2) What does society at large say about the obligations of science? First, let us find out what the contemporary philosophers of science say should be the broader social concerns of science. Here, I quote from an essay on the new philosophy of science by philosopher Marjorie Grene of the University of California at Davis; she has published extensively on the philosophy of biology:

> Scientists are real people trying, like other creatures, to solve problems presented to them by some aspect of the world that specially concerns them. . . . To insist that scientists are real people facing real problems does not mean in this context that they are concerned with practical problems—how to cure cancer or clean up pollution; nor does world here mean environment in the environmentalist's

> sense—oil slicks or acid rain. The problems of scientists are questions about how something in the real world really works, whether it be the reproductive system of the aardvark, the origin of galaxies, or the production of a protein by a particular gene or gene complex. . . . The oddity of science is that it is a specifically theoretical inquiry about what is actually the case, not about what is useful or just aesthetically appealing. It is the attempt to orient oneself in one's own scientific environment that constitutes the scientist's activity as scientist. (Grene 1985, p. 2-3)[3]

That pretty much sums up the research scientist's preference, should he or she be in a position to have the freedom of choice—which many don't.

Turning to the second question, we need to realize that the relationship of science to society has certainly changed over the past half century. In 1940, science was not generally viewed by the public as having an obligation to improve the state of the nation or its security. Scientific discoveries were favorably received simply as additions to knowledge of the universe, while human benefits arising from discoveries were gratefully accepted as a bonus of research. Alamogordo changed all that in 1945. In 1990, science is heavily pressured from public and government alike to save the nation and the world from a host of burgeoning and seemingly insoluble problems in almost every conceivable category. Because the science machine is driven by energy (funding) from the same sources that are making the demands for its product, science has been forced to trade its freedom of choice for servitude to the short-range goals of the society that encloses it. True, there are many interesting scientific puzzles to be solved on the road to finding a vaccine for AIDS or an answer to the problem of acidified lakes now barren of trout. The catch is that many of these interesting puzzles are clearly side branches that must be regretfully passed by in order to reach the society-oriented (society-financed) goal. Always there is the public background outcry expressed through the media: "Find that cure!"

The new philosophy of science feeds beautifully into the changing public image of scientists and their product. If science generates only fallible results derived by use of subjective methods, then science differs in no fundamental way from political practice and sociopolitical ideologies or from ethical or religious formulations. That being the case, science must submit to being used and abused as are all other forms of knowledge.

Perhaps there is something more to be said that might differentiate science from sociopolitical ideology and its nonempirical classmates in the total spectrum of knowledge. Kenneth E. Boulding has given us some suggestions along this line by pointing out that science has "a very distinctive ethic of its own" (1984, p. 146). He says that it is a "four-fold way" having four essential components:

> The first is a high value on curiosity, which not all cultures possess. The second is a high value on veracity—that is, not telling lies—which many other cultures also do not possess. . . . Error is often pardonable, but lying is a sin that cannot be forgiven. (P. 146)

Boulding's third ethical principle "is the high value on the testing of images of the world against the external world that they are supposed to map." Ethics comes in when we compare sociopolitical ideology, ethics, or religion against the same standard; all of them can claim anything they choose to claim about what is good, right, or true without the slightest need that it be testable by observation of nature. "The fourth principle of scientific ethics is abstention from threat, embodied in the principle that people should be persuaded only by evidence and never by threat" (pp. 146-47). As Boulding points out, this ethical principle "is in striking contrast to the ethics of many religious organizations and of all political organizations."

Scientism—What Is It?

Continuing a bit further on the theme of how society views science, consider a negative attitude or antiscience sentiment that has resurfaced within some sectors of the public in the past two or three decades. We will encounter this attitude in later chapters on pseudoscience cults and the highly-touted "New Age" movement. There, we speculate on the causes of such negative attitudes.

Scientism is term you will encounter in the writings of those critical of science as a public institution. When the suffix *ism* is attached to a term that describes a field of knowledge, the connotation can be drawn that it is a belief field (although not always the case). In religion this meaning is obvious, as in deism, theism, Taoism, Judaism, or Buddhism. Correspondingly, in sociopolitical fields we find racism, socialism, communism, Leninism, or Nazism.

The meaning of "scientism" varies with context, and you should be aware of these variations. Here, I quote from Peter Angeles, *Dictionary of Philosophy*, 1981. From within the philosophy of science, the term can refer to the logical positivists' assertion that "science is the *only* method for obtaining knowledge" (p. 251). Later, the logical empiricists decided to insert the qualifier "reliable" in front of "knowledge." From outside the science community, "scientism" has a strongly pejorative connotation: "the unwarranted idolization of science as the sole authority of truth and source of knowledge" (p. 251).

Harvard University science historian I. Bernard Cohen traces scientism back to Henri Saint-Simon (1760-1825): ". . . although he was virtually ignorant of science, he wrote eloquently about the importance of science and he envisaged that scientists would play a crucial role in reordering society. . . . He even dreamed of a religion of science with scientists-priests and a physicist as analogue of the pope" (Cohen 1985, pp. 328-29). In describing Saint-Simon, Cohen uses the phrase "an early preacher of the cult of scientism" (p. 329). Saint-Simon was primarily occupied with sociopolitical ideology, and thus an outsider relative to the science of his time, even though he wrote at length about the nature of scientific revolutions.

Karl Popper also refers to scientism in his later major work, *Objective Knowledge* (1972, revised edition). The context is a discussion of the place of the humanities—especially human history—as a knowledge field. In complaining that his positivism is being denounced by those historians who disagree with him, Popper uses "positivistic" as synonymous with "scientistic" and "scientism" with "positivism" in citing those denunciations (p. 185). He then explains in a footnote: "The term 'scientism' meant originally 'the slavish imitation of the method and language of [natural] science', especially by social scientists." He states that the term was introduced in 1962 by F. A. Hayek, and that he (Popper) subsequently suggested use of the term "as a name for the aping of what is widely *mistaken* for the method of science" (p. 185).

This may seem like a lot of nit-picking over the meaning of a term, but it is important to get the right feel for the word; i.e., that it is used by those outside of science, is pejorative in application, and may be a false view of science.

With the coming of a new philosophy of science that admits to fallibility of science and recognizes a large element of subjectivity in the decision-making process in science, the validity of accusations of scientism has perhaps been effectively countered. At the same time, however, we should recognize that some scientists continue to invite this invective by their sweeping statements of what science can achieve in terms of cosmic knowledge and what it can do for society. As an example of such chauvinism, the distinguished English physicist Stephen Hawking has been criticized for his statement in *A Brief History of Time* that the eventual goal of science is to provide a unified theory that completely describes the universe we live in (1988, p. 13). Writing in *New Scientist*, Chris Beckett, a social worker, takes Hawking to task for mistakenly extrapolating a successful combination of relativity and quantum mechanics as a unifying theory applicable to all fields of human knowledge (Beckett 1990, pp. 60-61). Beckett feels sure that Hawking did not intend his pronouncement to have that universal connotation. Intent not withstanding, it is the kind of statement that invites accusations of scientism. The same lesson applies in the society/science interchange on the Darwinian theory of biological evolution. Public pronouncement by leading biologists and paleontologists that "evolution is a fact" is met by charges of scientism in which the accusers transform "evolution" and "evolutionary" into "evolutionism" and "evolutionistic," respectively.

Is Science a Monolith?

Our final topic in this chapter is on a question of *scientific pluralism*. To have pluralism science would be an aggregation of subject areas, each of which possesses some unique feature or combination of features not found in the others. Uniqueness would be found in classes of data used and methods of investigation and analysis. The antithesis would be called *holism*, the view that science is one monolithic body in terms of its methods of operation—one science, one method.

Marjorie Grene describes the older, holistic view expressed by logical positivism and logical empiricism:

> An aspect of the older orthodoxy, defended by many of its adherents, concerned a program for the "unity of science." If there is *the* scientific method, it was thought, there should also be, when science eventually matures sufficiently, *the* one body of science itself, with its one unified set of axioms, from which ideally all the empirical content of science can be deduced. (Grene 1985, p. 16)[3]

One important point to recall is that adherents to logical positivism and logical empiricism thought of genuine (bona fide) science as consisting entirely of physics. Their earlier writings considered organic evolution and other branches of historical science as outside the fold of science because, they argued, the propositions they contained were not even testable. It was a case of holism by fiat exclusion of anything that didn't fit.

Grene expresses the contrasting position of the new philosophy of science:

> If sciences are practices, on the contrary, not only is this one family of practices distinguished from others, but it comprises an interlocking network of disciplines, each impinging on, but different from, its neighbors. If one becomes a biochemist, one cannot at the same time and by the same training become an astronomer. A neurophysiologist has no competence in paleobotany. Nor is this a case of stultifying specialization. Expertise is essential to every practice. At the same time, neighboring disciplines do impinge on one another. . . .The point is that, on principle, there is not science as such; there are disciplines, mutually interrelated and interacting, that develop as distinctive scientific practices. (1985, pp. 16-17)[3]

Is there a suitable term we might use for the new philosophy of science to parallel "logical empiricism"? Grene suggests *comprehensive realism* (1985, p. 18). The new science is comprehensive in its scope, including many branches that the old view simply disregarded. The new science is realistic in the sense that it recognizes what scientists actually do in pursuing the answers to interesting questions about the real world. (Note in both terms the implication of "-ism" that they are belief systems!)

I can find no better way to close this chapter than with yet another quotation from Marjorie Grene's essay on the new philosophy of science. I like this passage because it reasserts the great strength science possesses and the tremendous body of reliable knowledge it has amassed.

> The option I am proposing, it cannot be too heavily stressed, does not constitute an abandonment of the reality of science as a collection of cognitive achievements or of the authority of science in its appropriate sphere, but an effort to see science as real and as authoritative because it is an important collection of human activities pursued out of motives we consider honorable and with results we respect and admire. Above all, it is a reasonable and rational pursuit, of which we can give a reasonable, and indeed a rational, account, even though—or just because—it is historical and social in its existence and its nature. It is also, and just as fundamentally, as we noticed first of all, rooted in the effort to make out through the use of one's perceptual systems the real structures of the world in which, bodily, we have our being. (1985, p. 18) [3]

This chapter brings to a close our description of science and the scientific method—"once over lightly." At this point, you may feel that you know as much about science as suits your needs. Our next chapter investigates some major problems and issues of science, selected to illustrate further how scientists view the universe and its laws; this is information important to all serious science majors. You may, however, wish to skip to Chapters 7 and 8, which are rather easy-going and enjoyable accounts of a dubious form of knowledge known as "pseudo-science." Everyone today needs to be able to recognize pseudoscience, no matter how skillfully it is disguised.

Credits

3. From Marjorie Grene, "Perception, Interpretation, and the Sciences." Pp. 1-20 in David J. Depew and Bruce H. Weber, *Evolution at the Crossroads: The New Biology and the New Philosophy of Science*. Copyright © 1985 by the MIT Press. Used by permission.

CHAPTER 6

Determinism, Randomness, Chaos, and Quantum Mechanics

In this chapter we consider an important topic in science—determinism versus randomness—that illustrates how a fundamental historical difference in viewpoints was resolved by the rise of a new paradigm. We also examine some of the philosophical implications of quantum mechanics and its application to modern cosmology. As these accounts unfold you will gain a better idea of how scientists think and work to develop not only new empirical knowledge but also a set of theoretical models to accommodate that knowledge.

The Deterministic Model

One of the controversies in science most difficult to explain in simple language has smoldered for decades. It concerns the basic manner in which natural systems operate or, at least, how they are imagined to operate. Suppose that we had in our possession a book containing all physical laws needed to explain anything and everything in the universe. This volume would represent the ultimate achievement of epistemological reduction, discussed in Chapter 4. The question before us now is: How useful are these laws in describing and explaining systems of matter and energy that involve countless individual parts or particles, each of which is engaged in its own form of activity and in interactions with other parts or particles? The book of laws would be extremely useful in solving single-body problems, such as the trajectory of an intercontinental ballistic missile. Knowing with considerable accuracy the initial conditions at the instant of launch, we could use the laws to predict with a high level of accuracy the point of impact. But can the laws be applied to make an accurate description of all the trajectories of, say, 100 billion gas atoms in a closed container at a given instant, and a prediction of their trajectories at a future instant?

Let us approach the problem of the gas atoms in a container, using a similar but much simpler mechanical system for demonstration purposes.

It is a billiard table (a carom billiards table with no pockets) and a set of billiard balls. Let us use six balls for this experiment. Place the balls in more or less random positions over the table. Let six players step forward, each holding a cue and taking up a position to strike a ball. Aiming in any convenient direction will be allowed. At a given signal, the six balls are struck in unison. The balls collide with each other and rebound from the cushioned edge of the table. Because of frictional resistance with the cloth and the air, the balls quickly slow down and finally come to rest. To cope with this difficulty, we must imagine that there is no friction in the system, so all the energy stays in the form of kinetic energy, with no energy loss to the outside. It is a closed energy system following the instant of impact.

The physicist now asks this question: Knowing exactly the starting position of each ball on the board and its initial velocity (speed and direction), can we predict the exact position of each ball after a lapse of exactly ten seconds? or one hundred seconds? A physicist who is a *determinist* will answer yes. The laws of motion govern every move made by every ball; it is only a matter of plugging the initial positions and velocities into a set of mathematical equations and solving for the desired elapsed time. The system activity is precisely determined at all points in time by this unique set of conditions and the laws of motion. The same answer could be given for a real billiard-table situation, provided that the exact values of frictional resistance and all other disturbing effects could be entered into the calculation. The system we have envisioned is a *deterministic system*; the way it is viewed as a scientific phenomenon is called *determinism*.

Determinism in empirical science was first expressed in the early 1800s by the French astronomer and mathematician Marquis de Laplace. This was about 130 years after Newton had formulated the laws of gravitation and motion. Other scientists had in the meantime developed these laws and applied them to many common mechanical phenomena. As translated from the French, Laplace wrote in 1812:

> Let us imagine an Intelligence who would know at a given instant of time all forces acting in nature and the position of all things of which the world consists; let us assume further that this Intelligence would be capable of subjecting all these data to mathematical analysis. Then it could derive a result which would embrace in one and the same formula the motions of the largest bodies in the universe and of the slightest atoms. Nothing would be uncertain for this Intelligence. The past and the future would be present to its eyes. (Wartofsky 1968, p. 298)

On this cosmological scale, a deterministic solution would be far from practical. It works fairly well for orbiting objects in the solar system and allows us to predict eclipses and tides far in advance and with a remarkable degree of accuracy. It is not, however, practical in analyzing most ordinary natural systems we can observe quite closely on our planet. For example, in the billiard table case, the surface of the table has many tiny bumps and hollows we can scarcely see or feel, but these affect the motions of the balls. These disturbing effects are distributed more or less at random over the surface of the table. Given plenty of time and some fancy instrumental equipment, a scientist could measure all the irregularities and enter their effects into the equations. Even so, the corrective effort would be incomplete and some errors would remain to upset final conditions. The effect of the small initial errors tends to grow cumulatively as the experiment continues to run, so prediction would be subject to greater error as time passes (see Smart 1979, p. 652).

Uncertainty and Probability

Strict determinism encounters insuperable difficulties in actual scientific research when an investigator must actually measure real quantities of mass, length, or time, and their combinations, such as density, force, velocity, or acceleration. If these measurements are not perfectly accurate—and, of course, they cannot be perfect—the future states in which they play a role cannot be accurately predicted. Errors of measurement are compounded, and the further we carry the prediction into the future, the worse the errors become. (This idea is developed in our later section on chaos theory.) Measurement errors are of two general sorts. First there are the errors of observation of a single entity. Each of ten students asked to measure the distance between the same two boundary posts, using the same meter stick, will come up with ten different values. Second, there is a natural variation in the magnitude of a specified property measured in all individuals of a given class. For example, what is the weight of a ripe apple? If we weight every ripe apple from a particular tree, no two weights will be exactly the same. Our commonsense solution might be to take the average of the measurements in each case, and this decision would probably be wise because the individual measurements seem to cluster about some middle value. Let us follow through an imagined example of determining a representative middle value that describes the outcome of a repeated experiment.

Imagine that a rifle is securely locked into position on a test bench in a long straight tunnel. The rifle is fired repeatedly at a target 500 meters distant. The target is a sheet of graph paper—a grid of horizontal and vertical lines. We don't know where the bull's-eye should lie—that is something to be ascertained from the results and not in advance. Because the bullets make rather large holes in the target, it will soon be badly shredded in an area where the bullets tend to concentrate, so we use a fresh sheet of graph paper each time a bullet is to be fired. The exact position of the bullet hole is determined in terms of X and Y coordinates on the graph field and this information is fed into computer storage. On a computer screen the shots are displayed as sharp points as they accumulate. The points are at first scattered in a heterogeneous pattern, but as the number of points increases, the points show clustering about a common center. The flight of each bullet is subject to aberrations from several sources. The bullets are not exactly uniform in mass, internal density distribution, and external form. Some, because of their shapes, will veer to the right or left, and up or down. The powder charge is not exactly the same in each shell, so the muzzle velocity of the bullet varies from one to the next. The Coriolis force that deflects to the right in the northern hemisphere increases in proportion to the speed of the bullet, so variations in speed will cause deflections in the left-to-right direction. Air in the test tunnel is in turbulent motion, with innumerable eddies that form and dissolve. The eddies act as if they were forces exerted on the bullet as it travels, pushing or pulling in all directions.

For each bullet fired, the various aberrations lead to a single final effect, which determines the point at which the bullet impacts the target. With no such imperfections and disturbing forces whatsoever, all bullets would arrive at exactly the same point. With respect to that ideal point—the bull's-eye—we can describe the magnitude and direction of the aberrational effects. The direction of the effect is a radial line drawn from the bull's-eye (like the spoke of a wheel); the magnitude of the effect is the radial distance out from the ideal point. In theory, the most probable location for the individual bullet hole is exactly on the bull's-eye. The probability that the bullet hole will lie at a given distance from the bull's-eye diminishes with increasing distance from the bull's-eye. This probability distribution will begin to show on the computer screen as more and more shots are recorded. The clustering around a central region will become more smoothly graded and the cloud of points will show an outward thinning in density. There will, however, appear occasionally a rather isolated shot as far out as the screen permits. An important concept to keep in mind is that we can never predict the coordinate location of the next shot to be fired.

Ultimately, the cloud of points in the central region becomes so dense that it appears solid. We must locate this central point of highest density, because it will be needed as the common center of a set of concentric circles we will draw to indicate levels of equal probability in deviations of the bullet holes. To help overcome this difficulty, the computer has been programmed to give us a three-dimensional picture of the points it has in storage. The two-dimensional graph serves as the horizontal base on which to erect perpendicular columns whose heights are proportional to the number of points falling within each square on the graph. We now have before us the image of a mound or hill, approximately circular at the base. The summit consists of a nest of four rectangular columns of nearly equal height.

Our next step is to use a grid with smaller squares and increase the number of shots fired in the experiment. The hill then becomes smoother in surface form and its summit point is more clearly defined. Finally, we go through the process of fitting an ideal smooth, continuous surface to the hill of tiny columns. This ideal envelope appears to have the shape of a bell—the Liberty Bell in Philadelphia, for example. Its side steepens upward to a middle region of maximum steepness, then lessens in steepness toward a summit that is broadly rounded. The base of the bell tapers outward to what seems to be zero thickness, but actually it extends outward to approach infinity in all directions.

What we have derived here is a particular kind of *probability distribution.* The central summit point of the bell is the average position, or *statistical mean* of the probability distribution; it is "the most likely position." The position coordinate (x,y) for each bullet that was fired is called a *statistical variate*; the total number of such coordinates makes up the *statistical sample*. We could never fire enough shots to produce the perfect bell form, but it can be arrived at by mathematics. In theory, the shots we fired came from a statistical population of possible shots whose number approaches infinity. Another concept of mathematical statistics is most important in our experiment. Each shot must be thought of as having been drawn at random; meaning that from the box of shells containing all the bullets we fired, each shell has an equal probability of being drawn out of the box on each draw. For those of you who are way ahead of me, let me add that the shells would all have to be identified by a number; each shell as it is drawn would have to be listed in serial order by that number, then returned to the box and mixed in well in preparation for the next draw. You could achieve the same randomness more easily by drawing the serial numbers at random, using a *table of random numbers* readily available to all scientists. The principle is that each variate drawn to become part of a sample of variates has exactly the same probability of being drawn as every other variate in the sample. Thus we have a *random sample* of variates.

The next thing we can ask our computer to do is to slice the bell surface with a vertical plane passing through the summit point, or mean. Let the computer display this surface as a simple curve. What we have here is a special kind of probability distribution known as the Gaussian distribution, or the *normal curve of error*. Figure 6.1 shows the Gaussian curve. It is said that the curve was formulated by one Abraham de Moivre in 1733, based on probabilities encountered in games of chance. It later took the name of Karl Gauss, the mathematician, who derived the curve in an attempt to idealize the distribution of measurement errors. In science, when a controlled experiment involves numerous repetitions to arrive at a measurement of some dimensional property—such as mass, length, or time, or a product of those dimensions—the individual measurements (variates) tend to fall to the left or right of the mean value with a frequency in numbers proportional to the height of the Gaussian curve.

The great importance of the Gaussian distribution (along with certain other mathematical formulations of probability distributions) lies in its ability to describe the natural variation of things that are measured by scientists. For example, the body weights of many individual organisms of a single species at a given age or stage in development can be sampled and fitted with a Gaussian curve. This gives us a general model of the relationship of individuals to the population average—extremes are rare, near-average values are common. Examples of the kinds of objects that are measured in science and treated in this manner are almost without limit. In geology, measurements are made of the dimensions of crystal grains in a rock or of sand grains in a beach or dune. Use of the Gaussian curve enables an investigator to make estimates of the reliability of scientific statements that serve as tentative conclusions based on measurements. Reliability is evaluated in terms of the percentage probability of being in error when a statement is asserted to be true.

The Null Hypothesis

To get a better idea of how the Gaussian model of distribution can be used in testing a scientific hypothesis, consider the following imaginary case. A physical anthropologist is studying two isolated groups of aboriginal

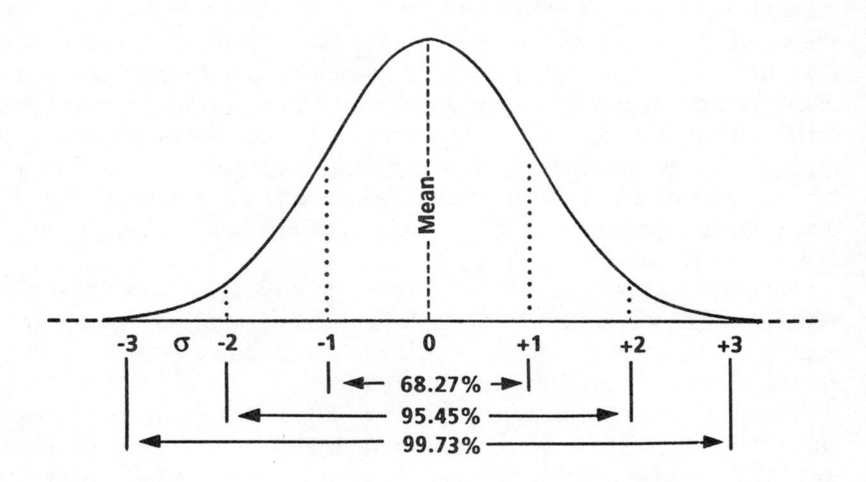

Figure 6.1 The Gaussian curve, or normal curve of error. The scale shows standard deviation (σ) from the mean, with percentage of area beneath the curve within outer limits of one, two, and three standard deviations.

humans separated from each other and from contacts with other populations by mountain barriers. The question under investigation is whether the two groups have important physical differences that would suggest that they migrated to their present sites from different source areas. The conservative scientific hypothesis reads thus: "There is no difference in the physical characteristics of the two groups." This is known as the *null hypothesis*. The anthropologist now takes measurements of various dimensions of individuals in each group. Suppose that one of these is the girth of the skull. The sample size is limited by the small total population of humans in each group. A Gaussian curve is fitted to each sample. When the two curves are drawn on the same graph, it is easy to see that although their two means differ by a small amount, the two curves overlap a great deal. This means that the range of variation within each group is large, larger in fact than the difference between the two means. A rigorous statistical test is performed and leads to the statement that, in adopting the null hypothesis, the chances of being in error in holding to it are very small—say, less than one percent. In view of this outcome, the wise course of action is to accept the null hypothesis, at least for the time being.

Figure 6.2 may help to explain the use of the null hypothesis. It is the same "ladder of excellence" shown in Figure 2.1, but now the probability scale is reversed. Going up the ladder, the probability decreases by powers of ten; this is the probability that the scientific hypothesis is false when it is asserted as being true. Arrows point to a probability of 1 percent, which is a rather conservative number to use in deciding whether to accept or reject the scientific hypothesis. If, from the sample data, the test of significance of differences in the two means had yielded a probability smaller than 1 percent, we would have rejected the null statistical hypothesis and accepted the scientific hypothesis (i.e., the two populations of aborigines are actually different). As it happened, the probability was much larger than 1 percent and we were forced to adopt the opposite conclusion. The use of the null hypothesis in the test may seem an unnecessary complication and a waste of time, but it is an application of Occam's razor. It keeps saying to us: "Don't read into the observations a conclusion that complicates things in nature, unless there is a compelling reason to do so."

No scientific hypothesis can be either proven or disproven by mathematical statistics rigorously carried out on a sound mathematical basis. All that you will get out of it is a statement of odds of being in error when stating your hypothesis or proposition. Forget the politicians' favorite slogan: "You can prove anything with statistics." In empirical science you can't prove anything with mathematical statistics.

Have we drifted far from the notion of deterministic versus nondeterministic systems? Not too far, I think. The hypothetical examples of bullet holes in a target and skull girths of aborigines deal with indeterminate solutions. As for the bullet holes, we can also adopt a rather strict deterministic approach. Each time a bullet is fired at the target, its course is fully determined by the forces that act upon it. Through every millimeter of its flight, the bullet moves in conformance with the laws of mechanics. In that sense, there is nothing at all "chancy" in the behavior of a particular bullet on a particular flight. All that we really need is the services of Laplace's Intelligence to work out the necessary equations to satisfy all the initial conditions and the transient conditions en route. It would be far too difficult for humans to actually do this job.

In practical terms, then, the deterministic approach simply has no scientific value, because to subject it to testing is close to being impossible. This conclusion would probably be looked on favorably by physicist Percy W. Bridgman, a Harvard professor for many years and a Nobel laureate as well. Bridgman expressed a rather practical or hard-nosed view of what is

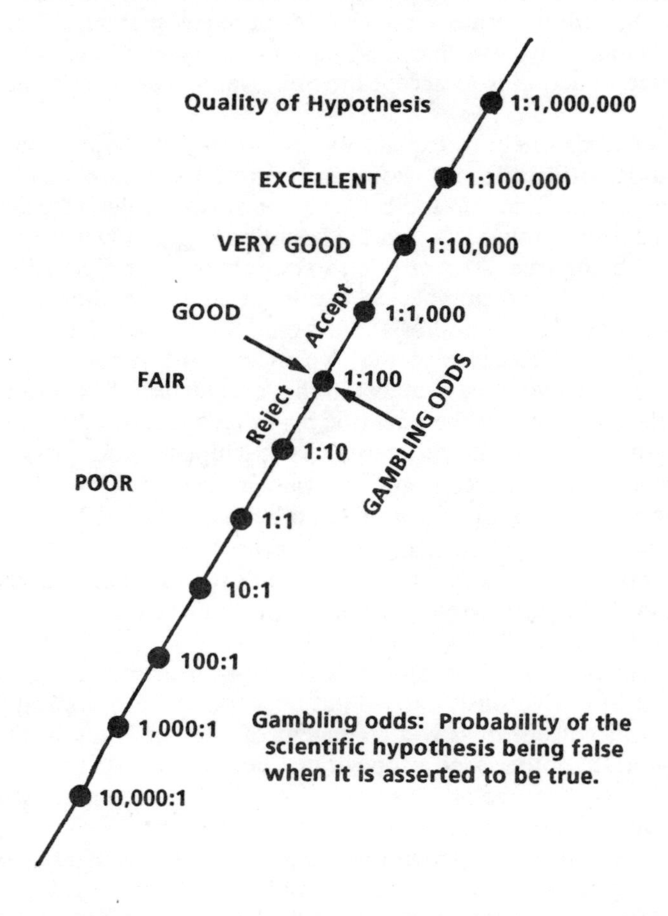

Figure 6.2 The quality of a hypothesis, set in terms of gambling odds.

useful or meaningful in scientific endeavor. His "operational method" in physics considers that a hypothesis has meaning only if it can be tested by accepted scientific procedure (Bridgman 1936, p. 10). Deterministic models of complex systems have essentially the status of untestable hypotheses—in the absence, that is, of Laplace's Intelligence.

The New View of Probability

This section may get me in trouble with a lot of science researchers who apply mathematical statistics in attempts to judge the quality of their propositions. In the previous section on the null hypothesis, we introduced the principle of applying tests of significance to sample data and reaching objective decisions as to whether the hypothesis should be rejected or accepted. Let's summarize briefly the steps involved in this type of quantitative (measurement) investigation:

(1) The objects (variates) to be measured are selected at random from the available population of variates. Randomness of selection is absolutely essential to assure total objectivity. The selection process must be purely mechanical in the sense that the observer is prevented from selecting according to any preconception as to what the outcome should be.

(2) The distribution of the sample variates must be "fitted" to an appropriate theoretical distribution function, such as the Gaussian normal curve. The data may require routine reprocessing (such as transformation into the logarithmic form).

(3) The appropriate statistical test is applied. In our example of skull girths, it would probably be the t-test of significance of difference in the means of the two samples. Reference to prepared tables presents a choice of probabilities to be considered; one is selected.

(4) The probability, P, yielded by the test is compared to the previously selected value that determines whether the decision shall be to accept or reject the null hypothesis.

(5) Finally, the decision to accept or reject is transformed in concept from that of probability (chance) to one of absoluteness. Another way of saying this might be: "Probabilistic" is transformed to "non-probabilitistic."

Adherents to the schools of logical positivism and logical empiricism embraced the five steps as part of their program of asserted complete objectivity in corroboration and/or falsification of hypotheses. They described their method as one of *inductive probability*, determined by logic alone, and completely free of subjective attitudes. The method is essentially an algorithm—a foolproof set of instructions requiring no judgment in application. But, of course, the fourth step is one of judgment, used in setting the decisive level of probability.

The transformation in Step 5 is where the philosophical problem lies. The investigator has switched from the concept of fallibility to that of infallibility. This is a very grave move to make in terms of epistemology, because the scientist has taken a statement that is uncertain knowledge and relabeled it as certain (true, genuine) knowledge. This fifth step appears arbitrary to the point of being irrational.

The English mathematician R. A. Fisher is usually given credit as the founder and developer of the prevailing methods of statistical testing for science researchers. His first work appeared in 1922 and others followed through the 1930s, '40s, and '50s. First published in 1925, the Seventeenth edition of his best known work, *Statistical Methods for Research Workers*, appeared in 1970. You will find Fisher's methods and others based on the same principles in use in almost any current scientific journal you open, regardless of what the subject area is. Because "everybody does it," the method of inductive probability is both the old way and the current way. There is, however, a challenger of this established way; it fits neatly into the new philosophy of science and requires a modification in the traditional way.

As we stated in Chapter 5, the new philosophy of science (NPS) asserts that because truth and falsity are unattainable ideals, science must rely on application of informed judgment using the rational process. Thus NPS recognizes the fourth and fifth steps as part of its program. Confronted by an infinite spectrum of probabilities, there is no realistic alternative but to use informed judgment. NPS says that subjectivity is no longer a dirty word in science; subjectivity is accepted and utilized as an ally rather than a despised foe. Recall our emphasis in Chapter 5 on the role of prior knowledge in making informed judgments by the rational process.

The "new" probability theory has early origins. Back in 1763, an English clergyman, Thomas Bayes, introduced the role of prior knowledge in probability calculation in a memoir published posthumously by the Royal Society of London. His mathematical equation, now known as *Bayes's Theorem*, is the basis of what is currently known as the Bayesian theory of probability. The French mathematician Laplace completed the formulation of the theorem.

We can approach the essential idea of Bayes's Theorem with simple examples, avoiding mathematical statements. Consider the roulette wheel, or better still, its relative, the wheel-of-fortune, that rotates in a vertical plane. Suppose that ours has stops at digits ranging from 00 to 99; 100 in all. The probability, P, of any of one of the 100 numbers being drawn is the same: 1 in 100, or P = 0.01. The assumption that the wheel is mechanically perfect represents the scientist's assumption of a theoretical probability distribution that models an infinitely large population. No other assumption is reasonable where the scientist has no information to the contrary, and this is the case in many investigations of entirely new subjects about which the investigator brings along no previous knowledge of those subjects. In the case of our example of two isolated aboriginal tribes, the anthropologist never knew they existed before and had no prior knowledge of them.

Back to the wheel-of-fortune, what if the player has been tipped off that the wheel is "rigged" to change the betting odds? A weight may have been hidden at one point on the wheel, skewing the odds in favor of the numbers near that location. Knowledge of such details is prior knowledge. If you elected to stay and play, a much more realistic set of odds would be available to you to avoid being wiped out by, say, using just your credit-card numbers in sequence. Using the actual odds would make you a Bayesian fan. Bayesians claim they are consistent winners.

Back to the aborigines, suppose the scientist recognizes that the language spoken by tribe A has many similarities to the language spoken in a nearby accessible region A, whereas the language of tribe B is closely related to that of inhabitants of region B. Next, add prior knowledge consisting of well-established published evidence that skull girths in region A are significantly greater than in region B. Reason would dictate that the difference in means of the two tribal skull girths should be safely judged significant (null hypothesis rejected) at a much larger value of the cutoff probability than the strict rules usually call for.

A different example, bringing us closer to Bayes's Theorem, is given by science philosopher Keith M. Parsons:

> The same statement can have different probabilities; it all depends on what other statements we take to be true when we are considering the statement in question. For instance, the statement
>
> A: Billy Bob can quote from Geothe's *Faust*
>
> is highly improbable if all we know about Billy Bob is the statement
>
> B: Billy Bob comes from Boondocks County, and 99.9 percent of the people in Boondocks County cannot quote from Geothe's *Faust*.
>
> However, A is much more probable if all we know about Billy Bob is the statement
>
> C: Billy Bob is a German major at Princeton and 98 percent of German majors at Princeton can quote from Geothe's *Faust.*
>
> (Parsons 1989, p. 69)

To carry the analysis further it should be explained that "probability" means "probability that the statement is true (or false)." For A, the required probability in the absence of any prior knowledge is 50 percent [P(A) = 0.5]. Next we say "the probability of A given B is only one in a thousand" [P(A/B) = 0.001]. For A given C, however, the probability is 98 percent [P(A/C) = 0.98]. What we have done is to determine *conditional probability*, meaning "probability conditioned by prior knowledge." Commonly, prior knowledge in this context is called "relevant background knowledge," including all pertinent knowledge we have available on the subject (Parsons 1989, p. 69).

While Bayes's Theorem is rigorous mathematically, there are obvious practical difficulties in making the necessary calculations in specific cases. Rounding up all the relevant bits and pieces of your prior knowledge—some of them positive and some negative—and casting all in a single number may simply border on the impossible. Nevertheless, your experience as a scientist may greatly aid you in evaluating the results of an experiment. More testing may be called for than was first planned. If your prior knowledge was largely accurate, its effect would be to move the experimental results steadily toward your anticipated conclusion.

The case in favor of practicing the Bayesian approach to scientific reasoning has been strongly supported in a recent book by English authors Colin Howson and Peter Urbach; both are Lecturers at the London School of Economics—Howson in logic and Urbach in philosophy (Howson and Urbach 1989). These authors point out that critics who find fault with the Bayesian approach rely largely on the argument that it is heavily subjective (which obviously it must be in practice), but they contrast this subjectivity with what they regard as pure objectivity of the traditional (Fisher) system. The supposed logic of this argument is that mixing subjectivity with objectivity wipes out the objectivity. Bayesian supporters counter with the observation that the traditional method also requires subjectivity in its final stages. This is essentially what we have already stated in foregoing paragraphs to be the case. What the Bayesian method does is to mix or amalgamate subjectivity from two sources. Howson and Urbach criticize the "classical" methods of logical positivism and logical empiricism, saying they wish to show that

> classical methods are really quite unsuccessful, despite their influence amongst philosophers and scientists, and that their pre-eminence is undeserved. Indeed, we shall argue that the ideal of total objectivity is unattainable and that classical methods, which pose as guardians of that ideal, in fact violate it at every turn; virtually none of those methods can be applied without a generous helping of personal judgment and arbitarary assumption. (1989, p. 11)

Much the same argument is made by James O. Berger and Donald A. Berry; both university professors specializing in mathematical statistics

(Purdue and University of Minnesota, respectively). The title of their article in *American Scientist* (1988) is provocative: "Statistical Analysis and the Illusion of Objectivity." Gleaned from their text are the following comments:

> We will discuss whether or not is it possible to arrive at an objective conclusion based on data from an experiment. We grant that objective data can be obtained, but we will argue that reaching sensible conclusions from statistical analysis of these data may require subjective input. (p. 159)

> Bayesian statistics treats subjectivity with respect by placing it in the open and under the control of the consumer. (p. 163)

> Statistical analysis plays a central role in scientific inquiry. The adoption of today's statistical methods has led to enormous improvements in the understanding of experimental evidence. But common usage of statistics seems to have become fossilized, mainly because of the view that the standard statistics is *the* objective way to analyze data. Discarding this notion, and indeed embracing the need for subjectivity through Bayesian analysis, can lead to more flexible, powerful, and understandable analysis of data. (p. 165)

The Stochastic System and Natural Laws

At this point we can no longer defer the introduction of a horrendous new term: *stochastic.* It is an adjective we can use to replace "indeterminate" as the alternative choice to "deterministic." Think of "stochastic" as meaning "related to chance"; it is a system activity that involves chance variations in some parts of that system. A *stochastic system* has built into it a random process continually at work. Gambling machines illustrate the *stochastic process* very nicely. The roulette wheel, if it is not tilted or rigged in some way, delivers its winning or losing numbers in a random sequence. Drawing the numbers of military draftees or lottery tickets from a large goldfish bowl is a stochastic process, for it exploits randomness.

The stochastic nature of empirical science investigations lies in two basic sources of randomness in the sequence of numbers it delivers: (1) errors of observation (measurement), and (2) natural variations in the individual objects or properties that make up the variates of the statistical sample. Scientific investigation must deal in stochastic models or hypotheses because there is no practical alternative. This operational method in no way substitutes "laws of chance" for "laws of physics." Laws of physics can, by the deterministic process, explain fully why the roulette wheel stopped on a particular number on a particular turn of the wheel. No goddess of chance put her finger on the wheel to make it stop at a point other than where it would have stopped solely because of frictional loss of energy.

I place so much stress on this subject because there prevails a misconception to the effect that randomness explains physical phenomena, and since that is the case (the argument goes), we have no further need of the basic laws of physical science. The misconception is the notion that when we have identified random variations in natural phenomena, we have therefore also provided a full explanation of what we observe. What we have actually provided should be described as a statistical explanation.

The Stochastic Model and Prediction

Consider the suggestion that a purely stochastic system always has a stochastic beginning, in which initial quantities, such as energy, momentum, speed, acceleration, and direction, arise at random from a population that is described by a probabilistic model. Now, take a second look back at the deterministic model and ask this question: does it have a stochastic requirement somewhere in the system operation?

Is it not likely that the conditions we specify at a given instant of time arise by chance in the first instance, if we could trace the state of the system back to an earlier point in time? In the case of the billiard balls, the zero point in time was the instant that all six balls were struck by the six different players. The balls had been placed in a random pattern to start with and it can be assumed that the directions in which the players aimed were also randomly distributed over a 360-degree range.

Let us now apply the stochastic model to the more general case of atoms of a gas in a closed container, perfectly insulated from its surroundings. The atoms of the gas are assumed to be composed of only one element (nitrogen or helium, for example). The individual atom is visualized as a perfect elastic sphere; all the spheres are exactly alike (as in our example of the billiard balls). These elastic spheres are continually in motion, flying through the vacuum in straight-line paths at high speeds, perhaps on the order of 5 kilometers per second (about 10,000 miles per hour). A particular atom will strike either another atom or the wall of the container and will change course or rebound, as do the billiard balls. The paths of free flight are, on the average, of great length compared with the diameter of the individual atom. Weak forces of attraction between the atoms can be disregarded. Constant bombardment of the walls of the vessel by atoms accounts for the pressure the gas exerts on those walls.

The atoms of a gas cannot actually be observed as flying objects; they are much too small and move much too fast. To apply a deterministic model to the entire container of gas is ruled out as a practical procedure because of the impossibility of actually measuring the initial and sequential positions and velocities of all (or any) of the atoms in the container. This impracticality in no way detracts from the validity of the mechanical laws that govern the activities of each and every atom. The only physical variables that can actually be measured in this experiment apply to the entire population of atoms treated as a uniform substance. These variables are the temperature of the gas and the pressure that the gas exerts on the container walls. Temperature can be read from a thermometer that penetrates the wall of the container; pressure can be read from a gauge that is connected with the gas chamber by means of a small tube. Limited to these observations, we turn to a *stochastic model* that will relate what the gas atoms are doing collectively to the observed temperature and pressure of the gas. In so doing, we are relating Newtonian laws governing motion of each gas molecule to the laws of bulk behavior of an ideal gas.

The stochastic model views the gas atoms as individual variates in a statistical population. The speeds of the atoms at any given instant can range from as low as zero to very high values. Thus there is a statistical population of speeds. This being the case, there must be an average speed that characterizes the entire population of speeds. We can also envision a probability distribution of speeds; it will be a peaked curve, perhaps something like the Gaussian curve. Physicists think of it as being sharply peaked, so that the speed of most of the atoms is not far from the average speed. Physicists use a relatively simple mathematical equation to demonstrate to their satisfaction that average speed of the population is related to the temperature of the gas. The equation requires that the mass of the individual atom be known; it has, in fact, been established independently. Without going into further details, it can be said that it is a relatively easy task to establish the statistical mean speed of the gas atoms. The actual speeds of individual atoms can be estimated only in terms of probability. If individual atom speeds could be sampled at random, the probable speed of a molecule would be indicated by the probability distribution. There is no practical way that an actual speed in kilometers per second can be assigned to a particular atom drawn at random.

Under the stochastic model of behavior of gas atoms, the exact positions in space of individual atoms at a given instant cannot be stated, but the distribution is assumed to be described by a particular probability distribution. The average distance separating one atom from its neighbors can be calculated from the number of atoms in the container and the volume of that container. Here, again, an ideal probability curve could be assumed to predict the likelihood of any two atoms being separated by a certain distance. As for the travel direction in space of a single atom at a given instant, that can be assumed to follow a random model—all directions are equally probable when individual travel paths are randomly sampled from the population.

In the stochastic model of behavior of gases there is provision for events of extremely low probability to occur. The stochastic model we have outlined describes the most probable state of the system in terms of energy. But because the probability curves on which it is based have "tails" extending toward infinity in either direction (or approaching zero), extreme values are possible. There is always a possibility, no matter how remote, that some very strange arrangement of the gas atoms will occur for at least an instant of time. For example, there is a remote possibility that we might observe an instant in which all the atoms spontaneously became crowded together in one corner of the container. Just don't wait around for this remarkable event to occur.

It may distress some readers that scientists must resort to stochastic models to analyze natural systems. The rules of stochastic procedures say that no statement can be proved either correct or incorrect; therefore that absolute truth is unobtainable. The same rules present a scientist with a scale of betting odds. The scientist must select in advance the particular odds that will lead to tentative rejection or retention of the hypothesis being tested. Can you think of a better way to investigate complex natural phenomena?

For an appropriate comment with which to conclude this section, I defer to Alan Lightman, at the time a contributing editor to the popular journal *Science 83*, published by the American Association for the Advancement of Science. In a brief essay on probability in science, "Weighing the Odds," he wrote:

> Most people, I suspect, have a deepseated reluctance to welcome probabilities into their private lives. At least since the Greeks, mankind has harbored a passion for knowing some things with certainty. Probabilities, by definition, shimmer in a mist of uncertainty. Einstein contributed little to science in the last three decades of his life, in large part because he could never accept the probabilistic nature of the emerging quantum physics. "God does not play dice," he insisted. (1983, p. 22)

Thermodynamics, Entropy, and Universal Decay

Randomness and the stochastic process are heavily involved in one of the fundamental subjects of physical science—thermodynamics and its laws. *Thermodynamics*, the study of the relationships between heat and work, had its beginnings during the Industrial Revolution. It was a period in which the coal-fired steam-engine powered textile factories and mine pumps. The first practical steam engine had appeared in 1689, and those that followed it made use of the principle that a chamber or cylinder filled with steam could be cooled suddenly by a water jet to produce a vacuum, which then forced a piston to move under atmospheric pressure. The Newcomen engine of 1705 was one of these and it was capable of lifting water from great depths in mines. Most persons know of James Watt's steam engines of the late 1700s; they represented major improvements in efficiency. Scientific study of such machines was undertaken not only to improve their efficiency, but also to prove (if possible) that perpetual-motion machines are an impossibility. Thermodynamics did that job effectively but seems never to have succeeded in completely quashing public interest in such miraculous machines. The honor of being the founder of thermodynamics has been given to Sadi Carnot, a French physicist whose 1824 treatise on heat contained the description of an ideal heat engine that makes use of expanding gas to do work. The Carnot cycle, carried out by this imaginary engine, leads to the conclusion that no heat engine can convert into mechanical work all of the heat energy supplied to it.

In its classical early phase, thermodynamics took no account of the atomic nature of the gases, liquids, and solids with which the subject dealt. Bringing into consideration the atomic (kinetic) theory of gases was the accomplishment of Ludwig Boltzmann (1844-1906), an Austrian physicist. A younger contemporary of Charles Darwin, whose theory of evolution he warmly supported, Boltzmann produced a statistical analysis of the behavior of gas atoms, putting thermodynamics into a new framework, much as we find it today. Presentations of thermodynamics and its laws to students of physics invariably draw on the atomic theory of gases, which is what we shall do here.

Kinds of Systems

Classical thermodynamics is usually presented in terms of the *isolated system*, a kind of system quite different from the open system described in Chapter 4 in our discussion of the cell as a biological system. In outline form, we can classify all systems as follows:

I. Isolated systems framed in formal logic and pure mathematics; neither energy nor matter can be transported across the system boundary (systems of classical thermodynamics).

II. Open and closed systems framed in empirical science and identifiable in nature. They are of two kinds:

  A. Open systems in which matter or energy or both can be transported across the system boundaries (examples: living cell, thunderstorm, river system).

B. Closed systems in which matter cannot be transported across system boundaries, but the matter inside can be driven by external energy sources (example: refrigerator).

The isolated system usually presented in classical thermodynamics is unreal in every aspect. It is typically pictured as a cylinder whose walls provide perfect insulation (impossible), containing a piston that moves in frictionless contact with the cylinder walls (impossible), and in which the motion of the piston and the changes in pressure or temperature of the enclosed gas take place at a rate approaching zero (impossible). To make this impossible machine "do something," heat is somehow injected into the cylinder or removed from it, or a weight has to be inserted on top of the piston (or removed from that position); yet these operations are forbidden by the very premise of total isolation. No wonder the average person finds it almost impossible to understand the laws of thermodynamics based on such a strange machine!

Class II systems deal in reality. They can be observed in nature and are often replicated and controlled in model form in the laboratory. Their inputs and outputs of energy and matter can be easily imagined in terms of sensory input; i.e., these systems fall within the range of human experience or analogs of human experience. Their operation can be understood without using the formal terminology of the laws of thermo-dynamics, even though they conform with those laws. Nevertheless, we need to review those laws and know something of the kinetic theory of gases to understand their meaning.

The Laws of Thermodynamics

The first law of thermodynamics can be stated in various ways. In classical terms, it is stated as follows: The quantity of heat added to an isolated system is always equal to the increase of energy within the system. If it bothers you that you are not allowed to add heat to an isolated system in the first place, try this statement: The total quantity of energy within an isolated system remains constant. The important message is that energy can be neither created nor destroyed; thus the first law is a way of stating the principle of conservation of energy. Now, the only possible truly isolated system of which we can conceive is the universe itself, provided that the universe is finite in extent. This leads to another way to phrase the first law: The energy inventory of the universe is constant.

To take into account that matter (mass) can be changed into energy and vice versa in such phenomena as radioactive decay and matter/antimatter interactions, the statement of the first law needs to be broadened, thus: The total quantity of energy and matter existing in the universe is constant. So stated, it is commonly known as the law of conservation of energy and matter. This definition is universally accepted. Unfortunately, the first law is really of no help in understanding the second law.

To understand the second law, we need to take into account principles of behavior of gas atoms in a closed container, stated earlier in this chapter. Gas atoms (and gas molecules) are constantly moving at high speed. Atoms are visualized as perfect elastic spheres, the distance separating the atoms being very great in comparison with the atom diameter, but frequent collisions occur. Atoms following a free path must impact the wall of the container, from which they will rebound at the same angle as that at which they strike the wall. The atoms thus obey the Newtonian laws of motion. Impacts of atoms on the container wall provide the outward pressure of the gas. It is assumed that the gas consists entirely of single, pointlike atoms of the same element (i.e., the gas is monatomic and homogeneous), and that weak interactive forces such as gravitation that may also be in action are small enough to be neglected. Also assumed is that the gas behaves according to the well-known gas laws, the names of which are familiar to students of physics (Boyle's law, Charles's law); these relate temperature, pressure, and volume of the gas. The heat present in the gas is in the form of the kinetic energy (energy of motion) of the gas atoms. With this brief introduction, we are ready to tackle the statistical aspect of the atom motions.

As explained earlier, within the gas the distribution of the atoms in space is perfectly random in the statistical sense. The travel paths of the atoms are uniformly distributed throughout all possible directions. However, the travel speeds of the atoms are not all the same. The range of speeds is from zero to very high values. Since it would be impossible to measure individual speeds of large numbers of atoms, an average, or mean, value of speed is used to represent the gas as a whole. A constant average speed is associated with a constant gas temperature. Temperature is always referred to the absolute scale, degrees Kelvin. At zero K, called "absolute zero," the atoms have no motion whatsoever and therefore the matter has no kinetic energy. The average kinetic energy of the gas, dependent on the average speed, is always proportional to the absolute temperature. Recall that the total population of speed values is described by a probability distribution that is strongly peaked, the summit of the peak being the mean value. Values fall off rapidly in both directions.

Randomness, in this stochastic view of thermodynamics, is described as a state of *disorder*; it is the most probable state of the system. In contrast, *order*, as used here, means an initial or special physical state of matter possessed of a large quantity of stored energy capable of being subsequently dispersed. Generally, order is considered a less probable state, as compared with disorder within the same system. The second law requires that spontaneous changes in an isolated system always go in the direction from order to disorder. Can this statement be challenged, or is it always true? (When using formal logic, a conclusion can be declared true or false.)

As we are interested in the probability aspect of order and disorder, I offer the following example from the writings of physicist Philipp Frank (1946, pp. 24-26). Let us imagine a container of monatomic gas subdivided into a large number of small space-compartments, or cells, all of the same volume. (The imagined cell walls have no effect on motions of the atoms.) Writes Frank, if we could trace and record the path of a single atom over a time of years or centuries, we could establish the specific fraction of the total time that the atom spends in one particular cell—that would be the "dwelling time," expressed as a ratio. The dwelling time, *p*, represents the probability that the atom will be found in the particular cell. For a system of 100,000 cells, p = 1/100,000. A second gas atom also has the same probability, p, of being in that same cell, i.e., 1/100,000. Now let us assume that the total number of atoms in the container, *N*, is one million. We can calculate the time that we can expect to find all the atoms in the same cell; it is equal to p raised to the Nth power times the total observation time, *T*:

$$T\,(p^{1{,}000{,}000}) = T\,(1/100{,}000)^{1{,}000{,}000} = T\,(1/10^{5{,}000{,}000})$$

If T is, say, a billion years, the average dwelling time during which all atoms are in one cell "will be much less than the billionth of the billionth part of a second" (1946, p. 25). He adds: "Moreover, if we are lucky observers of such an event, we can bet that it will soon disappear and not reappear for a billion generations to come."

Perhaps easier to grasp is a similar example based on a deck of playing cards. We start with a new deck in which there is perfect order—two through ace in each of the four suits. Repeated shuffling quickly destroys that initial order and replaces it with a random distribution that represents disorder. The random distributions we continue to obtain are rarely exactly the same, but the level of disorder represented by the randomness is overwhelmingly the more probable. The probability of recovering the initial ordered distribution with any one deal is about 10^{-68}.

Now comes the most difficult concept in thermodynamics. The term *entropy* is a mathematical concept related to laws of probability, but we can best handle it here by relating it to order and disorder. The dictionary definition of the word typically reads: "A measure of the unavailable energy in an isolated system." We can simply say that the greater the disorder in the system, the greater is the entropy. This relationship allows us to state the *second law of thermodynamics* as given by Professor Kenneth Atkins: "When a system containing a large number of particles is left to itself, it assumes a state with maximum entropy, that is, it becomes as disordered as possible" (Atkins 1972, p. 100). The state referred to is the *equilibrium state*, which is the final stable condition.

The only possibility we have of a truly isolated system is the entire universe (if it is finite in extent). When it originated, we suppose, the degree of order was enormously high and the entropy low (to the point of being virtually nonexistent). In the billions of years that have elapsed since the Big Bang with which it started, the total entropy has been increasing and so has the degree of disorder. But the system is still changing and equilibrium has by no means been reached. It is this very state of change—a state of flux—that allows local and temporary reversals of the entropy change from positive to negative to occur, accompanied by corresponding local changes in the direction toward increasing order. Recall from Chapter 4 that such reversals are the rule in living organisims on our planet. Let us turn next to consider that possibility on a cosmic scale.

A schematic view of the universe is shown in Figure 6.3. Within the limits of the universe are many open subsystems within which entropy change can be negative to the accompaniment of the construction of

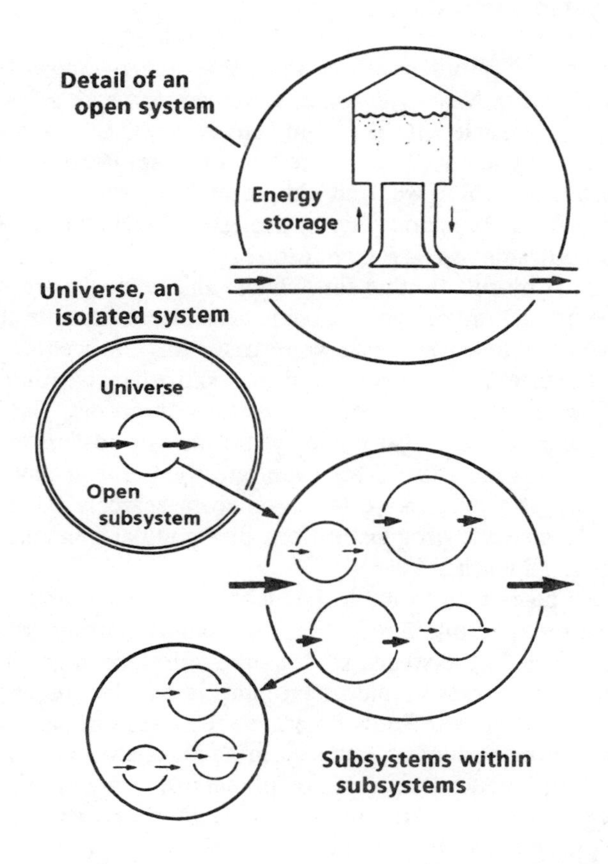

Figure 6.3 The universe visualized as an isolated system within which are nested sets of open subsystems. (A. N. Strahler.)

highly ordered states of matter. The largest of the open subsystems are galaxy clusters and galaxies; within them are smaller subsystems—luminous gas clouds and stars. Many stars have planetary systems, and within each solar system the individual planets and other orbiting objects form another set of subsystems. Each planet, in turn, includes numerous subsystems. Associated with a typical energy subsystem is a subsystem of matter that may be either closed or open. Subsystems of all types and scales continue to be formed. After the growth stage of decreasing entropy is completed, the system may enter into a steady state, and this may be followed by a decay state in which system entropy increases along with increasing disorder, until the system itself goes out of existence. Meantime, the universe as a whole proceeds on its general course toward increasing entropy and increasing disorder.

Chaotic Systems in Nature

Open almost any popular science magazine today—*Scientific American*, *American Scientist*, *New Scientist*, or *Discover*, to name a few—and there will likely be an article with the word "chaos" in its title. Do we have here a new concept in science? It sounds like the final death knell of classical determinism, but then we find the term "deterministic chaos" being applied to it by a Princeton physics professor (Anderson 1990, p. 9). Isn't that a self-contradiction—an oxymoron?

Let us start by substituting for "chaos" the term *chaotic system*. It can perhaps be easiest understood through a historical example that is usually credited with its discovery. Everyone castigates the weather forecasters, including those most highly trained and skilled in weather science and equipped with the finest of powerful computers. While weather forecasting has improved in recent years for periods up to, say, two to three days in advance, those for even one week ahead have shown little improvement, and long-range forecasts seem scarcely better today than those pseudoscientific prognostications in a popular almanac, published a whole year in advance.

Our story takes us into the M.I.T. laboratory of research meteorologist Edward Lorenz, about 1960. He was experimenting with computer modeling of weather systems, such as the spiralling highs and lows that move from west to east in middle latitudes under the guidance of sinuous jet streams. His plan was firmly deterministic, just as in our example of the billiard balls struck in unison and sent into collision courses. It works quite well when applied to the orbits of planets, moons, and orbiting space vehicles. Measure as accurately as possible the intial values of a few key dynamic parameters that largely control the motion of the objects comprising the system. Let the computer take over and predict the values of these parameters as far into the future as time and funds will allow. The

French mathematician Jules Henri Poincaré had stated very clearly in 1903 what can be expected of such a procedure. Notice that he builds upon Laplace's idea of a supreme Intelligence:

> A very small cause which escapes our notice determines a considerable effect that we cannot fail to see, and then we say that the effect is due to chance. If we knew exactly the laws of nature and the situation of the universe at the initial moment, we could predict exactly the situation of that same universe at a succeeding moment. But even if it were the case that the natural laws had no longer any secret for us, we could still only know the initial situation *approximately*. If that enabled us to predict the succeeding situation with *the same approximation*, that is all we require, and we should say that the phenomenon had been predicted, that it is governed by laws. But it is not always so; it may happen that small differences in the initial conditions produce very great ones in the final phenomena. A small error in the former will produce an enormous error in the latter. Prediction becomes impossible, and we have the fortuitous phenomenon. (Quoted in Crutchfield, et al. 1986, p. 48.)

For the model weather systems Lorenz was studying, the results were mixed. Lorenz worked with three rather simple differential equations of motion that could be used to simulate cycles of change; these formed simple deterministic systems (Tsonis and Elsner 1989, p 17). Because the cycles were generated by a computer, if exactly the same numbers were fed into it on each trial, its output should have been perfectly repeated each time. Lorenz discovered for himself that when the starting numbers were altered by extremely small percentages (resulting from rounding off to fewer decimal places), the pattern of change departed rapidly and radically from that of the original set of numbers. Unpredicted behavior of a system, in this sense, can be described as "chaotic." The chaotic effect is characterized by the key phrase "sensitivity to initial conditions." Note further that "chaotic systems are sensitive at every point in their motion" (Crutchfield, et al. 1986, p. 48).

Lorenz's finding contravened the going conservative theory in scientific research that very small errors in measurement of the initial values would lead only to very small errors in prediction. It had been assumed that those errors would conform with a random probability distribution and that the causes of the errors were external. That has been at the heart of stochastic theory right along. The enormous philosophical significance of Lorenz's discovery lies in the dreadful thought that no matter how small the observational errors of scientific measurement are made, the future course of a natural process within a chaotic system is not, for practical purposes, predictable. In broader terms, then, long-range temporal prediction is (in many cases, at least) impossible. Lorenz may have been the first scientist to feel the impact of this depressing conclusion. James Gleick, whose popular book 1987 book *Chaos* is well worth reading, comments on Lorenz's reaction to his initial discovery of the chaotic phenomenon:

> But for reasons of mathematical intuition that his colleagues would begin to understand only later, Lorenz felt a jolt: something was philosophically out of joint. The practical import could be staggering. Although his equations were gross parodies of the earth's weather, he had faith that they captured the essence of the real atmosphere. That first day, he decided that long-range weather forecasting was doomed. (Gleick 1987, p. 17.)

Gleick also quotes Lorenz as saying:

> The average person, seeing that we can predict tides pretty well a few months ahead would say, why can't we do the same thing with the atmosphere, it's just a different fluid system, the laws are about as complicated. But I realized that any physical system that behaved nonperiodically would be unpredictable. (Gleick, p. 18.)

The popular handle "butterfly effect" is often applied to the phenomenon of sensitive dependence on initial conditions used in weather forecasting. It alludes to the quip that "a butterfly stirring the air today in Peking can transform storm systems next month in New York" (Gleick, p. 8). There is more to the butterfly effect than simply a degeneration of a simple deterministic beginning into a chaotic system appearing to be totally random. Closer studies showed signs of order appearing and reappearing as the computer continued its output. Generally similar stretches of the same weather pattern recur, then dissolve, then appear again. In three-dimensional graphic plots, these cycles appear as figures resembling the scribblings we often make on a pad as we talk on the phone or hear a class lecture. For example, we may create multiple superposed circles, ellipses, or figure-eights. Once attracted to a particular doodle design, we keep doing it over and over, but never in exactly the same way. Our figure-eight design simulates a chaotic system that flips from one mode to another. We can also make several circles on the left side before switching to the right side, where we make some more circles and switch back again. The resulting graphic output of a computer can resemble paired butterfly wings or an owl's eyes. In Lorenz's computer diagram, shown here as Figure 6.4, the lines never repeat exactly the same trajectory, nor do they touch one another. Visually, we perceive of the system as being "attracted" to one (or two) geometrical modes, and this has led to the term "Lorenz attractor" or "strange attractor" for the repetitive activity of the system. Perhaps a better description of "attractor" would be that of a mathematical region within which the coordinates of the equation variables are confined or largely restricted.

Chaotic systems have been recognized in many branches of natural science. In biology a good example comes from the behavior of populations of animals within an ecosystem. In the arctic ecosystem, for example, populations of mammals such as the lemming and snowshoe hare show cycles of a few years' duration in which rapid expansion is followed by collapse to only a few individuals—a "boom or bust" rhythm. Ecologists use a linear equation to model this form of oscillating population density:

$$x_{n+1} = k \bullet x\,(1 - x_n)$$

where x is actual population divided by the minimum possible value, x_n is this year's population; x_{n+1} is next year's population, and k is a proportionality factor.

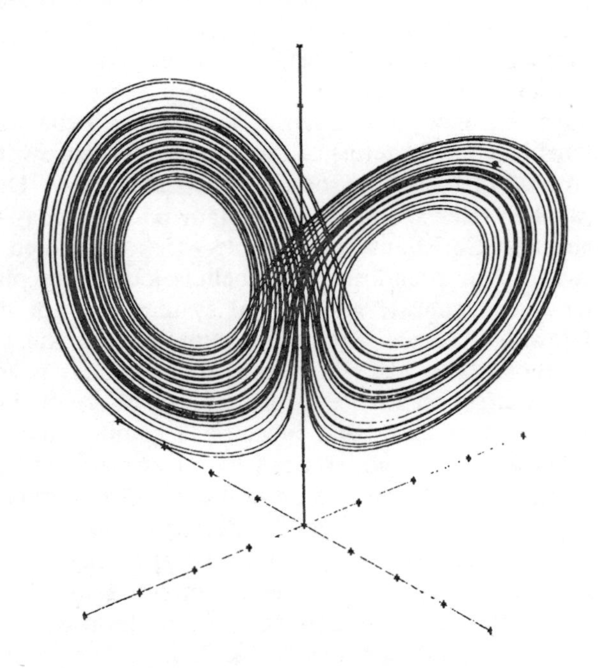

Figure 6.4 This computer-generated image, often likened to an owl's mask, is known as "The Lorenz Attractor." It shows in three dimensions simultaneously the output of three differential equations that together describe the behavior of a fluid layer heated from below so as to produce rising convection currents. The image is formed by a moving single point forming a single continuous line that never occupies the same path in space and never intersects itself. A dot (upper right) marks the end of the line. (Reproduced by permission of Dr. James P. Crutchfield.)

Simply stated, next-year's population is calculated from the current year's value by introducing a variable proportionality factor, k, that can be selected for growth rate. (For details see Gleick 1987, p. 70, and Polkinghorne 1991, p. 223.) Biologist Robert May investigated the properties of this equation by running it on a computer with different values of k to simulate periods of many years duration. To be very brief, a low value of k resulted in the population declining to zero (extinction); for a k value of 2 the population fluctuates for several years, then becomes steady; at 3.2, the population oscillates indefinitely between two extremes. Once k rises above 3.5, a very odd change occurs: a twofold cycle gives way to a fourfold cycle, giving way next to an eightfold cycle, in turn to a sixteenfold cycle, etc. Plotted on a graph, this doubling progression resembles a branching tree; it belongs to a class of mathematical forms known as *fractals*. Above a k of 3.58, the population varies in a completely random fashion that can be accurately described as chaotic. In that mode, the behavior is extremely sensitive to very small differences in the initial value. To summarize, the simple predictive equation can generate long-term system behavior ranging from completely predictable to apparently random (completely unpredictable), with intermediate forms in which those two extremes alternate.

Where does this new knowledge of the strange behaviors of some natural systems leave us in terms of the pursuit of science with the stated goal of improving the reliability of scientific statements? Does this new knowledge weaken science and its laws? Many persons today—New Agers and fundamentalist Christians, for example—seek to weaken and demean science to make their pseudoscientific beliefs look more plausible. The new findings about chaotic and fractal systems do not diminish the integrity of laws of science, for at no point in even the most chaotic behavior are those laws suspended, nor are new ones needed in their places. The total effect on science is, instead, one of greatly increasing the levels of system complexity that must be accounted for. Many more special cases and deviant modes need to be incorporated into various natural systems. What does appear to be a downgrading factor is a decrease in the confidence that can be placed in simplistic assumptions attached to probability models used in statistical testing of significance of scientific data. Each dynamic system, expressed as a mathematical equation, or combination of equations, must be thoroughly investigated for signs of "Jekyll/Hyde" behavior shifts. Perhaps, for example, some of the prevalent forms of prediction in economics will need to be thrown out or redesigned. Scientists may need to adopt the slogan "back to the drawing board" and to take a more critical and skeptical stance generally. All this increased difficulty and complexity in carrying out scientific research fits well into our description of the new philosophy of science, given in Chapter 5.

A number of prominent scientists and mathematicians have gone so far as to pronounce chaos theory a major paradigm shift that has superseded quantum mechanics. Others see it as only as an important advance that has affected most major branches of science, but not as a revolution. Philosopher of science Steven Toulmin sees chaos theory as having given us "extra intellectual weapons, but not an entirely new world view" (Pool 1989, p. 28).

Philosophical Implications of Quantum Mechanics

What is *quantum mechanics*? Perhaps no major area of science is more foreign to the general public, although most persons know that it is an investigation of the fundamental, bottom-line nature of matter and energy that make up the universe. It would be impossible to sketch even the barest outlines of so vast a field as quantum mechanics (QM) in the few pages available here. I limit my introduction to a few salient principles, concepts, and ideas that can perhaps help to put the uninitiated reader in a receptive frame of mind to understand the major philosophical problems of QM I have chosen to discuss.

Most persons know that quantum mechanics (QM) concerns itself with particles of various sorts, all of them much too small to be directly observed even with our most powerful electron microscopes. Most of us would claim some degree of familiarity with the concept of a class of discrete units of "solid" matter known as atoms. These we visualize as having a dense central sphere, or nucleus, surrounded by a sort of blurry "cloud" of extremely small, fast-moving electrons, and also that the nucleus of the atom consists of two kinds of massive particles—protons and neutrons. Atoms set firmly in fixed geometrical patterns in solid crystals can be identified as rather blurred dots on electron microscope photos. Seeing such photos, we may feel satisfied that we have accomplished at least indirect observation only a step beyond our sense of direct vision, and we are prepared to accept what we have observed as empirical reality—something "real" or "concrete" existing "out there." Electrons lie far beyond the limits of direct "viewing" by any optical devices. They were first identified in 1897 as streams of charged particles. That identification was based on the physical effects they produced. They were soon named "electrons," an appropriate move because each electron is the bearer of an electrical charge.

In quantum mechanics (QM) "a particle is a thing that can be detected at a particular point in space, and that has certain measurable properties such as mass, electric charge, and angular momentum (or spin)" (Rothman 1988, p. 44). It may be helpful to know that the electron is assigned a mass of 1, called "one electron mass" and the masses of other particles are given accordingly in the same unit. For example, the enormously massive tauon "weighs in" at 3,535 electron masses. Other particles have no mass, a condition designated as "zero mass," but this applies only when the particle is at rest. To get some sense of scale, consider that the diameter of an electron is said to be on the order of 10^{-13} cm, which can be written as a simple fraction with the denominator of 1 followed by 13 zeroes. For our discussion of philosophical aspects of QM we will make use only of the electron and another fundamental particle, the photon.

The fundamental particles with which QM deals include forms of energy as well as of matter (mass). Of course, mass and energy have been virtually equated through Einstein's familiar formula, $E = mc^2$. (Energy, E, is equal to the product of mass, m, times the square of the speed of light, c^2.) Physicist Kenneth R. Atkins describes the relationship as follows: "Mass and energy can be considered to be different manifestations of the same physical quantity. A quantity of energy E . . . can always be considered to have a mass E/c^2. Conversely, the mass m of a material object can be considered to be the quivalent to an amount of energy equal to mc^2" (Atkins, et al. 1978, p. 107).

The idea of a flow of energy consisting of particles may be difficult to accept. We can sense various forms of energy directly—as heat, light, pressure, or the sensation of acceleration. Take light, for example; our eyes detect no moving particles in light beam. In contrast, under the wave theory of light, it is not difficult to accept a continuous beam as a carrier of energy from a light source to our eyes. To visualize light as a stream of particles bombarding our retinas is not easy to take. We don't usually think of particles being able to pass through a thick plate glass window and then through the lens of the eye.

QM is often said to have made its entrance in 1900, when Max Planck put forward his quantum theory of light and of electromagnetic radiation (EMR) in general. It simply held that EMR consists of discrete packages of energy, each bearing a *quantum* of energy. We now call these packages *photons* and think of them as particles. At the same time, the photon is also described as "wave packet" that can be visualized as a very short "burst" of waves traveling together as a unit. These energy packets can be thought of as separated from another. Thus a wave packet has a one-on-one correspondence with a photon. A photon has zero mass when at rest, and in that sense would seem to consist of pure energy. On the other hand, a photon always travels at the speed of light, and by this motion is endowed with a certain quantity of mass.

Many other kinds fundamental particles are included within the scope of QM—some 100 kinds are well documented. All but a few of these are extremely short-lived—on the order of a millionth or a billionth of a second. They disappear by decaying to form other particles, which in turn decay to give rise to still others. Obviously, these short-lived particles were extremely difficult to identify.

Particles come in a hierarchy. One major class consists of the *hadrons*, of which the neutron and proton of the atomic nucleus are long-lived examples. Another major class consists of the *leptons,* of which there are six varieties; one of these is the electron. One of the others you may have heard about is the elusive *neutrino*—elusive because has no rest mass and no charge. Hadrons, it is now known, are in turn composed of smaller particles known as *quarks*. Physicists seem to think that quarks are the true elementary particles, indivisible and lacking in any internal structure. To sum up, "all ordinary matter in the universe is made from just the two lightest leptons (the electron and its neutrino) and the two lightest quarks" (Davies 1984, p. 90).

We need to give thought to the reality of particles well beyond the range of our senses, and therefore requiring special mechanical means of indirect detection and measurement. This is a major concern of philosophy, embodied in the age-old question: What is reality? Are particles real, even if we can't sense them? Science requires that the existence of all particles of the quantum realm be established through use of proven lawful physical phenomena, and that the observational procedures be repeatable to extremely high levels of accuracy. This was expressed by Ernst Mach and was a tenet of logical positivism since its earliest days. The phenomena we have in mind are interactions between and among particles in what we can call the "microrealm" and between those particles and larger objects of the "macrorealm." Thus the concept of *fundamental interaction* is our guide (Rothman 1988, p. 39).

Long before the dawn of QM, science faced the problem of establishing the existence of molecules as a physical reality. Here the microscope came in handy to observe the physical effect of energetic, fast-moving molecules (too small to be observed) upon specks of matter suspended in a fluid. This Brownian motion, recognized as early as 1827, was easily judged the cause of the seemingly random trajectories of the relatively huge suspended masses. Albert Einstein in 1905 deduced an equation describing the Brownian motion. As physicist Milton Rothman states it: ". . . the fantasy of molecules has been *realized*—turned into reality" (p. 41). Next came the identification of various kinds of particles associated with radiation phenomena, such as cathode emission and natural radioactivity. In 1897, cathode rays were demonstrated to consist of electrically charged particles, which came to be appropriately named "electrons." It was not long before the beta particles radiating from uranium were identified as also being electrons, an advance made possible by measuring both the mass and electrical charge of particles of both sources and finding them to be identical. Rothman describes the cloud chamber, invented in 1911, in which an invisible particle generates a highly visible "vapor trail" of condensed water molecules. Rothman writes:

> Even though the particles themselves are invisible to the naked eye, their effect on the environment is so dramatic and immediate that

> their reality forcibly impresses itself on any observer's consciousness. Instrumentation transforms fundamental particles from abstract concepts into real things. We need no longer depend on the deceptive naked eye. (P. 43)

One of the early problems of atomic physics was that of predicting exactly where, at a given instant, a particular electron orbiting the atomic nucleus will be located and how fast it will be traveling. Physicist Werner Heisenberg attempted to solve this problem by experimental means. In 1927 he came to a rather startling conclusion: it is simply not possible to measure precisely both the location and the velocity. (Strictly speaking, "momentum" should be substituted for "velocity.") Either quantity could be measured accurately, but not both, because the very act of precise measurement of one quantity disturbs the system, preventing the other quantity from also being precisely measured. The solution to the problem is said to be *indeterminate* (i.e., it cannot be determined). The principle involved here is called the *Heisenberg uncertainty principle* (Atkins 1972, pp. 251-53).

From this established principle there emerged what is now known as the *Copenhagen interpretation* of QM: a particle does not possess specific properties until those properties are measured. Does this sound weird, or just specious? Even though I have never measured the height of the desk my computer rests on, am I to conclude that therefore the desk does not possess the quality of height? But how can I get the tape measure and determine its height, if it does not already possess that property, i.e., if the table is unmeasurable in the first place?

Victor Stenger, a university professor of physics and astronomy, has presented much the same response to the Copenhagen interpretation:

> Extending this idea to all other physical quantities, we conclude that they become real only upon being measured. Now it may seem as if I am adopting ancient Hindu idealism, with everything in our heads after all. As I have noted, this view is also a tenet of the New Age: Reality is what you want it to be. The fact that reality rarely is what you want it to be is the best evidence that world beyond our heads does indeed exist. (1990, p. 217)

Stenger restates the Copenhagen interpretation when he adds: "As I will show, however, observations force upon us the conclusion that the variables we measure have no meaning beyond their actual measurement" (p. 217). He cites Einstein's rejection of the Copenhagen interpretation in favor of the conventional view of scientists that objects have external reality. Actually, Stenger claims, "quantum mechanics does not reject the existence of external objective reality":

> Conventional quantum mechanics simply says that the physical quantities we use to describe reality are meaningless except as they result from a carefully prescribed, but nonetheless arbitrary measurement procedure. This is the view that was questioned by Einstein in 1935, but is now confirmed by experiment. (P. 217)

The Problem of Interaction

But there are much more serious problems of reality to be faced, and with which the above statements do not deal. The electrons and protons that leave trails or are picked up by laboratory amplifier/counters can be emitted *spontaneously* from their source materials. When this is the case, the receivers are passive investigators. Imagine a different situation suggested by the following scenario. Driving in the pitch darkness of a moonlight night across a desert plain in the American Southwest, my headlights reveal distant pairs of reddish lights. They are the eyes of jackrabbits staring back at me while motionless on their haunches. Some of the energy of my light beam has been reflected back to me. The animals have stopped whatever they were doing and are now riveted to the headlights. They have reacted physically to a beam of energy in a manner that is not normally part of their nocturnal activity. To identify them I have had to initiate an input of energy (photons) into the target. I will never know what a jackrabbit looks like going about its normal business. To observe the animal, I have changed its state of being, so to speak. We have here the detection of an interaction, and the interaction has changed the target itself.

Even in the case of passive reception of particles from a spontaneous emission source, the particle itself is changed, or even obliterated by its impact with the detector. If not transformed, the particle is relieved of part of its kinetic energy or its electrical charge. This interaction changes both of the interacting components. In order to gain information about the particle, I have had to change or destroy it. "At the atomic and subatomic level, photons can wreak havoc with the system being observed" (Stenger 1990, p. 216). "Havoc" is also a description of what takes place in an "atom-smashing" accelerator. Particle collisions at enormously high energy levels destroy the initial states of the colliding components, transforming those initial particles into other kinds of particles. The problem here is that reality of the initial particles is not directly observed *prior* to the collision. Those of us who are on the outside of QM must assume that the identifications of all the many particles on the list have been established in one way or another to the satisfaction of the community of QM physicists.

British physicist Paul Davies provides a deeper philosophical meaning for the interaction process relative to the the Heisenberg uncertainty principle that lies at the heart of all quantum physics (1984, pp. 39-40). First, he emphasizes that the microrealm of quantum particles and wave packets and the macrorealm of classical physics "are intimately interwoven" (p. 39). "Only the system *as a whole* gives concrete expression to microscopic reality. The big and the small coexist. One does not subsume wholly the other, nor does the other wholly 'explain' the one" (p. 39). He then notes that neurophysiology, now in tune with naturalistic organic evolution, has accomplished a reduction of mental activity to purely physical processes—a system of electrochemical impulses that are only the motions of electrons and ions. But then he leaps forward to a rather astounding proposition: The human mind, he suggests, is not merely an autonomous agent that investigates nature, but an agent necessary to the very existence of nature. Read his own words and judge for yourself:

> The new physics, by contrast, restores mind to a central position in nature. The quantum theory, as it is usually interpreted, is meaningless without introducing an observer of some sort. The act of observation in quantum physics is not just an incidental feature, a means of accessing information already existing in the external world: the observer enters the subatomic reality in a fundamental way and the equations of quantum physics explicitly enclose the act of observation in their description. An observation brings about a distinct transformation in the physical situation. (Pp. 39-40)

Davies concludes: "Commonsense may have collapsed in the face of the new physics, but the universe that is being uncovered by these advances has found once more a place for man in the great scheme of things" (p. 40). Surely, Davies has confused an artifact (perceptional knowledge) with an external reality that has been present without humans and their brains for billions of years. He has also introduced a category of human knowledge, suspiciously religious in tone, that lies in a belief field having no place in science. We shall develop this "anthropic principle" in Chapter 14, where it belongs—with religion.

Togetherness in Separation: The Nonlocality Problem

Our second philosophical topic in quantum mechanics concerns "the weird nature of quantum reality," as Paul Davies calls it (1984, p. 47). It is commonly described informally as "action-at-a-distance," "togetherness in separation," and more formally, as "nonlocality of phenomena." The subject requires much more of an in-depth treatment than we can give it here. Again, our limited objective is to "get the hang of it."

Albert Einstein was never able to accept the Copenhagen interpretation of an indeterminate solution to the problem of measuring both the location of a particle and its momentum. Nor was he able to accept the conclusion that a wave packet (wave function) is intrinsically incomprehensible to the human mind as a form of reality. Earlier in this chapter (quotation from Alan Lightman) Einstein's famous statement "God does not play dice" was quoted. He carried on a long debate with Niels Bohr, spokesperson for the Copenhagen interpretation. It culminated in 1935, when Einstein and two associates—Boris Podolsky and Nathan Rosen—proposed a showdown of sorts. It was a thought experiment that, if implemented, would decide once and for all whether a particle can have both a position and a momentum at the same time. Referred to as the "EPR," the experiment was to be "a scheme in which it appears that both these quantities can, in principle at least, be measured to any desired degree of accuracy" (Davies, 1984, p. 43). The strategy was to use two particles. These "twins" could be made first to interact, then to fly off in different directions, becoming widely separated in space. The momentum of particle 1 would be measured, and it would be reasonable to assume that this measurement would also hold for particle 2. Simultaneously, the position of distant particle 2 would be measured. Thus both measurements would have been accomplished on what was virtually the same particle. The EPR assumption had to be made that neither particle could influence the other, once they had separated. It was further assumed that no "influence" (signal) can travel faster than the speed of light, thereby negating the total independence of the two particles. Thus, for two photons moving apart in exactly opposite directions, particle 1 could not send out a signal notifying particle 2 not to respond.

A memorable step toward implementing the EPR experiment came in 1960 when John Bell formulated and proved a theorem showing that such an experiment was theoretically possible. Known as Bell's Theorem (or "Bell's inequality"), it provided a model for experimental determination of not only position and momentum, but also a property of particles known in QM as "spin." At that time, experimental technology was incapable of achieving the level of accuracy needed to perform the experiment, and it was not until 1981 that Alain Aspect and a group of corroborators, working in France, were successful. They used two photons emitted simultaneously by a single atom.

Details of Aspect's experiment are fully and clearly described by Milton Rothman at a level most college science majors can easily understand (1988, pp. 75-81). We can only give the barest suggestion of what it was all about with the help of a simple diagram (Figure 6.5). From a central source the two photons are simultaneously emitted in opposite directions—left (A) and right (B)—both traveling at the speed of light. Detectors at equal distances from the source recorded the simultaneous arrivals of the photons. Suppose, now, that two switches are inserted, as shown, one in each of the paths. Switch A can divert photon A to an alternative detector, A′; switch B can divert photon B to detector B′.

Either switch can be thrown while the photon is midway between the source and the switch. Commonsense would lead us to say that switching photon A to detector A′ could have no effect whatsoever on the path of photon B, which would complete its trip straight to detector B. But that is not what the Aspect group found to be the case. Instead, photon B simultaneously changed course and ended up in detector B′. What's more, the operation could be run in reverse: throw switch B and photon A responds in the same way, ending up in detector A′. That's the "spooky part"! It might be explained if the throw of one switch caused the photon to send a signal to its counterpart, instructing it, too, to divert course accordingly. But that would require us to postulate travel of an information signal faster than the speed of light; a postulate forbidden by all physicists, no matter what their views of QM theory.

We are left with a mystery: the "mysterious action-at-a-distance" that Einstein refused to accept. Followers of Niels Bohr and the Copenhagen interpretation consider the mysterious action-at-a-distance to be compatible with their theory of QM, and their position seems to represent the current consensus. The following statement by Milton Rothman may help those of us outside the priveleged area of QM to cope with what is now that "accepted" or "majority" opinion:

> Human language describes macroscopic objects well, but breaks down when we attempt to use it to describe the microscopic. Expressions like "the location of a particle" or "the location of the energy in a particle" or the "polarization of a photon" are quite simply without meaning. A particle may be detected anywhere within a large volume of space; the detection of a photon may depend on the state of another, distant photon; and even the term "wave packet" may not mean what we naively think it to mean when we write about it in textbooks. Clearly, these expressions are metaphors for things that can really only be described mathematically. (1988, p. 81.)

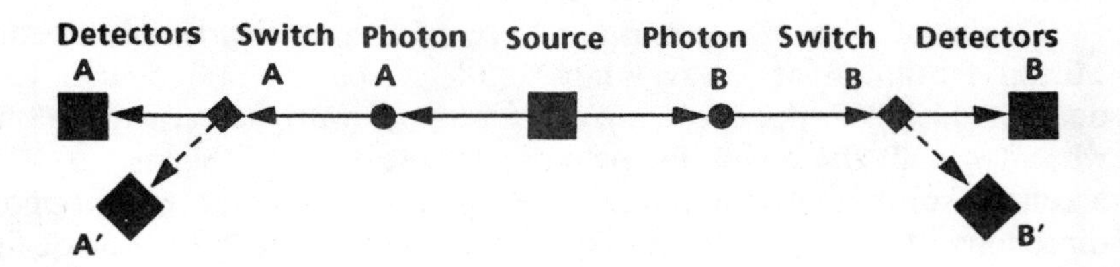

Figure 6.5 Schematic diagram to illustrate the concept underlying the Aspect experiment.

Accepting the above consensus and perhaps reluctantly abandoning Einstein and his commonsense view of QM leaves us with a fundamental question in the philosophy of science. We expressed it in Chapter 4 in connection with evolutionary biology and the recognition of a possibly unique emergentism setting that field apart from reductionistic physics. Must science be forced to conform to a single, all-pervasive system with a uniform set of laws; i.e., a *holistic* view of the universe? Or must we manage with a science partitioned into two divisions, one of which has certain special laws of its own, not required of the other? Perhaps we have no choice but to leave those questions unanswered. The history of science shows that more than once the paradigm of one epoch has given way to a new and unanticipated paradigm. The QM paradigm is described as being today in the "mopping up" stage. Perhaps there will follow a stage in which serious new problems begin to appear and grow in persistence.

This chapter may have been difficult going for many readers. Much of the subject matter is intrinsically strange, because we cannot easily visualize the entities being described in the framework of our endowed senses. Moreover, areas such as uncertainty, probability, chaos, and quantum mechanics are couched in the language of mathematics that many find difficult to grasp. Our next two chapters, in contrast, deal with familiar topics encountered in popular writing, both factual and fictional, and in news media reports. Pseudoscience is all around us all the time and we need to sharpen our awareness of its insidious nature.

CHAPTER 7

Pseudoscience: I. Case Studies

We devote two chapters to pseudoscience. In this chapter, four examples are presented to help us discern the criteria for recognition of pseudoscience. Chapter 8 summarizes the criteria of distinction and considers the historical and sociological factors in the phenomenon of pseudoscience.

Pseudoscience—A Checklist

Pseudoscience is information that is promulgated as science but fails to meet the qualifications for admission to empirical science; it is false science—not science at all. Don't confuse pseudoscience with science fiction; they can sound almost alike in places. Science fiction is a legitimate branch of literature, dealing in fantasy in the established tradition of such famous writers as Jules Verne. Science fiction has its devotees, among them many excellent scientists; pseudoscience has its cults, whose followers can get pretty nasty.

Listed below in random order are some topics that may or may not qualify as being partly or wholly in the category of pseudoscience. How many of them are familiar to you? Place a "P" in front of those you recognize as pseudoscience, and an "S" in front of those that belong in science. I have enough sense not to offer my own ratings—I just couldn't cope with the calumny that would be aimed at my head! For some answers, see Martin Gardner's excellent books: *Fads and Fallacies in the Name of Science* (1957) and *Science: Good, Bad, and Bogus* (1981).

Bermuda Triangle
Big Foot (Sasquatch); Abominable Snowman (Yeti); Loch Ness Monster
Flat Earthers; Koreshanity; Hollow Earthers; Atlanteans (Lost Atlantis); Lemurians
Shroud of Turin
Jupiter Effect
Pyramid Power; Pyramidology

Astrology
Lysenkoism
Cold fusion
UFOs, Flying Saucers, Extra Terrestrial Intelligence
Worlds in Collision (Velikovsky)
Chariots of the Gods (von Däniken)
Exobiology
Dianetics; Biorhythms
Homeopathy; Naturopathy; Osteopathy; Chiropractic
Acupuncture
Crystal power
Laetrile; Krebiozen

Here's another one to add to the list: A new and rather startling hypothesis has been making its way into geology and biology. An asteroid impacted the earth with great violence some 65 million years ago. The effects of this explosive impact were lethal to many forms of life and caused mass extinctions of whole species, families, and even some entire orders of animals. On land, the dinosaurs were completely wiped out; in the shallow seas the ammonites all but disappeared. This scenario is now being studied and debated in reputable science journals and books by some of the best paleontologists and biochemists in our universities. The science community regards the asteroid hypothesis as good science, even though it is on the frontier of science. The scenario is as weird as almost any that could be found in science-fiction magazines and movies. On what grounds is it science and not pseudoscience? I hope to give the criteria by which you can reach your own conclusions.

Our first case study is introduced to illustrate the transition zone between bona fide science and questionable, or suspect science; i.e., a theory or hypothesis that lies in a possible fringe zone between science and pseudoscience.

Parapsychology

Parapsychology, abbreviated to *Psi* (the 23rd letter of the Greek alphabet), is a comparatively recent term coined to describe a wide range of phenomena that are purported to involve direct interaction of one human mind with that of another, or with matter, in ways that do not seem to have explanations in terms of physical laws as we now accept them. Claimed by a group of investigative scientists as a field of science, parapsychology remains extremely controversial. The problem in any discussion of parapsychology is to sort out from among a wide variety of activities, interpretations, and opinions, that content which is genuinely scientific in method (if not complete in theory and supporting evidence) from that which is transempirical or simply religious in nature and from that which is simple charlatanism.

The following list of varied psychic phenomena can be gleaned from recent books on parapsychology:

Automatism (automatic writing, Ouija-board activity)
Clairvoyance
Dermography
Dowsing, divining
Extrasensory perception (ESP)
Haunting
Kirlian photography (aura photography)
Levitation (biogravitation)
Materialization
Mediumship
Metal bending
Out-of-body experience (OBE) (disembodiment, astral travel)
Poltergeist activity (recurrent spontaneous psychokinesis)
Possession by spirits
Precognition
Psychic healing, injury, and surgery
Psychic photography (psychophotography)
Psychokinesis (PK)
Psychometry
Reincarnation
Spirit recording
Stigmata
Survival (including death-bed phenomena)
Teleportation
Xenoglossy; xenography

The above phenomena, as listed in alphabetical order, make a mixed bag and the list serves only to test your memory. Surely, almost everyone who is exposed to the entertainment and news media has heard of many of these items.

Psychic Phenomena and Darwinism

To make some kind of order out of the entire package of so-called psychic phenomena, it will be helpful to review the historical background in which these terms arose. In this review I follow Alfred Douglas's historical account in his book *Extra-Sensory Powers: A Century of Psychical Research* (Douglas 1977, pp. 12-17). The book is written for the general reader and contains fascinating accounts of the most celebrated psychical cases and claims, but at the same time it is objective in its approach and appraisals.

Prior to about 1800 in Europe and the Americas, psychic phenomena were closely tied in with the dominant theistic Christian religion and the paraphernalia connected with it. Psychic phenomena were often recognized as being promulgated by God or the Devil in the struggle for possession of the human soul. Belief in the hereafter made easy the interpretation of spiritualistic phenomena as an interchange between the living and the dead. With the rise of naturalistic and materialistic views of the world of nature and the place of humanity in that world, Christianity was challenged and threatened. Materialistic science emerged and developed as a dominant force through the 1800s. Physical and natural science combined to form a pervasive influence that involved all areas of human activity. Charles Darwin's theory of organic evolution seemed to reduce biology, including the origin of the human species, to a mechanistic level devoid of spiritual values.

By the 1800s, in England, a group of persons known as Spiritualists organized themselves into the Society for Psychical Research (SPR) for the purpose of investigating spiritualistic phenomena claimed by its practitioners to be bona fide forms of communication with spirits of deceased persons. According to Douglas, the SPR attracted to its membership some prominent scientists who "were attracted to psychical research partly in the hope that it might provide ammunition with which to fight the growing materialism of their day. Many of them came from pious Christian backgrounds and they were profoundly disturbed by the implications of the Newtonian 'clockwork universe' which held no place for spiritual forces and which, they feared, might not only undermine religious faith but also public morality" (1977, p. 12). Among the prominent individuals named by Douglas were philosopher Henry Sidgwick and distinguished physical scientists Lord Rayleigh and William Crookes. The SPR supported research on spiritualism, which had become a popular movement in England. Its practice included clairvoyance, telepathy, occult diagnosis and healing, and contact with spirits of the dead (p. 13). The research was not altogether satisfactory in outcome. In Douglas's words:

> The early hopes of the founders of the SPR were not easily realized. The claims of the Spiritualists were not readily authenticated, although much of the phenomena studied was inexplicable, and there was ample evidence of fraud and gullibility. (P. 13)

The American Society for Psychical Research was founded in 1884 and worked along similar lines to those of the SPR. Continued research by both societies produced what seemed to be evidence of the reality of psychic phenomena, but there emerged no conclusive evidence of survival after death. This early period of psychical investigation based on case studies of spiritualist mediums ended about 1930; it was to be superseded by laboratory studies conducted according to strict rules of empirical science and evaluated by rigorous tests of mathematical statistics.

Extrasensory Perception and Psychokinesis

A middle phase of parapsychology of the 1930s and '40s was dominated by the work of J. B. Rhine, his wife, Louisa E. Rhine, and their associates, largely carried out at Duke University, North Carolina. Both Rhines held doctoral degrees in science and were trained in the structuring of biological experiments and the use of statistical tests to determine significance levels of test results. This is not the place for a review of their experiments, but for those entirely unfamiliar with their work a brief description is in order.

The Rhines used a test subject (usually a university student) who was asked to focus attention on a "target" isolated from any contact with the subject through the known forms of sensory perception. Many of the Duke experiments were conducted with a set of specially designed cards (Zener cards), each imprinted with one of five different pictorial symbols. A deck of 25 cards, five cards for each symbol, constituted the target matter. In a typical test, an operator would turn up cards in succession from a shuffled deck; the isolated subject was asked to identify the symbol on each card. (Other procedures were used as well.) On a simple probability basis, a subject with no ability to perceive the sequence of the deck would, on the average with random order of the cards, obtain by chance a distribution of correspondences (called "hits") commensurate with the population distribution (in this case, five hits per run of one deck of 25 cards). Some subjects were able to identify the symbol correctly more frequently than the random model called for. A null hypothesis was proposed and the probability that such a result could occur by chance alone was evaluated. The result was judged as significant or not significant at a predetermined acceptance level. Repeating the experiment had the effect of increasing the size of the sample, and in this way the probability of rejecting the null hypothesis when it is actually true could be reduced to very small values. Similar experiments were devised using dice and sequences of random numbers. Certain sensitive subjects were found to be highly successful in perceptive activity and scored probability levels far beyond the limits used generally in experimental science to establish the outcome of tests as "almost beyond question."

The Rhines (Rhine and Pratt 1957) used tests to determine the ability of a subject to perceive information in an extrasensory manner in three activity categories: *telepathy* (to perceive the thoughts of another), *clairvoyance* (to perceive information directly), and *precognition* (to perceive events before they happen). These three forms of perception were named *extrasensory perception* (*ESP*). A fourth mental activity, also included by the Rhines in parapsychological phenomena, is called *psychokinesis* (*PK*). Very different from the first three, PK is the ability of a subject to exert physical influence on the environment by mental processes alone. In this case, "influence" can involve changes that under ordinary circumstances would involve the application of physical forces. For example, a subject would be asked to influence the outcome of the throw of dice in such a way as to increase significantly the incidence of occurrence of higher-than-average totals, or, alternatively, of lower-than-average totals. The reason for including PK investigations with those of ESP is that in a particular test activity it is not always possible to establish which of the four mental activities is involved.

By about 1950, the major findings of the Rhines, their coworkers, and colleagues in parapsychological phenomena had been published in the *Journal of Parapsychology*. In 1957, the Parapsychological Association was formed; in 1969, it was granted affiliated status by the American Association for the Advancement of Science. One influential individual who urged this admission was anthropologist Margaret Mead; she commented: "The whole history of scientific advancement is full of scientists investigating phenomena that the establishment did not believe were there" (Bowles and Hynds 1978, p. 9). According to Carroll B. Nash, Director of the Parapsychology Laboratory of St. Joseph's University, the accumulated experimental data from such laboratory tests is now overwhelming:

> Most ESP experiments are no longer directed at proving extrasensory perception. This would be comparable to continuing the stockpiling of nuclear explosives when there are already enough to destroy the world's population. Most parapsychologists believe that the experimental evidence for ESP has long been sufficient for its acceptance and that perusal of the numerous experimental reports in scientific journals should convince any openminded skeptic of the existence of ESP. (Nash 1978, p. 47)

In the past three decades, parapsychological research has taken some new twists, such as a study of the relationship between parapsychological performance and altered states of consciousness. There has been a general return of attention to cases of spontaneous psychic phenomena, including poltergeist activity, premonitions, out-of-body travel, apparitions, and the power of spiritualist mediums.

The Scientific Community Reacts

Is parapsychological research to be accepted as a scientific activity, as Margaret Mead recommended and as the AAAS's acceptance of the Parapsychological Association seems to affirm? Based on our description of the essential components of science, the affirmative answer to this question is at least worthy of debate. We have stated that science at its best consists of both description and explanation. A third form of content—logical prediction—is usually given as a requirement for scientific hypotheses. Thus prediction, in the sense of testability, should be included, but in the temporal sense it can be dispensed with.

How does parapsychological research, such as that published in the *Journal of Parapsychology*, rate in each of the three essential requirements of science? Consider first the methods of experimentation leading to a description of the phenomena of ESP and PK. Are the results as positive as claimed by Rhine, Pratt, and Nash? The statistical evaluation methods they used have, despite some adverse criticism, stood up well because of conformity to testing methods used in the fields of psychology, biology, and other areas of natural science. But what of the striking "successes" reported in test scores of those rather rare individual test subjects found to be exceptionally sensitive to ESP?

Fortunately for science, there have been at least a few skeptics who did much more than simply pooh-pooh parapsychology as nonsense. They continued to probe the published descriptions of parapsychological tests, always on the lookout for flaws. With Occam's razor in mind, they searched for simpler answers to significant outcomes of tests. Two of these skeptics in particular have made extraordinary contributions by their unflagging probes of psi research. One is Martin Gardner, whose many books and innumerable magazine articles on fringe science and pseudoscience have appeared over a span of three decades. Two of his books were named earlier in this chapter (1957; 1981). The second contains reprints of his numerous magazine articles together with replies from authors he has criticized, and his rebuttals to their replies. A second skeptic, dealing particularly with ESP and PK, was C. E. M. Hansel, then a professor of psychology at the University of Sawnsea, England. Among his major works is *ESP and Parapsychology: A Critical Reevaluation* (1980). Even the briefest summary of the hundreds of pages of criticism produced by these authors would be too long and too detailed to include here. Since my purpose is to explain how science works, I will attempt only a few general statements about the nature of the criticisms leveled at experimental work on parapsychology.

First, we must take into account that for psychology as a whole, and for parapsychology in particular, the human mind is investigating itself. In no other area of science can subjectivity be so easily magnified into self-deceit. Second—and possibly in consequence of the first point—is what seems to be the total commitment of psi researchers to the belief that ESP exists. This is the method of the ruling hypothesis, against which Professor T. C. Chamberlin warned so eloquently. The self-inflicted blindness that plagues a scientist dedicated to a ruling hypothesis greatly increases the chances of making errors of observation, both deliberate and unintentional, and of overlooking alternative explanations that may to others seem obvious.

Much of the probing done by Hansel and Gardner into experimental efforts in ESP and PK has focused on alternative explanations for successful experiments that seemed on statistical grounds to point to the existence of psi phenomena. Alternative explanations that are naturalistic and acceptable to scientists generally are of two kinds: (1) information was unknowingly transmitted from one participant to another by sensory or mechanical means; (2) and/or deliberate fraud was practiced by participants to yield false data. Instances of the second kind have been exposed and subsequently recognized as fraudulent by coworkers in psi. A good example is a study conducted by Dr. S. G. Soal, a British researcher in parapsychology located at London University. The case has been analyzed in great detail by Professor Hansel under the title of the Soal-Goldney Experiment (Hansel 1980, pp. 141-67). A statistician, Betty Marwick, succeeded in detecting the manner in which numbers of the test series had been altered to generate test results with probabilities as small as 1 in 1070 that the outcome was due to chance alone. An assistant testified that she had observed Dr. Soal altering the figures on more than one occasion. Careful analysis by skeptics of the experimental procedures of numerous earlier tests, on which the claims for ESP and PK were based, disclosed numerous ways in which fraud could have been carried out, had the experimenters chosen to do so. Hansel has stated an important principle in assessing the validity of an experiment in psi: "An experiment that has any defect such that its result may be due to a cause other than ESP cannot provide proof of ESP" (p. 20).

Hansel writes very forcefully about the meaning of the extremely low probabilities observed in some ESP experiments—odds of 1 in 1070 in Soal's experiment, for example. Such odds are commonly "quoted as if they indicate that the experiment *proves* the existence of ESP" (p. 23). Hansel goes on to say that if there is even the smallest possibility of some other explanation, the results of the experiment support that other explanation as much as they support the ESP explanation. "The probability obtained in the experiment is that of the score having arisen by chance." Hansel concludes:

> To provide statistically overwhelming evidence for the occurrence of ESP in experiments of this nature required satisfaction of two conditions: (1) the scores achieved by the subject must be such as

> are very unlikely to arise by chance, and (2) the experimental conditions must be such that only ESP could account for them. (P. 23)
>
> The second condition is the more difficult to assess. The experimental conditions may be examined most meticulously and no flaw found, but there is no certainty that nothing has been missed. The conclusion that an experiment provides statistically overwhelming evidence for the appearance of ESP is misleading if no allowance is made for the possibility of error in the experimental setup. The incidence of trickery, deception, and error in psychical research is such that the probability of their occurrence is certainly far from insignificant. (P. 24)

In response to criticism such as that quoted above, attempts have been made to devise an electronic machine that can eliminate all possibility of fraudulent manipulation of data. One such attempt, made in 1962, was the construction by a member of the Air Force Cambridge Research Laboratories of a machine called VERITAC. It automatically selects digits at random, records the subject's guesses, and holds the results of the test locked inside until instructed to print out the data. In a large number of trials conducted with VERITAC in 1962 "there was no deviation from chance either for the entire group or for any individual" (Gardner 1981, p. 76). Similar tests, run with newer highly sophisticated electronic machinery in 1972 with financial support from NASA likewise showed no significant differences from the expected chance distribution (pp. 75-90).

Martin Gardner summarizes the conservative scientific view of parapsychology in these words:

> Science cannot absolutely rule out the possibility of anything, but it can assign low degrees of probability to unusual claims. In my view, which is the view of most psychologists, the classic psi experiments are more simply and plausibly explained in terms of unconscious experimenter bias, unconscious sensory cuing, fraud on the part of subjects eager to prove their psychic powers, and, on rare occasions (such as those recently disclosed about S. G. Soal), deliberate fraud on the part of respected investigators. (1981, p. 183)

Is an Explanation Required?

On the second attribute of a well-constructed hypothesis—explanatory power—we encounter a total blank from the experimenters. J. B. Rhine's position in this matter has always been clearly stated: there is no physical explanation. To quote from Rhine and Pratt:

> What, then, identifies a psychical phenomenon as parapsychical? It is an occurrence that has been shown by experimental investigation to be unexplainable wholly in terms of physical principles. It is, in fact, the manifestly nonphysical character of parapsychical phenomena that for the present constitutes their only general identifying feature and marks them off from the rest of general psychology. This does not, of course, alter the fact that the data of parapsychology are natural. As a matter of fact, our concept of what is "natural" is built up out of just such discoveries of science as they are made; accordingly it goes on growing, and will continue to do so, with each added bit of knowledge. It is now clear that, contrary to some of the limiting philosophies that currently prevail, nature extends beyond the domain of purely physical law. (1957, p. 6)

Despite what skeptical investigators have turned up in terms of alternative interpretations of experimental results, some may wish to accept parapsychology provisionally as an "incomplete" or "partial" science, pending the possibility that in due time investigation will at least allow a reasonable working hypothesis of the mechanism of ESP and PK to be formulated and tested. There is, however, a rather sticky point to be discussed in connection with Rhine's steadfast position, expressed in his own words. Take note of the words "beyond the domain of purely physical law." On the same page Rhine and Pratt state that in distinguishing parapsychical occurrences from physics "there is always some distinct point at which a completely physical interpretation is manifestly inadequate" And a few lines later: "we may now characterize it (parapsychology) as a branch of inquiry which deals with nonphysical personal operations or phenomena."

Physicist Milton A. Rothman, in his book *A Physicist's Guide to Skepticism*, makes some clarifying statements on the distinctions between "physical" and "nonphysical" phenomena (1988, pp. 116-17). First, he stresses the skeptical physicist's tenet: "The behavior of living matter (including thought and consciousness) can be explained entirely by the behavior of fundamental particles acting under the influence of the four forces, chiefly the electromagnetic" (p. 116). If explanations of ESP are to fit into this principle, the alleged phonomena "can be explained only by invoking new and presently unknown interactions that follow uniform and natural laws" (p. 116). He also adds the possibility that these new and unknown principles may operate within the inner structure of quantum theory. He points out that it would, however, be a serious error to interpret quantum theory as allowing instantaneous communication at a distance (p. 117). The "nonphysical" category of explanation clearly falls outside the domain of physics, and Rothman recognizes it as being in the supernatural realm along with religion. He offers some details of his reasons for concluding that "physical interactions cannot account for the claimed properties of most paranormal phenomena" (p. 117). The latter violate the principle of conservation of energy, the law that relates signal strength to distance (inverse-square law), and the principle of causality that requires a signal to be transmitted before it is received (pp. 117, 150-51). Another interesting observation Rothman makes concerns the possibility of the existence of nonphysical ways of communication under a dualistic system of mind and brain. He asks: How could the "nonphysical mind" interact with the electrons of the nervous system to start the working of the brain (p. 119)? I see this required interrealm communication as no more than an a priori assertion, advanced despite the objection from logic that it is ruled out by mutually exclusive definitions of the two realms.

The sense of "nonphysical" is extremely important from an ontological viewpoint. I see two interpretations before us: (1) The causative agents of the phenomena do not lie within the capability of observation of empirical science; they are not in the natural realm, but they lie instead in the transnatural (supernatural) realm; in that realm their existence lies in human belief, as do religious beliefs asserted to be of divine origin and known to humans only through revelation. (2) The causative agents of the observed phenomena have not yet been identified from among existing forms of matter and energy, but will eventually be so identified. When this happens the agents will take their places beside (or in addition to) known agents of empirical science. This fits a monistic ontological model that can be described as wholly materialistic or naturalistic.

The first possibility requires a dualistic ontology, similar to that in which science and religion (or science and ethics) are assigned to different realms (Ryzl 1970, p. 181). We see in that view a persistence of the dualistic mind-body model or soul-mind model that has been argued for centuries. It reminds us of Nobel Laureate George Wald's recently expressed belief in the existence of consciousness as a pervasive cosmic force apart from physical reality (see Chapter 1). Using only the words of Rhine and Pratt it is not easy to put aside such dualistic interpretations. If the dualistic model is to be adopted, there is no hope for parapsychology to attain scientific status.

The second interpretation seems clearly stated in these words by biochemist Milan Ryzl, who was at one time a coworker with Rhine at Duke University:

> The object of parapsychology is distinct from the objects of other sciences. Research has shown the existence of a definite group of natural phenomena controlled by laws that *are not based on any energetic influence known so far to us.* In other words, a new form of energy, so far unknown to physics, is responsible. This presumption helps us define the sphere of parapsychology—the study of phenomena not explainable by common energetic effects. (1970, p. 3)

This interpretation is both optimistic in outlook and the more parsimonious of the two alternatives. Moreover, it is in harmony with the history of empirical science, which shows a long series of discoveries of the causes of phenomena that could not have been suspected of existence at an earlier stage in science. Take, for example, the discovery in the Van Allen radiation belts, far out from the earth's surface, made up of electrons trapped in the magnetic field of the earth. Traveling at enormously high speeds, these electrons emit energy in the form of photons, producing the auroral displays of colored lights. This "synchrotron radiation," emitted by relativistic electrons traveling at speeds approaching that of light, has been identified in enormously vast and powerful magnetic fields within active galaxies. Humans observed and documented the aurora borealis as a visual display in a scientific manner for some centuries, while at the same time finding for it no physical causative agent whatsoever. Under certain circumstances we simply have no alternative but to take a wait-and-see attitude, perhaps hoping that we will live long enough for the discovery to be made.

My position seems to be in agreement with that expressed by Martin Gardner:

> The history of science swarms with observed phenomena that were genuine but had to wait for centuries until a good theory explained them. A lodestone's magnetism was sheer magic until the modern theory of magnetism was formulated, and even today no physicist knows why the acceleration of electrical charges inside atoms causes magnetic effects. It is not even known why electricity comes in units of positive and negative charge, or whether magnetic monopoles exist as theory seems to demand. Kepler correctly decided, on the basis of confirmable correlations, that the moon causes tides; but in the absence of a theory, even the great Galileo refused to believe it. One could add hundreds of other instances in which a phenomenon was authenticated long before a theory "explained" it. On this I find myself in full agreement with J. B. Rhine and other parapsychologists who regard the lack of a physical theory as no obstacle whatever to the acceptance of psi. (1981, p. 183)

To say that no acceptable physical explanation of parapsychological phenomena now exists does not mean that speculations have not been made as to the nature of the agent or agents involved. Quite to the contrary, speculations have long been available, but with evidence from observation entirely lacking. Putting aside explanations of religious nature, there are speculations as to the existence of some kind of medium or substance capable of conducting or conveying mental imagery from one mind to another. This medium has been thought of as a kind of halo enveloping the earth, permeating all matter, but being impossible of detection by any means now known to science. It can be speculated that the medium either carries a weak flux of energy or exerts energy on physical substances. Where a high level of energy is required, as for example in poltergeist pranks or in bending of metallic objects, it is speculated that weak energy can be focused into intense "beams," much as a magnifying glass focuses the solar rays to allow combustion temperatures to be applied to a surface.

Attempts have been made to build scientific hypotheses about the nature of a form of energy enabling ESP to take place (Ryzl 1970, pp. 183-84). About 1940, H. Burger, the discoverer of EEG waves (brain waves) proposed that there exists another type of energy, called *psychical energy*. He suggested that this energy is generated within the brain cortex and moves out into the environment, where it can enter the brain of another person. There, the energy is transformed into usable form. Among Russian parapsychologists, who seem to have been quite receptive to the idea of psychical energy, there was presumed to exist a *psychic field*, or *psi-field*, which has some physical similarities to an electromagnetic field. Precise as psi-field theory may be in its statement, no means to observe it—if it exists—have been found. As a hypothesis, it cannot be tested, at least at the present time. By this criterion it does not find acceptability as a viable scientific hypothesis.

Other speculative hypotheses invoke fluxes of elementary particles (see Gardner 1981, pp. 185-204, "Quantum Theory and Quack Theory"). One particle, the neutrino, is thought by physicists to penetrate matter with almost no resistance whatsoever. It has been suggested that neutrinos generated within the human body carry information to other locations. The hypothesis suffers from a weakness that might seem fatal: If neutrinos are rarely intercepted by atoms, they cannot effectively transfer energy to a receptor consisting of atoms. One imaginative individual has postulated the existence of a form of elementary particle, dubbed the "psitron." Emitted by various kinds of ordinary matter, psitrons form a psifield. Needless to say, the psitron has not yet been discovered by physicists.

Precognition—A Bit Too Far Out?

While waiting for science to provide a causative agent for alleged parapsychological phenomena, we should be pondering a very difficult and perhaps deeply disturbing concept that arises when the implications of human precognitive ability are faced. For the Duke University investigators, precognition is simply the perception of a future event by means of ESP. They attached rigorous qualifications to such perception:

> To qualify as a genuine instance of precognition an experience must refer to a coming event to an extent that is more than merely

> accidental; it must identify a future happening that could not have been inferred as about to occur; and finally, it must refer to an event that could not have been brought about as a consequence of the prediction. (Rhine and Pratt 1957, p. 10)

In the Duke laboratories, precognition was tested through the ability of a subject to predict the ordering of symbols on cards before the card deck was shuffled (or the order of a succession of dies before each die is tossed, or to predict a succession of random numbers before they are generated). If, despite what the skeptics have to say, we accept as highly likely the ability of some sensitive subjects to make such predictions in significant proportions, and if we can rule out to our satisfaction the possibility that the operator, knowing by ESP the subject's choices, was able to influence by the psychokinetic process the card distribution during shuffling, we must conclude that the human mind has the ability to eliminate time as a dimension. This means that beyond merely imagining what may take place at a future time, the mind moves ahead to that actual point in time where a real or actual state of matter and energy is in existence.

That strong objections, based on ingrained "common sense" attitudes, would greet claims of observed precognition is not surprising. Milan Ryzl describes the situation:

> Objection has also been raised to the possibility of precognition—a parapsychic transmission of information crossing the barrier of time. Such phenomenon contradicts our conception of causality—how can a future event, the cognized process, be the cause of a percept when it precedes it in time? Besides, precognition contradicts the concept of free will. If precognition exists then it would seem necessary to admit the absolute determinism of all processes in nature; that future events are "existing predetermined" already in the present time, without allowing any possibility of intervention by an individual's decision. This leads to a paradox: how would precognition work if a person related to the predicted event decided to do the opposite of what was predicted? (1970, p. 13)

Ryzl admits that, should acceptable proof be brought forward of the possibility to transmit information about a future time, we will be forced to change, at least in special cases, our present concepts of the space-time continuum (1970, p. 13).

Milton Rothman firmly rules out precognition on grounds that it violates the principle of causality by allowing a cause to come later in time than the effect (1988, p. 150). Precognition, he observes, "requires that information be sent from the future into the present. But transmitting a signal at a time later than the time of reception violates causality. There is no interaction that goes backward in time" (p. 150).

Philosophers, who seem to take little notice of parapsychological phenomena involving transfer of information, perk up their ears at the mere mention of precognition. There is a bone well worth chewing! The seeming illogical implications of precognition are too flagrant to be ignored. I offer no excuses for not reviewing here the philosophical arguments concerning precognition, except one that is most obvious—it is very difficult going for a scientist who is unskilled in the language of logic. I recommend to braver souls two collections of papers that cover the ground quite thoroughly: *Philosophical Dimensions of Parapsychology* (Wheatley and Edge 1976) and *Philosophy and Parapsychology* (Ludwig 1978).

Does everything that ever was and ever will be exist somewhere, waiting only for a human mind to call out a selected instant of reality and display it for observation, as one calls out information from storage in a computer and orders it to be displayed on a screen? Is there a deterministic universe at our fingertips? What a field day for science fiction writers!

The Verdict on Psi

An overview of the status of parapsychology from the skeptic's viewpoint is fully developed in a collection of twenty articles in a volume edited by Kendrick Frazier titled *Paranormal Borderlands of Science* (Frazier 1981). I recommend especially the first article, "Is Parapsychology a Science?" by Paul Kurtz. An emeritus of philosophy at the State University of New York at Buffalo, Dr. Kurtz serves as Chairman of the Committee for the Scientific Investigation of Claims of the Paranormal. I can do no better in offering a concluding statement than to quote some lines found at the close of his essay:

> All that a constructive skeptic asks of the parapsychologist is genuine confirmation of his findings and theories, no more and no less. I should make it clear that I am not denying the possible existence of psi phenomena, remote viewing, precognition, or PK. I am merely saying that, since these claims contravene a substantial body of existing scientific knowledge, in order for us to modify our basic principles—we must be prepared to do so—the evidence must be *extremely strong*. But that it *is*, remains highly questionable. In the last analysis, the only resolution of the impasse between parapsychologists and their critics will come from the *evidence* itself. (Kurtz 1981, p. 21)

This, then, is an invitation and a challenge to parapsychologists to bring their findings to the most thoroughgoing skeptics they can

> locate and have them examine their claims of the paranormal under the most stringent test conditions. If parapsychologists can convince the skeptics, then they will have satisfied an essential criterion of a genuine science: the ability to replicate hypotheses in any and all laboratories and under standard experimental conditions. Until they can do that, their claims will continue to be held suspect by a large body of scientists. (Kurtz 1981, p. 22)

Three Historical Scenarios to Consider

Our continued investigation of pseudoscience develops along the following lines. I have selected three major scenarios put forward just after World War II. All three are reconstructions of supposed events fixed in time; their content falls in the category of timebound knowledge. Whereas parapsychology lies in the category of timeless knowledge, what we now examine consists of assertions or allegations that particular events occurred at particular times; all introduce agents responsible for those events; in all, the agents are asserted to be extraterrestrial, but do not require the intervention of the God of contemporary theistic religions. All are based, in one way or another, on testimony given and recorded in a language by humans following sensory experience. In some cases the testimony is in the form of symbols on artifacts and in the nature of the artifacts themselves. The events are of a singular nature, not capable of being duplicated on demand for direct observation by either their authors or independent investigators.

Additional common characteristics of the three scenarios relate to their social impact. Each quickly caught the imagination of millions of persons because of rapid and widespread dissemination through the popular media. Large followings of believers were generated from among persons outside the scientific community, and cults have arisen in support of each scenario. In each case the scientific community reacted with skepticism and strongly expressed its disapproval. Within each cult, believers have attempted to draw support from the body of empirical science in an effort to claim that the phenomena described are explicable in a naturalistic manner and do not require supernatural intervention.

The three scenarios are (1) Immanuel Velikovsky's reconstruction of world history of the first and second millennia B.C., (2) Erich von Däniken's claims of visitations by extraterrestrial beings from the time of earliest recorded human history to modern times, (3) and alleged visitations by beings in spaceships from outer space in the period following World War II. The last item is closely tied to, but not to be equated with, the total content of cases of Unidentified Flying Objects (UFOs).

The word "scenario" is used with the implication that the succession of singular events contained in each may be in some degree fictional; indeed, the entire scenario may be an artifact. The scientific community views all three scenarios as being outside the realm of empirical science (i.e., they belong in pseudoscience). No one of the scenarios poses a serious threat to the security of the human race, because the first two deal with ancient history and the third is viewed by many of its believers as a benign phenomenon. In this light, the intensity of conflict between scientists and cultists seems all out of proportion to its importance in terms of content.

We cannot escape from the realization that the highly emotional exchanges between scientists and cultists in public debates are symptoms of a deeply rooted social malaise. The conflict between science and the forces of antiscience or cultism is one in which the stakes are much greater than for mere refutation or confirmation of the particular scenario. The malaise is not new but seems to have taken a new form adapted to an age of space technology.

My suggestion for your consideration is that during the first three decades of this century throughout the culture of western Europe and North America, highly dogmatic Christian beliefs were being abandoned by increasing numbers of persons. Some sectors of Protestantism, and to some extent Catholicism as well, became sufficiently liberalized to accept the phenomenon of organic evolution occurring through an enormously long period of geologic time; this trend was perhaps accelerated by the impact of the Scopes trial in 1925. Freethought was on the rise and secular humanists were advertising the merits of a purely rational approach to human problems, using science as a guide to a new ethics. Perhaps the population of agnostics, atheists, and those simply having little interest in religion was steadily growing larger. In the 1940s, science in the area of nuclear physics showed its enormous power to create destruction through the atomic bomb while developing its theoretical concepts far beyond any possibility of being understood by the vast majority of citizens.

Perhaps for vast numbers of persons the need was growing for something both new and irrational to believe in—something that, like a religion, required no formal education and did not tax the intellectual faculties, but at the same time carried an air of mystery and romance. The arrival on the scene shortly after World War II of Velikovsky's scenario and a rash of reports of flying saucers almost instantly filled that growing vacuum and, at the same time, polarized the science and antiscience factions as never before in the century. Not only was the vacuum filled, but there was provided a public spectacle in which the ivory tower of science was to be struck repeatedly with roughhewn two-by-fours, wielded by folk heroes. However one may wish to explain why these scenarios of pseudoscience took off like huge rockets, the evidence is clear that they did. A somewhat similar explanation of this resurgence of pseudoscience is given in John T. Omohundro's article, "Von Däniken's Chariots: a Primer in the Art of Cooked Science" (1981, pp. 307-309).

Some surveys and census statistics presented by William S. Bainbridge and Rodney Stark (1981) are supportive of this analysis. These authors zero in quite closely on the major group that has been most ready to accept pseudoscience. It appears from the results of questionnaires in 1963 and 1973 that the religious liberals and the irreligious persons were more likely to accept the "new superstitions," such as occultism and pseudoscience, than were fundamentalists (1981, p. 50). The latter group, professing in the questionnaire to be born-again Christians, dislike Far Eastern cults and occult literature and took a dim view of UFOs being spaceships from other worlds. They were, however, strongly opposed as a group to the theory of evolution (p. 48). Evolution was most strongly supported by those with no religious preference and those admitting to be agnostics and atheists. (Note that the years of these polls roughly span the Vietnam War, a period of great turmoil among the younger persons of our society, especially the college students.)

According to 1970 U.S. Census Bureau figures quoted by Bainbridge and Stark, there were about 115 million church members in a total population of 205 million, the average rate being 560 church members per thousand (p. 54). This left 90 million persons unaccounted for. Some were children or incompetent adults who would not have participated in a response to pseudoscience, and some may be excluded as holding other faiths that disfavor pseudoscience, but there still remained an enormous audience available for participation in the cultist response to pseudoscience (exclusive of creation "science"). Bainbridge and Stark make a significant comment on our possible misconception of the character of this susceptible group:

> It would be an equally great mistake to conclude that religious liberals and the irreligious possess superior minds of great rationality, to see them as modern personalities who have no need of the supernatural or any propensity to believe unscientific superstitions. On the contrary, they are much more likely to accept the new superstitions. It is the fundamentalists who appear most virtuous according to scientific standards when we examine the cults and pseudosciences proliferating in our society today. (1981, p. 50)

The three scenarios I have selected are presented, argued, and discussed in thousands of pages of published books and magazine articles. There is no way this mountain of information (and misinformation) can be adequately digested and covered here. Keeping in mind that we are interested in gaining insights into the workings of science, we must settle for a very brief statement of the contents of each scenario—just enough to provide some targets and examples as we try to identify the essential characteristics of pseudoscience and its modus operandi. The crucial questions we will examine lie mostly in the areas of philosophy, psychology, and sociology. To find answers to these questions, we must try to stand clear of the arguments themselves. In any case, neither side will concede anything to the other, regardless of the merits of the arguments.

Velikovsky's *Worlds in Collision*

Immanuel Velikovsky's first major work, *Worlds in Collision* published in 1950, is essentially a reconstruction of world history spanning a period that began when human myths and legends were first being recorded in writing. His references to events earlier than the beginning of the second millennium B.C. (2000 B.C.) are vague; they include a novalike explosion that caused the Flood of Noah. Let us begin this brief review with a catastrophic event occurring about 1450 B.C., when what is now the planet Venus made its first major encounter with Earth. Velikovsky states that this encounter was recorded by many ancient cultures, but is particularly noted for causing the plagues associated with the Exodus of the Hebrews from Egypt (Ransom 1976, p. 21). At that time Venus was a comet; it had been born of material ejected from the great outer planet Jupiter. The comet's trajectory was erratic for a period of 800 years, during which time it terrorized Earth by near-collisions. In Velikovsky's words, when a comet makes contact with a planet, "the planet slips from its axis, runs in disorder off its orbit, wanders rather erratically, and in the end is freed from the embrace of the comet" (1950, p. 156). The Jews during Exodus experienced unusual phenomena including a fall of red dust, which "turned rivers to blood"; falls of either hail or meteorites (or perhaps both); a fall of petroleum droplets from the sky; fall of vermin from the sky; a period of several days of darkness; and great earthquakes.

Current astronomical knowledge suggests that the dense core of a typical comet consists of a mixture of icy particles that may include substantial amounts of such compounds as water, ammonia, and methane, and particles of solid mineral matter similar in composition to materials found in meteorites that have reached the earth's surface. In popular parlance, the typical comet is a "dirty snowball." The diversity of a comet's composition can perhaps be held to account for the diversity of substances alleged to have been transferred from a cometary Venus to Earth.

Other phenomena associated with this first encounter with Venus included the parting of the Red Sea, gigantic tides, and an enormous electrical discharge between Venus and Earth (Ransom 1976, p. 30). Great geologic events occurred; they included the growth of innumerable great volcanoes, the rise of mountain ranges, and large-scale crustal rifting. Later, food of some sort also fell from the sky. For Hebrews in the desert, the food was manna; in other parts of the world it was described variously as ambrosia, heavenly bread, and honey (pp. 31-32). Following the encounter there was a period of several years of dense cloud cover and gloom (p. 34).

Meantime, comet Venus continued on a highly elliptical orbit that eventually brought it back to a second encounter with Earth. Ransom describes that encounter in these words:

> The second near encounter was not as close or as destructive as the first. The Earth was not engulfed in the extended atmosphere of Venus although numerous meteorites hit the Earth. Earthquakes were also common. (P. 35)

At this time, the Israelites had crossed over the Jordan River, entered the Promised Land, and encountered the city of Jericho. The collapse of that city may have been accomplished by one of the earthquakes referred to above. During the battle of Gibeon, the sun "stood still" (held its apparent position in the sky) for a whole day, enabling Joshua to complete the slaughter of the enemy at Beth-horon, begun by a rain of huge hailstones sent down by Yahweh (Josh. 10: 1-15). This astronomical event is interpreted by Velikovsky as a mechanical coupling effect with Venus, causing earth rotation to cease, but then to start up again to resume its previous rotational rate.

In what we might want to label as "Act 2" of the Velikovsky scenario, Mars walks on stage as the principal troublemaker. Venus, much greater in mass than Mars, severely disrupted the orbit of Mars, causing that planet on at least three occasions to pass close to Earth. The Mars encounters with Earth were not as violent as were the Venus encounters. The first came in 776 B.C., about the time of the founding of Rome. It produced violent earthquakes and altered the earth's orbit, increasing its path length. Instead of 360 days in each year, the number increased to the present 365 days (approximate). Earth's axis was tilted by an additional ten degrees. The final encounter occurred in 687 B.C. It produced a great "thunderbolt" or "blast from heaven" that Velikovsky claims obliterated the Assyrian army, led by Sennacherib, as it lay camped near Jerusalem. The angle of tilt of the earth's axis was restored to its value prior to 776 B.C.

In the closing scene of the Velikovsky account, Mars engaged Venus in a final dramatic event, in which Mars was thrown outward (with respect to the sun) into an orbit beyond that of Earth, while Venus became a planet and assumed a nearly circular orbit between Earth and Mercury. In that location, it became a bright planet seen only as an evening star and morning star. The end of the "battle of the gods" ushered in a new age of serenity for planet Earth.

Immanuel Velikovsky was born in Russia in 1895, obtained his medical degree from Moscow University in 1921, and later migrated to Palestine, where he became a general medical practitioner. After study in Vienna, he became a psychoanalyst. In 1939 he and his family migrated to the United States, settling in New York City. For nearly a decade thereafter he spent much of his time in the library of Columbia University, searching out ancient writings of both Western and Eastern cultures. *Worlds in Collision* was published in 1950 by the Macmillan Company. It immediately drew heavy fire from the scientific community, and in particular from astronomers. One of the latter was Harlow Shapley of Harvard University, held in great distinction for his pioneering work on galaxies. Summaries of Velikovsky's scenario flourished in popular magazines, including *Harper's*, *Collier's*, and *Reader's Digest*. Under a second publisher, Doubleday, *Worlds in Collision* was reprinted numerous times and was followed by four more Velikovsky books.

The ranks of Velikovsky supporters grew rapidly and his scenario quickly became the basis of a loosely organized, but highly vocal cult. In using the word "cult," I have taken my cue from Martin Gardner, who refers to "the rise of the Velikovsky cult" (1981, p. 382). Devotees of the cult have attempted to demonstrate the scientific validity of events in the Velikovsky scenario. See, for example, a volume of such articles titled *Velikovsky Reconsidered* (*Pensée* Editors 1976). It contains reprints of papers appearing prior to 1976, including those in *Pensée*, the magazine of the Student Academic Freedom Forum, published in Portland, Oregon. (See also Ransom 1976, *The Age of Velikovsky*.)

In the early 1970s, scientists concerned about the popular persistence of Velikovsky's claims began to organize a symposium within the American Association for the Advancement of Science. It was held in 1974 and included a paper delivered by Velikovsky himself. Among the scientists participating, perhaps the best known to the public was astronomer Carl Sagan, who had been highly vocal in opposing Velikovsky's versions of planetary behavior. The principal papers of that symposium (sans Velikovsky's contribution) were published in a 1977 volume entitled *Scientists Confront Velikovsky* (Goldsmith, 1977). Velikovsky died in 1979 at the age of eighty-four.

Although few persons on either side of the debate are aware of it, Velikovsky's scenario was preceded by at least three theories of comet-caused catastrophes (Gardner 1957, p. 41). One was proposed by William Whiston, a British clergyman and mathematician, in a book titled *New Theory of the Earth*, published in 1696. Whiston had the solar system formed from the tail of a comet. Visited somewhat later by a second comet, the earth experienced a great flood (Flood of Noah) caused by precipitation coming from water vapor in the comet's tail.

Of considerably greater importance relative to the Velikovsky scenario is a massive tome published in 1913 by Hans Hörbiger, a Viennese mining engineer, under the title of *Glazial-Kosmogonie* (Glacial Cosmogony). An English account of the major points of Hörbiger's scenario was written by a British follower, Hans Bellamy, and published in the early 1930s. The main point of the story is that a small stray planet was captured by Earth, to circle Earth at close range. As this moon orbited Earth, it drew closer and closer, exerting increasing gravitational pull on Earth and raising an enormous equatorial bulge of the oceans. Ultimately, Earth's attraction overcame the cohesion of the moon and it began to fall apart. Great chunks of ice broken from the moon's surface melted and the water was drawn to the earth as rain and hail. There followed a rain of rocks, as the moon disintegrated entirely. Relieved of the moon's attraction, the equatorial tidal bulge subsided and the ocean spread over the higher latitudes to produce the Flood of Noah. A second capture gave Earth its present moon in a catastrophic event that resulted in the Ice Age, with the growth of great continental ice sheets and the foundering of the continent of Atlantis. The capture is said to have occurred about 13,500 years ago. "Racial memories" of the event are "buried in our subconscious." The present moon is also spiraling closer to Earth, and a new doomsday scenario can be anticipated.

What is most interesting about the Hörbiger scenario is that it became increasingly popular among anti-intellectual persons in the German Nazi movement, perhaps in part because conventional German astronomers spoke out against the scenario with such vigor. According to Martin Gardner (1957, p. 37), the fantastic reconstruction of world history soon acquired millions of followers and became a cult known as WEL (initials of *Welt-Eis-Lehre*, translated as Cosmic-Ice-Theory); the term was coined by Hörbiger. Gardner cites Willy Ley, the well-known German rocket scientist, as saying the WEL "functioned almost like a political party. It issued leaflets, posters, and publicity handouts. Dozens of popular books were printed describing its views, and the cult maintained a monthly magazine. . . . Disciples often attended scientific meetings . . . to interrupt the speaker with shouts of 'Out with astronomical orthodoxy! Give us Hörbiger!' " (p. 37). Hörbiger even wrote to Ley: "Either you believe in me and learn, or you must be treated as an enemy." I dare say few Americans have even heard mention of Hörbiger and his WEL. I, for one, was totally unaware of them. Charles Fair states that "Hörbiger's disciples tried to ally the WEL movement with the Nazis, as their official scientific arm, but Hitler would have none of it!" (1974, p. 185)

Von Däniken's *Chariots of the Gods?*

Erich von Däniken's major works are *Chariots of the Gods?* (1969), *Gods from Outer Space* (1971), and *The Gold of the Gods* (1973). The main thesis he presents in *Chariots* is that during the period of early human civilizations, the earth was visited repeatedly by highly intelligent, technologically advanced individuals who resembled humans and were transported in spaceships from some unknown planet in outer space. In Von Däniken's text, these astronauts are called "gods," which is the way they were regarded by the earthlings who felt their impacts. The date of the earliest visitation is not made clear, but visits were taking place at the time of the earliest Egyptian pyramid builders, and possibly as far back as 45,000 years before the present.

Von Däniken postulates that at some prehistoric time the visiting gods found a population of apes or of hominids that predated *Homo sapiens*. They are not identified in the conventional anthropological sequence, but let us suppose they were members either of *H. erectus*, or of Neanderthal man (now considered a subspecies of *H sapiens*). The gods artificially inseminated females of this indigenous species and were successful in breeding *H. sapiens*. The gods were evidently experts in hybridization and may have understood the genetic code. Later they destroyed the remaining population of parent hominids.

The scientific and technical capabilities of the gods were particularly noteworthy in areas of navigation, metallurgy, and construction of stone buildings and monuments; they demonstrated this capability in many lands of the earth. Von Däniken hopes that modern humans will, as soon as feasible, make spaceships that can transport earthlings to some distant planet in another solar system, where they will find a primitive culture, for which they in turn will serve as "gods" and return the favor, so to speak.

Von Däniken's evidence is in the form of a collection of diverse artifacts. They include: an ancient world map (attributed to Piri Re'is) interpreted by von Däniken as an aerial view of the earth from a vantage point in space about over Cairo; lines on the Nazca plain of Peru, interpreted as landing strips made by pre-Inca inhabitants to facilitate the landing of spaceships; huge mountainside drawings that served as signals to spaceships; an ancient astronomical calendar at Tiahuanaco in Peru, and assorted other astronomical data (Sumerian), for which the knowledge was supplied from extraterrestrial sources; a large assortment of buildings, idols, and monuments consisting of hewn stones far too large to have been carved and transported by earthlings—localities include Tiahuanaco, Sacsahuaman, Baalbeck, Egypt (pyramids), Easter Islands (stone idols), and Guatemala and Yucatan (pyramids); cuneiform texts and tablets from Ur telling of "gods who rode in the heavens in ships" and "gods who came from the stars, possessed terrible weapons, and returned to the stars"; cave drawings from many localities depicting beings from outer space dressed in space suits and wearing goggles; miscellaneous smaller objects from Lebanon, Egypt, Iraq, Peru, China, India, and elsewhere that required manufacturing technologies far beyond the capabilities of the earthlings (made of platinum, aluminum, and nonrusting iron); biblical references interpreted as describing gods from outer space (giants, angels, Ezekiel's spaceship). The total package of so-called evidence can be interpreted as making the following argument: Because so many of the artifacts and technologies of ancient civilizations are so very far beyond the capabilities of earthlings of that period to make or perform, we can only conclude that extra-terrestrial beings must have come to earth to supply the necessary advanced knowledge and technology. Von Däniken seems to look with approval on the interpretation of certain sightings of unidentified flying objects (UFOs) in the present century as continuing visitations of gods from outer space (1969, p. 138), at least for surveillance of what is going on here on our planet.

Early in his first book, von Däniken gives a brief review of modern astronomy that shows his acceptance of the great age of the cosmos and the earth, and the appearance of humans about one million years ago. In this respect, he differs from Velikovsky, who confined his scenario to events within the time frame of literal interpretations of the book of Genesis. The result is that von Däniken immediately came under fire of the funda- mentalist creationists; whereas that group seems to have paid little serious attention to Velikovsky.

In von Däniken's third work, *The Gold of the Gods* (1973), we encounter a scenario quite unlike that of the first two books. Now the space visitors are seeking refuge after suffering defeat in a cosmic war. They build great tunnel systems for safety and set up a decoy on a fifth planet then orbiting in our solar system between Mars and Jupiter. The enemy space beings annihilate that planet, which is fragmented into the asteroids, and then leave the scene for good. Meantime, back on planet Earth, the newcomer "gods" reel from the effects of the planetary explosion, which has caused a great flood (Flood of Noah?). Recovering, they turn on Earth "monkeys," representing the hominoids of that time, and manipulate their genes through controlled mutations to create intelligent humans. But then the gods become dissatisfied with their product and begin to punish and kill the humans, who seek shelter in underground hideouts.

Does this flip-flop mean that the von Däniken cultists must drop the *Chariots* version and substitute the *Gold* version? Will there appear a third scenario to replace the second? Is von Däniken getting in a little Velikovsky planetary catastrophism to increase book sales? I'm really confused! One nice feature of this switch is that we need no longer debate the merits of the *Chariots* evidence. If we wait a little longer, we may not need to debate the *Gold* evidence.

Erich von Däniken was born in Switzerland in 1935. His early education in a Catholic school was not followed by advanced schooling in either astronomy or archaeology, subjects that caught his interest and that he attempted to link to religion (Story 1976, pp. 1-2). He became an avid amateur archaeologist and traveled to many sites of ancient civilizations. While employed on the staff of a Swiss hotel he wrote his first book, published by Econ-Verlag in 1968. It was serialized in a Swiss newspaper and attained great popularity that quickly spread to Germany, where within a year it was the number-one-selling book. The English translation, titled *Chariots of the Gods?* was published a year later in England, followed in 1969 by publication in the United States by G. P. Putnam's

Sons. A German-made film version of the book was widely viewed and led to production of an American TV show based on the book, airing in 1973. As a result, U.S. sales of the book skyrocketed. An interesting sidelight is that in 1966, the year in which von Däniken was writing his book, two prominent astronomers—I. S. Shklovski of the USSR and Carl Sagan of Cornell University—published a book titled *Intelligent Life in the Universe*. According to Ronald Story, "This book contained many ideas that were later expressed (although some in a distorted form) in *Chariots*; it may well have given von Däniken the brainstorm to provide the world with a new set of gods to worship, to replace the traditional deity, who was being murdered by the poison pens of contemporary theologians" (1976, p. 5).[1] Reference in the last phrase of the quotation is to the God-is-dead movement among theologians at that time. Story also cites earlier publications by other authors that contained much the same archaeological "evidence" used by von Däniken (1976, p. 5).

Von Däniken's chariots-of-the-gods scenario seems to have had a milder impact than Velikovsky's worlds-in-collision on both the general public and the scientific community. If a cult of believers has formed, it has not been very vocal in the English-speaking world, at least. Reaction has been strongly negative from the fundamentalist creationists, who regard it as an affront to the literal interpretation of the Old Testament. The idea of numerous "gods" from outer space replacing the functions of the one and only true and living God of all Creation is anathema. A denunciation from the creationists' side was not long in making its appearance. In 1972, *Crash Go the Chariots* appeared as a short book written by Clifford Wilson, Ph.D., "archaeologist, authority in psycholinguistics, education professor, and Bible scholar" (so the Foreword to his book reads). Goaded by von Däniken's put-down of archaelogists as a group, Wilson scathingly attacks the gods and their chariots. He puts things back into their traditional Judeo-Christian framework with one Supreme God at the controls.

I recommend the detailed critical analysis of von Däniken's scenario and interpretation of archaelogical materials in Ronald Story's *The Space Gods Revealed* (1976). It seems completely secular in tone and will appeal to scientists and humanists who might be turned off by the emotive style and *ad hominem* arguments that characterize Wilson's *Crash*. Story's book contains a Foreword written by Carl Sagan, offering some interesting speculations on the theological connotations of von Däniken's scenario. Sagan sees in popular acceptance of the questionable scholarship of von Däniken's books a profound yearning of vast numbers of people to embrace extraterrestrial beings for their superior support and guidance, a commodity no longer available through traditional religious beliefs. As we will see, this same yearning may have found satisfaction in belief in contemporary visitations by benign extraterrestrial beings arriving in flying saucers.

UFOs and UFOlogy

Of the three scenarios, that of *unidentified flying objects* (*UFOs*) is by far the most difficult to analyze—it has a literature of incredible proportions throughout which confusion, contradiction, and calumny have proliferated unchecked. For a definition of UFO, I quote from Allan Hendry's tightly organized *UFO Handbook*:

> UFO (Unidentified flying object): Any anomalous aerial phenomenon whose appearance and/or behavior cannot be ascribed to conventional objects or effects by the original witness(es) as well as by technical analysts who possess qualifications that the original observer(s) may lack. (Hendry 1979, p. 4)

Is there any American capable of reading newspapers or comprehending a television or radio program who has not been exposed to stories of UFO "sightings" in the past few years? Most stories tell of one or more individuals seeing strange objects in the form of luminous bodies—commonly cigarshaped, spherical, or disklike—and showing a wide variety of behaviors, including rapid travel, hovering, abrupt change in direction, change in shape and size, and sudden disappearance. The highly luminous forms, emitting light from an internal source and displaying various colors, are usually of indistinct outline or outer boundary; those that seem to reflect light brilliantly are often described as solid objects of metallic luster with sharp outlines. Various kinds of sounds can accompany the antics of the object. A few of the objects, when close to the ground, exert physical effects on things in the vicinity; for example, they may cause automobile engines to misfunction, radios to emit static, or TV pictures to break up. In a very few cases, disturbances of the ground surface, including burning or charring of vegetation, are reported. Finally, in a very few reports some sort of living creatures are said to have emerged from the object—in this case a "flying saucer" or "space vehicle" of some kind—and even to have spoken to the observer(s). In the most bizarre reports, the observers are kidnapped and later released.

According to Hendry, a 1978 Gallup poll stated that 9 percent of adult Americans "believe they have seen a UFO" (1979, p. 1). This extrapolates to about 13 million sightings. Upon careful scrutiny by persons trained in UFO evaluation, a large proportion of the reported incidents—as high as 90 percent—can be resolved as explained phenomena and relabeled as IFOs (Identified Flying Objects). For example, the object has turned out to be an aircraft or an astronomical object. Some reports are hoaxes. Also under consideration is the possibility that what is reported as an object is a kind of hallucination, originating within the mind. Of those events remaining in the UFO category, only a few are so strongly supported in terms of the quality of the witnesses and their allegations that organizations engaged in serious study of the phenomena accept them as real physical phenomena worthy of careful further analysis.

The enormity and intensity of public interest in UFOs, together with a sustained interest over some forty years, results in a great diversity in groups of interested persons and a vast literature. I have read a statement that several hundred books have been published on the subject. Lists of organizations devoted to UFOs suggest that they are international in distribution and number at least in the dozens. Two major reasons for the magnitude of the social phenomenon are apparent. First, there was concern that the UFOs included surveillance tools of the Soviet Union and were thus a threat to national security, both because of the information they gathered and the manifest existence of advanced Soviet technologies as yet unknown to the Western democracies. Second has been the concern that some of the UFOs are space vehicles bearing life forms from other solar systems and visiting our planet for reasons unknown.

Because of the first type of concern, UFOs attracted the attention of the U.S. military establishment, the Central Intelligence Agency (CIA), and numbers of politicians. Because of the second type of concern, UFOs gave rise to vivid stories in the news media and attracted an enormous following of "believers" who form a cult of quasi-religious nature, organized into numerous societies for the culture of "UFOria." This movement is now worldwide. Charles Fair, in his 1974 book titled *The New Nonsense*, divided the cultists into two groups (1974, p. 137). The *angel school* views the visitors from outer space as benign and desirous of heading off the great nuclear holocaust or Armageddon. The visitors' superior intelligence may include the ability to save earthlings from this fate. The *devil school* sees the space visitors as evil and threatening, with the power to wreak a catastrophe on our planet.

A small handful of qualified scientists has taken a strong interest in the UFO problem and considers it worthy of serious scientific study. Fair states that these individuals can be divided into two groups (pp. 140-41). The skeptics are hardnosed individuals who tend to work from a strong bias against any unconventional proposal unsupported by physical evidence. An example is Edward U. Condon, a distinguished physicist who, as we shall see, headed up a committee to investigate UFOs. Another, but quite different example is Carl Sagan, noted astronomer, who has evaluated the possibility of contacting *extraterrestrial intelligence (ETI)*. A third example, also an astronomer, is David H. Menzel of Harvard University; he offered an alternative physical explanation for one class of UFOs. I should also include Philip J. Klass, a graduate electrical engineer who proposed his own scientific hypothesis, based on principles of plasma physics, to explain a large proportion of UFOs (Klass 1968).

The second group consists of a very few qualified scientists who are either believers in the possibility that some UFOs are manifestations of

ETI, or are neutral on that issue to the extent of being agnostic. One of the most puzzling of these (in my view) is James E. McDonald (deceased) of the Institute of Atmospheric Physics at the University of Arizona. His statements can be interpreted as indicating a certain measure of belief that ETI has reached our planet through UFOs and thus created a situation of grave scientific concern. A second person whose activities and statements suggest the possibility of belief is Dr. J. Allen Hynek (deceased), a former Chairman of the Department of Astronomy and Director of the Dearborn Observatory of Northwestern University. He became Director of the Center for UFO Studies and for some years was a scientific consultant to the Air Force in its program of documentation of UFO sightings. For many years a skeptic, Hynek became less sure of that position. He stated that from his long experience, most reported UFO cases were misrepresentations but a "relatively small residue of UFO reports . . . were so well attested and so compellingly strange that the chances were overwhelmingly great that they could not be ascribed to collective misidentification, hoax, or hallucination" (Hendry 1979, pp. ix-x). He goes no further than to leave open the possibility that UFO cases may be manifestations of ETI.

The beginning of the UFO era is usually placed in June 1947, with the first publicized sighting of an aerial object and its being dubbed by a newspaper reporter as a "flying saucer." Equally strange luminous objects, called "foo fighters," had been the subject of numerous accounts by military aircraft pilots in World War II, but those incidents did not precipitate a public reaction. Reports of strange aerial objects go back to the age of early hot-air balloons and to dirigibles of World War I vintage; some of these were described in lurid terms in the newspapers. Evidently, the immediate post-WW II period had generated a national psychology unusually favorable for a mass response in the United States to the UFO phenomenon. Tensions with the USSR were increasing with the start of a nuclear-weapons race and there were many rumors of Soviet spying. Perhaps because several UFO sightings occurred near the Pentagon in Washington, D.C., and many other sightings were described by trained aircraft pilots, including personnel of the U.S. Air Force, the military establishment began to take the phenomenon seriously as a possible security threat. This concern led to the setting up in 1947 of Project Blue Book, a registration and documentation of UFO sightings maintained by the Air Force in Washington and continued until 1969. The material was classified and closed to public access, perhaps contributing to the early charges from the public sector that the Air Force was withholding secret evidence that could prove that Soviet spying on the nation or extraterrestrial visitations were actually taking place.

In 1966, the Congress stepped into the investigatory picture. The House Armed Services Committee directed the Pentagon to set up a civilian inquiry into the UFO phenomenon to be based in part on a review and restudy of the accumulated data of Project Blue Book. The result was funding of a national committee to study UFOs. The committee was set up to do its work at the University of Colorado and requested to submit its report to the National Science Foundation, which, in turn, was to release it to the public. The committee was headed by Dr. Edward U. Condon, at one time a professor at Princeton University, a past president of the American Association for the Advancement of Science, and widely known for his research in nuclear physics and his role in developing the atomic bomb. Later he served as Director of the National Bureau of Standards and in advisory positions related to atomic-energy development and nuclear-weapons testing. With a staff of scientists that included both physical scientists and social scientists, Condon's committee set to work to reexamine and evaluate UFO cases in the Blue Book file.

The Condon Committee released its report in 1969 (see Condon 1969). The report concluded that no independent physical evidence had been found to substantiate the allegations made in the case studies they had examined; that further scientific study of the matter was not justified because no threat could be implied. A possible exception to the first part of that statement lies in the interpretation of three small, metallic fragments alleged to have been collected from a shallow-water site off the coast of Ubatuba, Brazil, where a UFO was alleged to have fallen, exploded, and burned. Analysis done under the committee's direction showed the metal to be magnesium. Neutron activation analysis of the metal showed an elemental composition that has been interpreted by some to mean that the material is not of earthly origin. (It could be of meteoritic origin.) Details of the actual collection and handling of the material are confused, and its direct connection with the alleged UFO incident is much in doubt (Fair 1974, pp. 149-50). In December 1969, a symposium sponsored by the American Association for the Advancement of Science aired the diverse opinions of a number of scientists on the significance of UFOs (Sagan and Page 1972). One of the opinions expressed was that the Condon Committee report should not be accepted as final, and that further scientific study of UFOs was both desirable and important.

In considering the status of UFOlogy in terms of whether it is science or pseudoscience it is essential to be quite precise as to exactly what we are talking about. Hendry reminds us: "We only get to study reports of UFOs—not the UFOs themselves" (1979, pp. 6-7). The reports consist only of allegations. Hendry puts it this way: "We would be dishonest with ourselves if we considered the reports as anything but allegations; this is unfortunate, but necessary, if we are to treat UFOlogy as a science" (p. 7).

It is generally agreed by both skeptics and believers that to date no physical object identifiable as a "flying saucer" or "spaceship" has been retrieved or captured for direct physical and chemical examination. Empirical science is reluctant to admit allegations as evidence in support of a working hypothesis that demands interpretation beyond sensory perception itself.

A particular sighting of a UFO, even if corroborated by several witnesses, is a singular timebound event and cannot be repeated on demand. What can be done as acceptable science is to classify and catalog the percepts on record and to rate their reliability according to some reasonable uniform standards, and from those data reconstruct a general description of a phenomenon (or several different phenomena) about which some pertinent questions can be asked. It would be foolish to rule a priori that everything claimed to be perceived by humans is nonreal, for that would be to deny the existence of all commonsense knowledge essential to the conduct of human activities. What we can work with is a body of persistent, repetitious allegations that various objects were observed to have certain properties (such as shape, size, color, or texture), that they changed those properties in certain ways, engaged in certain types of motions, and that they emitted, besides visual light, some other form of electromagnetic radiation or some kind of sound. In making the tentative assumption that these perceptions arise from real phenomena, it is quite permissible to formulate one or several hypotheses that could explain the typical or common perceptions through application of accepted laws and principles of empirical science. In such a program we would be simply carrying on a form of descriptive science typical of natural sciences in their formative periods.

If it be agreed by scientists that some common ground of observational reality exists within many UFO reports, hypotheses offered in explanation can be subjected to testing through the derivation of deduced consequences, which in turn may lead to the discovery of previously unnoticed relationships. For example, a certain class of statements might prove to be significantly associated with a particular time of day or night, or with a particular environmental setting. If an investigation of UFOs is to be taken seriously by scientists, the hypothesis will conform as closely as possible with the principle of parsimony. This means that where possible, the explanation makes use of natural phenomena already accepted through extensive prior observation and experimentation.

For example, a hypothesis requiring that the UFO is of extraterrestrial origin will be given relatively low status if another hypothesis can explain the same phenomenon in terms of well-known terrestrial phenomena. A case in point would be the plasma hypothesis advanced by Philip J. Klass, who suggests that many of the perceived qualities and behavior patterns of UFOs are what might be expected of masses of strongly ionized air containing highly excited free electrons capable of emitting various colors of light and other forms of electromagnetic radiation (Klass 1968). A commonplace example of this effect is the fluorescent light from gas-filled glass tubes through which an electrical current is passed. Related phenomena in nature include the corona (St. Elmo's fire), seen to emanate from solid objects on the ground or from ships at sea, and ball lightning, both of which are occasionally observed when intense fields of atmospheric electricity are present. Visible plasma effects are also known to occur on or near high-tension power lines. What Klass has done is to attempt to explain at least a large proportion of luminous UFOs as plasma phenomena. Numerous deduced consequences of this hypothesis are capable of being tested, and Klass has endeavored to do just that.

One might even wish to entertain as a working hypothesis from the field of psychology that most UFOs are hallucinations arising within the human brain. There might be difficulties in reconciling the allegations that the same hallucination occurred simultaneously in the brains of two or more observers, but the magnitude of those difficulties would be small compared with a hypothesis of UFOs as spaceships directed by ETI or occupied by extraterrestrial creatures. The ETI hypothesis has the fatal weakness of merely citing as "supporting evidence" the assertions of the hypothesis itself. The ETI hypothesis is thus not now capable of being dealt with by the methods of empirical science because no evidence independent of the hypothesis itself can be brought to bear on it. But, at the same time, its mere statement can be labeled as pseudoscience only when persons who advance it use in their argument unacceptable claims for supporting evidence. For example, the allegation that small creatures with large heads and spindly bodies emerged from the UFO cannot be accepted as evidence of ETI, because those little guys (or gals) are simply an extension of the hypothesis itself! The argument that "a particular UFO could not possibly be an artifact of earthlings, because it represents a technology far beyond the capability of earthlings" fails because that assertion is also an elaboration of the hypothesis itself. It also fails on grounds of logic because to claim that another hypothesis cannot explain the alleged phenomenon in no way adds value to the ETI hypothesis. Persons who use such specious arguments are practicing pseudoscience; they are abusing a hypothesis that might at some future time become amenable to treatment by acceptable methods of scientific investigations, deriving support from naturalistic lines of evidence not presently available.

As an example of an investigatory treatment of UFOs that I suggest as worthy of being described as science, I would cite for consideration the activities of the Center for UFO Studies, under the directorship of J. Allen Hynek. From his personal agnostic position, Hynek set up criteria for objective reporting of UFO incidents with a view to extracting the maximum possible information from the observers, while at the same time screening out misidentifications, hoaxes, and hallucinations. To this end, the Center for UFO Studies prepared an observer's handbook to guide in recording information and to recommend tools that may be useful in hardening up the information (Hendry 1979, p. ix).

With the three chosen scenarios in mind, we must in our next chapter attempt to decide whether they deal in science, pseudoscience, or some of both. To reach such a decision we need to probe into the general characteristics of pseudoscience and to set down some criteria of recognition to guide us.

Credit

1. From Ronald Story, *The Space Gods Revealed: A Close Look at the Theories of Erik von Däniken.* Harper & Row, Publishers, Inc., New York. Copyright © 1976 by Ronald Story. Used by permission of the publisher.

CHAPTER 8

Pseudoscience: II. The Phenomenon Examined

To prepare ourselves for passing judgment on whether Velikovsky, von Däniken, and the priests of the ETI cult have dealt in pseudoscience, we will find it helpful to compare their methods of obtaining and processing information with those used by researchers in mainstream science. Do Velikovsky and von Däniken share the same set of characteristics with the mainstream scientists? Do they practice in the same manner? Do they share the same set of norms?

Applying the Norms of Mainstream Science

Among the collected papers emerging from the AAAS Velikovsky symposium of 1974 (Goldsmith 1977) was that by sociologist Norman W. Storer, titled "The Sociological Context of the Velikovsky Controversy" (1977): We referred to Storer's text in Chapter 5, where we reviewed Robert K. Merton's four "norms" that are central to the ethics of science (1973).

Recall that the first norm was that the value attached to a scientific statement must in no way be connected with the personal characteristics of the scientist who makes that statement. Merton called this principle "universalism," meaning that the strength or weakness of a hypothesis proposed by a scientist must be considered strictly on its scientific content and supporting evidence. It should make no difference whatsoever that the scientist is of a certain race, religion, sex, age, political affiliation, or the like. If the principle were extended to include disregard of professional qualifications, we would need to give unbiased and serious attention to a scientific statement made by anyone, scientist or not, and including Velikovsky and von Däniken. I doubt that Merton intended such a sweeping inclusion.

Second is the principle of communality: that findings made by one scientist must be shared freely and openly with the entire scientific community. This norm doesn't help us to distinguish mainstream scientists from the practitioners of pseudoscience; the latter publish compulsively and could not be restrained from doing so.

Third was the point that scientists must practice organized skepticism. We noted that the most important part of the policing action occurs through peer reviews of articles submitted to scientific journals. The scrutiny of one's work by colleagues is a feature wholly lacking in the publication of pseudoscience literature. Velikovsky, von Däniken, and their publishers' editors never sought critical reviews from scientists familiar with those areas of astronomy, geology, and archaeology that form the skeletal structure of their scenarios. If those authors had submitted their manuscripts to scientific journals, rejection notices would have been swift in coming. It looks, then, as if authors of pseudoscientific material shy away from the scientific community. Instead, they seek support in the nonscience community, and particularly from those persons having little higher education in any field of knowledge. *Prima facie* evidence of this audience selection lies in the fact that pseudoscience is published by those same publishing houses (or divisions within a publishing house) that handle fiction, science fiction, and the more sensational forms of biography and autobiography. You would not find a *Worlds in Collision* or a *Chariots of the Gods?* on a publisher's list of scientific textbooks and monographs.

The fourth norm recognized by Merton is disinterestedness: a scientist's research should not be guided by desire for personal rewards such as private economic gain and glory in the eyes of the nonscientific public. For the pseudoscientists, the norm of disinterestedness is simply not there, and no public shame is to be incurred from violating such a norm. Emphasis is on rolling up the royalty earnings and fees from book sales, TV/motion-picture adaptations, and lectures to lay audiences, on public exposure through media interviews, and on receiving expressions of adulation from fan clubs within the cult. Pseudoscience is big business and very little else!

Pseudoscientists as Exoheretics

In Chapter 6, we noted that the driving force behind mainstream scientific research and publication is the acquisition of professional recognition (Storer 1977, p. 31). For the producers of pseudoscience, professional recognition within the mainstream science community is nonexistent. They are excluded in a very firm manner. Exclusion is then seized upon by the pseudoscientist and associated cultists as an opportunity to indulge in paranoia, a subject we will come to shortly.

Writing in a foreword to the AAAS Velikovsky symposium volume,

Isaac Asimov has identified and described two kinds of scientific heretics, distinguished in terms of their relationship to the scientific community we have just examined (Goldsmith 1977, pp. 8-15). One is the *endoheretic*, the other, the *exoheretic*. Endoheretics arise within the scientific community of which they are a part; exoheretics arise from outside the scientific community. As examples of endoheretics, Asimov cites Galileo and Charles Darwin. Both were well qualified as scientists according to the prevailing standards; both were opposed by orthodox members of the science community, who attempted to discipline them and bring them into line with the orthodox views. (In Galileo's time, Asimov points out, both scientific and religious orthodoxies were one and the same, so that pressure to recant came from the Inquisition.) In the long run their hypotheses won out. A striking example of an endoheretic of modern times was Alfred Wegener, the German geophysicist who proposed the drifting apart of the continents to the accompaniment of a widening of the intervening oceans. Geologists of the time persecuted Wegener mercilessly, and it was not until the 1960s—long after his death—that his hypothesis (greatly revised) prevailed.

Asimov suggests that there may have been at least fifty endoheretics whose hypotheses failed to every one whose hypothesis was eventually accepted. The losers simply dropped out of sight and the public knew little or nothing of their attempts. The successful endoheretics won out because the self-correcting methods of the scientific community, directed by norms described in earlier paragraphs, continually sorted the wheat from the chaff, retaining that which had value; but the process may be painfully slow—painful, particularly, to the endoheretic.

The exoheretic is an outsider to science and is not schooled in practice of the norms of the scientific community, or in the methodology and language of science. The manner in which the exoheretic attacks some area of orthodox science strikes those relatively few scientists who take note as being strange, unintelligible, or otherwise deviant, often to the point of causing amusement or derision. Knowing that this will happen, the exoheretic directs expression to the general public. Those persons educated primarily in the arts and humanities will receive the strange ideas as worthy expressions of the human imagination and will respond to the romantic overtones. The exoheretic finds public response deeply satisfying as well as lucrative. The public in turn will turn against the scientists and goad them into denunciations, if possible. Goading was not needed for Harlow Shapley, the Harvard astronomer; he attacked Velikovsky savagely even before *Worlds* was published, basing his information on prepublication reviews. Velikovsky cultists have made a big thing of these early attacks, which were successful in forcing the original publisher to turn the book over to another house. Seen in retrospect by members of the science community today, the concerted attack on Velikovsky was a tactical blunder, for it quickly brought the public to Velikovsky's side, making him a martyr, while doing nothing to cause his readers to reject the bizarre astronomical scenario.

You might want to question the classification of Velikovsky as an exoheretic, since he had obtained a medical degree and practiced psychoanalysis. That area of knowledge, however, is normative science and quite unrelated to the planetary science on which *Worlds* depends. Velikovsky's competence as a biblical scholar seems not to have been questioned. As to von Däniken, there seems to be little contest over his candidacy for exoheretical status. He admits to being an amateur. When it comes to UFOlogy and the question of ETI, both exo- and endo- types have gotten into the act; they must be sorted out (as we have done) according to their levels of skepticism or belief, as well as their professional qualifications and the quality of their writing.

Martin Gardner has set down what he regards as characteristics that most pseudoscientists share (1957, pp. 8-15). "First, and most important is that cranks work in almost total isolation of their colleagues" (p. 8). This certainly describes both Velikovsky and von Däniken. They not only worked in isolation from the scientific community, but from other persons of any description. It is said that Velikovsky "opened and closed the doors of the Columbia University Library" for several years, probing into the literature of the Fertile Crescent in biblical times. Von Däniken wrote in the solitude of his quarters in a Swiss hotel. Some successful scientists have worked for years as loners in the sense of having secured no support from colleagues, but that is another matter; they had full access to the scientific community and its body of empirical knowledge.

Second, the self-inflicted isolation of the pseudoscientist goes hand-in-hand with paranoia, which manifests itself in a sense of personal greatness, enabling him or her to stand firm in defiance of the recognized scientists. Gardner lists "five ways in which the sincere pseudoscientist's paranoid tendencies are exhibited."

> (1) He considers himself a genius. (2) He regards his colleagues, without exception, as ignorant blockheads. (3) He believes himself unjustly persecuted and discriminated against. (4) He has strong compulsions to focus his attacks on the greatest scientists and the best-established theories. (5) He often has a tendency to write in a complex jargon, in many cases making use of terms and phrases he himself has coined. (Pp. 12-14)

I'm not so sure all five of these criteria can be detected in Velikovsky and von Däniken—that is something best left to the psychiatrists. The second, third, and fourth criteria are clearly visible in the writings of the more outspoken supporters of both authors, and especially virulent in the cultist authors who are believers in UFOs with ETI connections.

Science and Pseudoscience as Cognitive Fields

In recent years philosophers of science have become increasingly aware that a single criterion for distinguishing between science and pseudoscience cannot be fully effective. This problem has been treated by Professor Mario Bunge of McGill University, whose analysis of cognitive fields we introduced early in Chapter 1. He has made an in-depth study of epistemology, presented in two major works: *Exploring the World* (1983a) and *Understanding the World* (1983b). In a 1984 paper in *The Skeptical Inquirer*, titled "What is Pseudoscience?" Bunge singles out the problem of comparing the two cognitive fields mentioned in Chapter 1: (a) belief fields that include religions, political ideologies, and pseudoscience; (b) research fields that include humanities, mathematics, and the pure and applied sciences. His opening paragraph reads as follows:

> Most philosophers have attempted to characterize science, and correspondingly pseudoscience, by a single feature. Some have chosen consensus as the mark of science, others empirical content, or success, or refutability, or the use of the scientific method, or what have you. Every one of these simplistic attempts has failed. Science is far too complex an object to be characterizable by a single trait—and the same holds for pseudoscience. Just as we must check a number of properties in addition to color and brilliance in order to make sure that a chunk of metal is not fake gold, so we must examine a number of features of a field of knowledge to ascertain whether it is scientific. (1984, p. 36)[1]

Bunge begins his analysis by listing ten essential characteristics, or components, of a cognitive field; they are all nouns—"things" that every cognitive field possesses. I list them in the order and with the names used by Bunge, but with some minor rephrasing of his explanations (pp. 38-39):

C: Cognitive community. The purveyors of the knowledge category; in this case, the scientists or the pseudoscientists themselves.

S: Society hosting the cognitive community. The society of which the cognitive community is a part.

G: General outlook. The world-view, or philosophy of the cognitive community.

D: Domain. The universe of discourse of the cognitive field; the objects of study or inquiry of the particular field (i.e., the subject matter itself).

F: Formal background. The logical and mathematical tools employable by the cognitive field.

B: Specific background. The set of presuppositions about the domain (D) borrowed from other fields of knowledge.

P: Problematics. The set of problems the cognitive field may handle or treat.

K: Specific fund of knowledge. The knowledge accumulated by the cognitive field.

A: Aims. The goals of the cognitive community in cultivating its cognitive field.

M: Methodics. The collection of methods utilizable in the cognitive field.

Figure 8.1 is my sketch of the ten characteristics or components, attempting to show their relationships to one another. The cognitive community (people) has a particular domain of reality in which it operates and within which lie its problems, its accumulated knowledge, and its aims. Drawn into the domain from outside it are essential inputs in the form of methods or information. This diagram may be going beyond or straying from what Bunge had in mind, but it may help in getting some coherent mental picture of the composition and structure of a cognitive field.

Bunge's next step is to describe and define the concept of science in terms of twelve conditions, all of which must be satisfied. If any cognitive field fails to satisfy all twelve conditions, it will be judged as *nonscientific.* Examples of nonscientific fields named by Bunge are theology and literary criticism. Nothing in such a designation implies the nonscientific field to be of lesser intellectual rank or value than science. Now comes the crucial point, which is "any cognitive field that, though nonscientific, is advertised as scientific will be said to be *pseudoscientific*" (p. 39). In other words, a claim put forward by its adherents that a belief field is science is a fraudulent claim; it is a misrepresentation of the real nature of that belief field.

So let us turn to the twelve conditions set down by Bunge (pp. 38-39). In places I have edited or paraphrased parts of his statements, but for the most part they are verbatim.

1. Every one of the ten components of the cognitive field changes, however slowly, as a result of inquiry in the same field as well as in related fields, particularly those supplying the formal background (F) and the specified background (B).

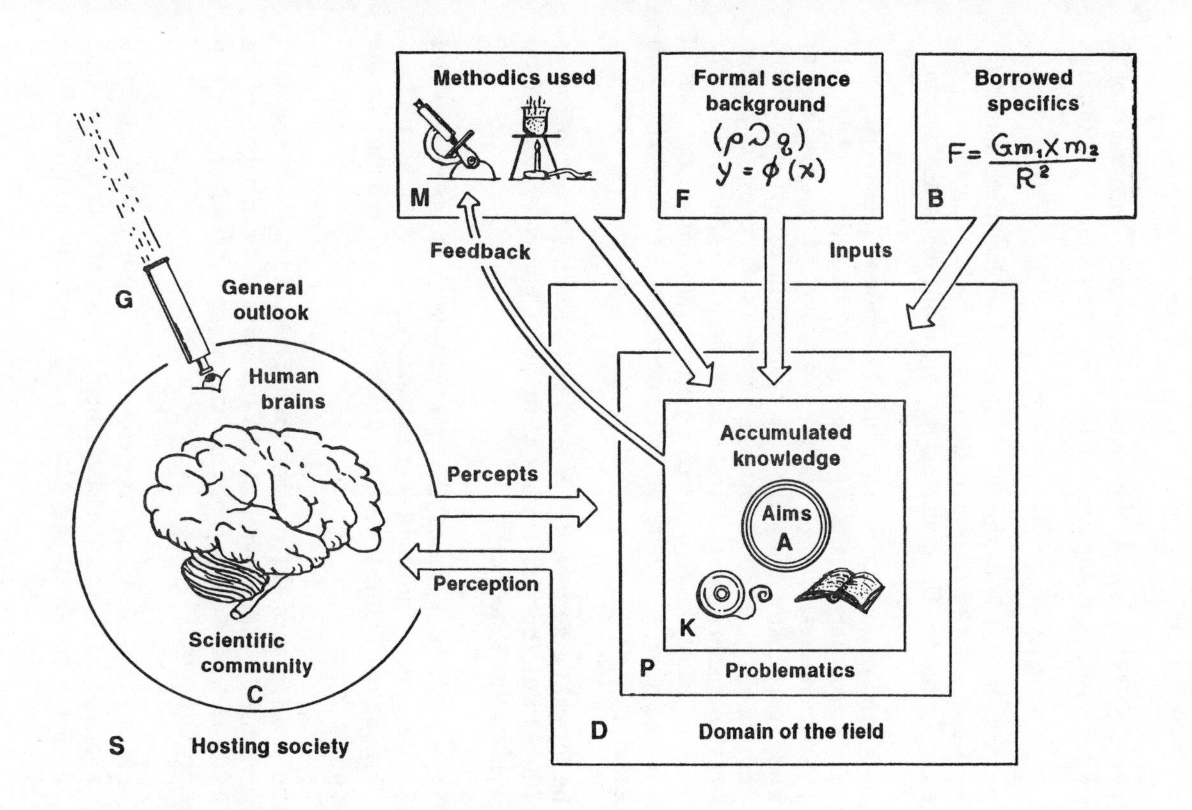

Figure 8.1 A fanciful presentation of the components of a cognitive field—with apologies to philosopher Mario Bunge. (A. N. Strahler.)

2. The research community (C) of the cognitive field is a system composed of persons who have received a specialized training, hold strong information links among themselves, and initiate or continue a tradition of inquiry.

3. The society (S) that hosts the cognitive community encourages or at least tolerates the activities of the ten components (A through S).

4. The domain (D) is composed exclusively of (certified and putatively) real entities (rather than, say, freely floating ideas) past, present, or future. The domain is entirely in the mechanistic realm.

5. The general outlook or philosophical background consists of (a) an ontology according to which the real world is composed of lawfully changing concrete things (rather than, say, of unchanging, or lawless, or ghostly things); (b) a realistic theory of knowledge (rather than, say, an idealistic or a conventionalist one); (c) a value system enshrining clarity, exactness, depth, consistency, and truth; (d) the ethos of the free search for truth (rather than, say, that of the bound quest for utility or for consensus or for conformity with dogma).

6. The formal background (F) is a collection of up-to-date logical or mathematical theories (rather than being empty or formed by obsolete formal theories).

7. The specific background (B) is a collection of up-to-date and reasonably well confirmed (yet not incorrigible) data, hypotheses, and theories obtained in other fields of inquiry relevant to the cognitive field.

8. The problematics (P) consists exclusively of cognitive problems concerning the nature (in particular the laws) of the members of the domain (D), as well as problems concerning other components of the cognitive field.

9. The fund of knowledge (K) is a collection of up-to-date and testable (though not final) theories, hypotheses, and data compatible with those of the borrowed specifics (B) and obtained in the cognitive field at previous times.

10. The aims (A) include discovering or using the laws of the domain of the field, systematizing (into theories) hypotheses about the domain, and refining methods in the methodics used (M).

11. The methodics (M) contains exclusively scrutable (checkable, analyzable, criticizable) and justifiable (explainable) procedures.

12. The cognitive field is a component of a wider cognitive field, i.e., there is at least one other (contiguous) research field such that (a) the general outlooks, formal backgrounds, specific backgrounds, funds of knowledge, aims, and methodics of the two fields have nonempty overlaps, and (b) either the domain of one field is included in that of the other, or each member of the domain of one of them is a component of a system belonging to the other domain.

Professor Bunge turns next to the set of conditions that will identify a cognitive field as pseudoscience (pp. 39-40). Point by point, it corresponds with the list we have just presented. In reviewing this list, I try to be as brief as possible in quoting or paraphrasing Bunge, but adding comments in parentheses relating to the two historical scenarios we have reviewed, namely, Velikovsky's *Worlds* and von Däniken's *Chariots*.

1. Little change occurs in the components in the course of time; when it happens, it is forced by controversy and outside pressure rather than by the results of research. (Velikovsky held to a single scenario; von Däniken switched to a radically different one after the first was widely received; no changes were forced by research findings.)

2. There is no research community as such. Instead, the cognitive community consists of believers, who, although calling themselves scientists, conduct no scientific research. (Velikovsky's library study and von Däniken's travels to archeological sites may have in some sense been research into history, but neither conducted research on the astronomical or space-travel phenomena at the heart of the historical scenario.)

3. The host society (S) supports the cognitive community for practical reasons (because it is good business) or tolerates it while relegating it beyond the border of its official culture. (Most certainly, *Worlds* and *Chariots* were good business for publishers, and press, and cultist writers, while at the same time providing great entertainment and amusement for a large sector of the public.)

4. The domain (D) teems with unreal or at least not certifiably real entities, such as astral influences, disembodied thoughts, superegos, and the like. (*Worlds* and *Chariots* certainly rest on astral influences; the "Space gods" were superegos. None of the actors,

whether aberrant comets, planets gyrating out of control, or space gods are certifiably real in the roles claimed for them.)

5. The general outlook (G) includes either (a) an ontology countenancing immaterial (nonmaterial) entities or processes, such as disembodied spirits, or (b) an epistemology making room for arguments from authority or for paranormal modes of cognition accessible only to the initiates or to those trained to interpret certain canonical texts, or (c) a value system that does not enshrine clarity, exactness, depth, consistency, or truth, or (d) an ethos that, far from facilitating the free search for truth, recommends the staunch sense of dogma, including deception if need be. (This is a long list, but the points are all antithetical to those listed for science. Perhaps the final point applies to *Worlds* and *Chariots*, both of which are delivered by their authors as dogma to be staunchly defended against the inquiries of science.)

6. The formal background (F) is usually modest. Logic is not always respected, and mathematical modeling is the exception rather than the rule. (This condition scarcely applies to the historical, or timebound, scenarios of *Worlds* and *Chariots*, but is important in other areas dealing with timeless knowledge, such as parapsychology.)

7. The specific background (B) is small or nil; a pseudoscience learns little or nothing from other cognitive fields. Likewise it contributes little or nothing to the development of other cognitive fields. (Surely this condition applies to all three scenarios we have reviewed in Chapter 7.)

8. The problematics (P) includes many more practical problems concerning human life (in particular how to feel better and influence other people) than cognitive problems. (This condition does not apply to *Worlds* or *Chariots*, but may apply to certain activities in parapsychology.)

9. The fund of knowledge (K) is practically stagnant and contains numerous untestable or even false hypotheses in conflict with well-confirmed scientific hypotheses. And it contains no universal and well-confirmed hypotheses. (This condition certainly applies to all three scenarios we reviewed in Chapter 7.)

10. The aims (A) of the members of the cognitive community are often practical rather than cognitive, in consonance with its problematics. They do not include the typical goals of scientific

research, namely, the finding of laws or their use to understand and predict facts. (This condition does not apply to the historical scenarios, but is certainly descriptive of extrasensory perception research, which has yielded no laws and nothing in the way of a testable explanation.)

11. The methodics (M) contains procedures that are neither checkable by alternative (in particular scientific) procedures nor justifiable by well-confirmed theories. In particular, criticism is not welcomed by pseudoscientists. (This condition would relate to extrasensory perception research, which has followed only a single mode of investigation for which no explanatory theory exists.)

12. No other field of knowledge, except possibly another pseudoscience, overlaps with the stated cognitive field. This means that the field is isolated and free from control of other cognitive fields. (The historical scenarios are, of course, isolated as unique conceptions.)

Finally, Bunge presents us with a table comparing the attitudes and activities of scientists and pseudoscientists (p. 41). If you have found the foregoing analysis of diagnostic conditions a bit too heavy, perhaps this checklist will be rewarding in its brevity and directness. (See Table 8.1)

In appraising Bunge's analysis, keep in mind that it fits best into physical science. The historical sciences, such as geology, paleontology, and evolutionary biology, have some special qualities that must be taken into consideration, as explained in Chapter 4. For example, formal science (mathematics and logic) may not be directly used in unraveling the historical sequence of events through geologic time. Appropriate adjustments can, however, be made to accommodate the historical (timebound) aspects of science.

Protoscience

Always at the outer limits of the expanding frontier of scientific knowledge there lies a nebulous zone in which tentative or suggestive scientific statements are made, but for which supporting evidence is lacking. This is the zone of *protoscience*, also described as *emerging science* (Bunge 1984, p. 44).

Perhaps a good example is the recent suggestion of certain astronomers that there may exist a companion star to our own Sun. Now dubbed "Nemesis" by some (and just "George" by others), the companion is a very small dead (or almost dead) dwarf star revealing nothing of itself through electromagnetic radiation or other measurable effects. Nemesis

Table 8.1 Attitudes and Activities of Scientists and Pseudoscientists

Typical Attitudes and Activities	*Scientist*	*Pseudoscientist*
Admits own ignorance, hence need for more research	Yes	No
Finds own field difficult and full of holes	Yes	No
Advances by posing and solving new problems	Yes	No
Welcomes new hypotheses and methods	Yes	No
Proposes and tries out new hypotheses	Yes	Optional
Attempts to find or apply new laws	Yes	No
Cherishes the unity of science	Yes	No
Relies on logic	Yes	Optional
Uses mathematics	Yes	Optional
Gathers or uses data, particularly quantitative ones	Yes	Optional
Looks for counterexamples	Yes	No
Invents or applies objective checking procedures	Yes	Optional
Settles disputes by experimentation or computation	Yes	No
Falls back consistently on authority	No	Yes
Suppresses or distorts unfavorable data	No	Yes
Updates own information	Yes	No
Seeks critical comments from others	Yes	No
Writes papers that can be understood by anyone	No	Yes
Is likely to achieve instant celebrity	No	Yes

From Mario Bunge, *The Skeptical Inquirer,* vol. 9, no. 1, p. 41,Table 1.

orbits far out beyond the solar system, but in an eccentric orbit that occasionally brings it into the region where comets or cometary materials reside. This region in itself is a hypothetical zone lying beyond the most distant of our known planets (Pluto). No one would have any reason to conjure up Nemesis were it not a possible mechanism for having a large number of cometary masses thrown out of their home region and sent hurtling toward the sun, on which course they can collide with our Earth and produce devastating impacts. Such impacts in the past might have been the cause of mass extinctions, such as that which killed off the dinosaurs and many other animal groups at the close of the Cretaceous Period, some 65 million years ago. The possibility that Nemesis exists and has played its disruptive game is pure conjecture, but one that can hardly be classed as pseudoscience. In this case, not even the phenomenon itself has been observed but, as conceived, it agrees with our knowledge of the classes of stars (binary stars, dwarf stars) to which it is assigned; its orbit is postulated in accordance with laws of motion and gravitation. The speculation is conducted within the cognitive community of scientists and is discussed in their journals and debated just as is any scientific hypothesis. The entire idea may eventually be shelved or dropped by those who proposed it. Perhaps the time will come when a companion star of this kind is actually located and its orbit computed. A telescopic search for Nemesis by scientists of the Lawrence Berkeley Laboratory was begun in 1984, so we are not dealing here with pseudoscience.

Pseudoscience often consists of speculations that might seem at first glance to be allowable, but on further consideration must be assigned to pseudoscience. Bunge looks to parapsychology for examples. He would relegate clairvoyance, precognition, and psychokinesis to pseudoscience on grounds that they all conflict with physical laws. On the other hand, he thinks that the investigation of mental telepathy, which is thought transmission directly from one brain to another, may perhaps be protoscience. He argues that "if the thought transmission does exist, then it must be a physical process. So, if it were discovered, it would not confirm parapsychology, but would become a subject of ordinary scientific research. . . . Such discovery would be the coup de grace of parapsychology, just as the chemistry of Boyle finished off alchemy and Newtonian astronomy killed astrology" (Bunge 1984, pp. 44-45).[1]

Protoscience may also be distinguished from pseudoscience by using Asimov's criteria for distinguishing between endoheretics (within the scientific community) and exoheretics (outside the scientific community). In this context, protoscience can be said to be endoheresy (Bunge 1984, p. 45). Endoheresy is not only tolerated by scientists but welcomed when it comes from a scientist highly respected by colleagues.

So that you may be better armed to take part in controversies involving science versus pseudoscience, I call attention to another term, *fringe science*, that appears in writings on pseudoscience. Fringe science has been discussed by Professor Steven I. Dutch of the University of Wisconsin in Greenbay (1982, pp. 6-13). It seems that J. S. Trefil, in a popular 1978 article in *Saturday Review* titled "A Consumer's Guide to Pseudoscience," classified scientific ideas into three categories: center, frontier, and fringe. The first two are genuine science, the third is often pseudoscience, but not necessarily so. From the definition that fringe science lies in "a region where ideas are highly speculative or weakly confirmed" (Dutch, p. 6). Dutch would place the Nemesis speculation in fringe science, but we have already argued for its correct place in science. Dutch would also place the Velikovsky and von Däniken speculations, along with creation "science," psychic phenomena, and the Loch Ness monster, in fringe science "because they are supported by little data, appear to be untestable or are in conflict with better and more conventional interpretations."

I sense confusion through ambiguity in the term "fringe science," since it seems to encompass both science and pseudoscience in a common zone of overlap. No such overlap needs to be recognized if we apply the list of distinguishing criteria laid down by Bunge.

How Pseudoscience Arguments Are Flawed

Flaws in arguments offered by pseudoscience have been exposed by Steven I. Dutch (1982). They include logical fallacies, often presented in subtle ways that can escape the attention of unwary readers.

We can start with perhaps the oldest logical fallacy known to orators—the *straw-man argument.* (In case some of these dummies were female, "straw-person" is to be preferred, but I defer to tradition here.) The strategy is to generate false assumptions or postulates, then show that they are indeed false. In other words, the pseudoscientist puts into the mouths of the scientists statements that no scientist has ever said or would have said. Actually, the pseudoscientists often pick up outmoded and discarded hypotheses that science has already repudiated on the basis of newer evidence. In such cases a repudiated hypothesis can be quoted from obsolete scientific articles, making it seem all the more respectable as genuine science. The false assumption is swallowed by the listener as a genuine current product of science, and is easily and quickly demolished by the orator, using evidence from science itself.

Another basic ploy, called by Dutch the *residue fallacy* (1982, p. 30), is for the pseudoscience advocate to point out that there are many things in nature that science cannot explain, and therefore that all explanations offered by scientists are inherently suspect. It is very easy to glean from the scientific literature statements to the effect that an observed phenomenon is not as yet explained or is poorly understood. Selective quotation is a favorite tool of the pseudoscientist. Frequently, the pseudoscientist deliberately ignores the existence of satisfactory, well-supported explanations readily available in the published scientific literature. After offering the selected statement, the false claim is then made that no explanations whatsoever are forthcoming from science. That assertion having been accepted by the gullible members of the public or the cult, there swiftly follows a fallacy of logic, the *non sequitur*, to the effect that because science has no explanation, the asserted scenario or hypothesis of pseudoscience must be valid. This is the well-known fallacy of "proof from ignorance."

Von Däniken repeatedly makes assertions that science has no explanation for this or that occurrence when, in fact, there do exist published articles containing carefully considered and reasonable explanations. Phrased as rhetorical questions—often many on a single page—his repeated calls for explanations are followed by exhortations to science to devote its full attention to the matter. In this way he tries to make fools of scientists and, of course, his audience loves it!

Dutch refers to a common practice of courtroom lawyers, which he calls the "stacked argument" (1982, p. 11). It uses a question so worded that you are given only two possible answers, either of which is incriminating. Example: Have you stopped distributing counterfeit bills? Answer yes or no! Dutch selects from von Däniken a question about the Great Pyramid; it is in the context of his general argument that humans are not physically capable of transporting and lifting great pieces of stone: "Who is so ingenuous as to believe the pyramid was nothing but the tomb of a king?" (von Däniken 1969, p. 11). You have no choice in answering but to choose either (a) "Yes, I'm a naive scientist, like all the rest," or (b) "Not I! Gods from outer space must have built the pyramid." The content of von Däniken's argument is *ad hominem*, which attacks the person of the opponent rather than the opponent's argument. Scientists have not always been above that sort of argument, but their bad acting does not make it acceptable.

Dutch describes another tactic of the pseudoscientists, which is to take aim at indirect scientific evidence and degrade its value (pp. 11-12). Geophysics, which studies the structure and composition of the earth's interior by analysis of seismic waves (earthquake waves), has presented strong evidence to support the hypothesis of a dense and probably metallic core, surrounded by a thick, rocky shell (the mantle). No human has ever seen this core, nor have samples of its material been extracted for laboratory examination. The geophysical evidence—especially from seismology—is nevertheless considered extremely good because it is based on laws of physics, has been experimentally repeated innumerable times, and is consistent with other categories of independent evidence (such as measurement of the earth's average density).

Perhaps you have read of the various "hollow earth" theories of pseudoscience, which assert that the earth's interior is hollow or has deep cavities and that the hollow space is inhabited by colonies of living creatures. In a modern version of one of these theories, its author will point out that scientific statements about what lies deep beneath the earth's surface are based only on conjectures, guesses, and suppositions, and not on any hard evidence. In making this assessment, the author has downgraded to mere speculation indirect evidence from highly sensitive and reliable scientific instruments interpreted in strict conformity to laws of physics. The practice is, of course, a form of attempted deceit, but it is effective in convincing an unwary and unsophisticated audience. What the pseudoscience author is really saying is: "My theory of a hollow Earth is just as good as your theory of a dense Earth without cavities, because no human has ever been down there to look at the core and get samples of it."

This brings us to a general logical fallacy prevalent in pseudoscience. Dutch identifies it as a form of *relativism*, which we discussed in Chapter 5. He describes it as follows:

> The general thrust of all of the fallacies above is to create a fog in which all theories appear to be on the same level, none more probable than any other, and the consensus among scientists is made to appear much weaker than is actually the case. Amid the seeming confusion, the fringe theorist argues that his theory is just as likely to be valid as any other, and that experts don't agree, so the non-scientist is free to choose whichever alternative looks best. (1982 p. 12)

The most devastating argument for segregating pseudoscience from science has not, I think, been brought forward strongly enough in the numerous writings of skeptics. In virtually every scenario of pseudoscience of which I am aware, the principle of parsimony is grossly violated. The pseudoscientists have postulated elaborate and complex events that are not necessary to explain the phenomena observed (or asserted to have been observed). In each case, a simpler explanation, reasonably well supported by the present state of empirical science, is available. For the Velikovsky case, the events described in ancient texts relating to the Flood of Noah and the exodus and the captivity in Babylon do not require a comet of unique dimensions and behavior to impact Earth and set off Mars into an orgy of destruction. All that is needed is an appreciation of the naturalistic manner in which humans, possessed of remarkable imaginative ability but of limited knowledge of the world of nature, were prone to construct myths and legends, some of which may have had a basis in real events that were not fully understood. Earthquakes, floods, tidal waves (tsunami), and volcanic explosions are examples of such events. As to the Velikovsky astronomy, the principle of parsimony suggests that, in the absence of any acceptable physical evidence to the contrary, we assume that the planets have maintained stable orbits for a vastly longer period of time than the few thousands of years Velikovsky asks for. Why was he compelled to read into vaguely worded, incomplete documents from ancient civilizations a sequence of catastrophic planetary phenomena? Perhaps he felt the same need to create myths that was felt by our ancestors in the dawn of civilization. Why should his mind work differently from theirs?

In addition to the principle of parsimony, but as a consequence of its abuse, the pseudoscientists share a common failing—they rely heavily on hearsay evidence, which is a body of allegations in oral or written language, passed from one person to another. As pointed out in the evaluation of UFOlogy, all that we have is a set of allegations, within which the distillate worthy of provisional admission as evidence is capable of a number of alternative physical explanations. Von Däniken, for example, bases his case on artifacts or what he claims to be artifacts, in addition to the texts and symbols of uncertain meanings in ancient languages. In this respect his scenario may demand more serious attention from science than those of Velikovsky and the UFOlogists. But, here again, those artifacts for which von Däniken asserts a requirement of space gods can be given reasonable alternative explanations commensurate with what engineering science can produce and, in fact, can reproduce. The quarrying, carving, transporting, and erecting of huge monoliths are operations capable of explanation in terms of human and animal power and the use of tools and machines available to humans who lived in those times.

Why Are Scientists Concerned about Pseudoscience?

If these and other forms of deceit and fallacious argument are so widely used and so easily exposed, why has not the scientific community quickly put down each pseudoscientific theory? Besides the obvious answer—that the believers within the cults of followers are totally oblivious to reasonable debate—there is another more serious. If scientists were to dissect every unsupported theory proposed in popular books and magazines, their entire working capacity would be filled and they would have no time for their research. To undertake a complete refutation of the Velikovsky scenario, for example, would require that a complete textbook be written, covering all the pertinent areas of astronomy, physics, chemistry, archaeology, and engineering. The scientific community has no obligation to perform this function; it is the obligation of the alleged pseudoscience author to document the scenario with high-quality evidence and present the whole package in scientific language with full referencing and in a form acceptable for publication, as in a scientific journal or monograph series—acceptable, that is, after searching review by scholars familiar with the areas of science involved in the scenario. Unless material of such quality is prepared by those who purport to be engaged in science, their work deserves only summary dismissal. That the AAAS should have sponsored symposia to focus attention upon the evidence for Velikovsky's assertions or for cultists' assertions of the existence of visitors from outer space is a sign of concern that the unchecked growth of cultism and its unwarranted attack upon science may undermine the strength of the nation.

Professor Bunge addresses our question in somewhat different language and perspective well worth our attention:

> Scientists and philosophers tend to treat superstition, pseudoscience, and even antiscience as harmless rubbish, or even as proper for mass consumption; they are far too busy with their own research to bother about such nonsense. This attitude is most unfortunate for the following reasons. First, superstition, pseudoscience, and antiscience are not rubbish that can be recycled into something useful; they are intellectual viruses that can attack anybody, layman or scientist, to the point of sickening an entire culture and turning it against scientific research. Second, the emergence and diffusion of superstition, pseudoscience, and antiscience are important psychosocial phenomena worth being investigated scientifically and perhaps even used as indicators of the state of health of a culture. Third, pseudoscience and antiscience are good test cases for any philosophy of science. Indeed, the worth of such philosophy can be gauged by its sensitivity to the differences between science and nonscience, high-grade and low-grade science, and living and dead science. (1984, p. 46)[1]

The Emotional Appeal of Pseudoscience

Why are the gods and their chariots needed? The answer may lie in psychology and sociology. Perhaps we need to pay more attention to how the mind of the author responded to the cultural environment and forces of history and less attention to questions of physical science. Ronald Story may have given us a clue to what is going on in his comments on von Däniken's "new mythology":

> Man's inability to rise to a high moral plane, even in times of dire need, is all too evident in the present world situation with its threat of atomic war, depletion of natural resources, and widespread hunger. Just as in the days of the contactees and the "Space Brothers," we are facing bad times and fears of the future—fears of an ever-advancing technology that seems to be running out of control.

> What could be more appealing than the modern Space Brothers—the ancient astronauts of von Däniken, who are godlike in their technical knowledge (which is so threatening but means so much to us) and in their wisdom (so we assume), and who could direct us in the use of advanced knowledge for the ultimate good of mankind? Since the gods may be our salvation, we want to believe in them, whether we realize it or not. By identifying with these gods, we both comfort our fears and flatter our egos. (1976, pp. 118)[2]
>
> [Note: The "Space Brothers" referred to are modern visitors from space in UFOlogy.]

The New Age Movement—What Is It?

Everyone who reads newspapers and views television has learned of the "New Age," and perhaps many are already swept up by its promoters to new heights of consciousness—or should we say "confusion"? Any succinct definition of the New Age belief system would prove inadequate and/or inaccurate. Seen as a world view, one large chunk of it is simply old-fashioned European-style spiritualism, disguised by new names but easily identified as pseudoscience. Another chunk is clearly religious mysticism imported directly from the Orient. Probably few New Agers could agree on any one description of their collective world view.

The New Age movement recently caught the attention of the Committee for the Scientific Investigation of Claims of the Paranormal (CSICOP), a group of humanistic skeptics that includes a wide spectrum of scientists and philosophers. In 1988 this body convened a national conference devoted to throwing some light on the New Age. One speaker after another arose to tell about a different facet of this movement, each viewing it from a different perspective. Here, I can do little more than compile a list of statements of what seemed to me to be applicable to the New Age movement, but only to suggest the variety of its contents. I take as my text an article by Lys Ann Shore in *The Skeptical Inquirer* (1989).

Paul Kurtz, chairman of CSICOP, opened with the remark that the much-touted New Age is simply "The Old Age repackaged." Indeed, much of the content of the New Age (NA) is reminiscent of the spiritualist movement a century ago. Like that movement, which sought to establish communications between living humans and souls of their long-departed loved ones, NA has a distinctly religious aspect. Speaker Robert Basil stated that the NA tries to hide its connection to religion by using a secular vocabulary that includes words such as holistic, synergy, and transformation. The religious aspect of the NA is nevertheless clear enough to the fundamentalist Christians, and they find the NA to be in conflict with biblical teachings (Shore 1989, p. 227).

Basil and another speaker, Maureen O'Hara, a humanistic psychologist, agreed that the NA "represents a large-scale rejection of science" (p. 227). In this connection, it is worth noting that the fundamentalist creation "scientists" accept a very large part of the content of modern science—that dealing with timeless knowledge—and many of its supporters are professionaly qualified in fields of science, such as physics, chemistry, and biology, along with the applied sciences (particularly engineering). If Basil is right in seeing the NA as "today's major alternative to American religious life," it does not follow that there is any clear or simple statement of what that alternative is. Vagueness of expression and lack of agreement to any simple dogma is evident in expressions by those identified as New Agers.

Basil noted that New Agers see reality as purely spiritual. In philosophy, this is a long-established position, known as spiritual monism, that denies the existence of what scientists refer to as the empirical realm. This idealistic aspect of the NA is a resurgence of a very ancient ontology (view of reality).

Speaker J. Gordon Melton observed that the NA shows strong Oriental influences and seems to have burgeoned at the same time—the mid-1960s—that an influx of Oriental immigrants introduced Eastern religious thought and the development of transpersonal psychology (Shore 1989, p. 228). Melton's analysis of the origin of the NA, stated by Shore, is as follows:

> The concept of personal transformation became the keystone of the New Age movement, whose chief message is "You can transform your drab, wretched existence." Personal transformation then becomes a model for social transformation, leading to the idea that the world will be changed if enough individuals change their lives. As a result, Melton pointed out, "the self assumes enormous importance in the New Age movement." (P. 228)

Another interesting aspect of the NA, stated by Jay Rosen, a professor of journalism, is its "belief in the equality of all opinions" (Shore, p. 228). The New Agers seem to attach no value to logical thinking, and they make no distinction between degrees of truth or falsity in knowledge. This amounts to the assertion that illogical and nonsensical statements have equal truth value with those logically and empirically based, and this is a totally anti-science position, appealing to relativism for its support.

Proponents of the NA movement emphasize the power of self-transformation to bring an individual to a position of power and fame in society. Actually, only a very few practitioners of the formula succeed in achieving any fame or power (Shore, p. 229).

Characteristic of the New Agers seems to be the susceptibility to fads, and they drift from one fad to another. One fad, "channeling," is a modern version of trance-mediumship, claiming contacts with "spirits," often specifically identified by name. None of that is new. Channeling becomes an opportunity for the medium (channeler) to charge fees that are willingly paid. Another type of fad to which the New Agers are susceptible lies in material products, illustrated in the fad for mineral crystals promoted as having mystical healing powers. The claims for such objects obviously fall into the area of pseudoscience. There's nothing new here—it's a form of exploitation as old as civilization itself.

Speaker Jay Rosen sees the NA as a "culture of narcissism," emphasizing egoism and egocentrism. In answer to the question "How can you recognize the New Ager?" Rosen replied: "Look for the combination of radical subjectivity accompanied by ferocious superficiality" (Shore, pp. 231-32). On that note, we close this attempt to describe what seems to be a very blurred and confused world view. In Part Two, we will have more to say about world views and their place in human knowledge.

An Unfinished World View

In these eight chapters of Part One, my attempt to describe empirical science and how it works has required that we observe science from a variety of vantage points and perspectives. Some insights will be new to many who are not scientists. To those who are established scientists I hope I have reminded them of some principles they may have neglected because of lifelong confinement in the tiny niches of their specialties. Before we turn in Part Two to an in-depth comparison of science with other knowledge fields, it may be well to let Professor Herbert Feigl state the guiding spirit of empirical science:

> Instead of presenting a finished account of the world, the genuine scientist keeps his unifying hypotheses open to revision and is always ready to modify or abandon them if evidence should render them doubtful. This self-corrective aspect of science has rightly been stressed as its most important characteristic and must always be kept in mind when we refer to the comprehensiveness or the unification achieved by the scientific account of the universe. It is a sign of one's maturity to be able to live with an unfinished world view. (1953, p. 13)

Credits

PART TWO

SCIENCE AND OTHER KNOWLEDGE FIELDS

Introduction

The logical positivists severely restricted the scope of what they regarded as genuine or bona fide knowledge to empirical science and the logical analysis of gratituitous propositions related to empirical science. Nothing else was allowed to count as knowledge. This left a vast area of products of the human mind to be placed in one great trash bin, labeled "Nothing of Value." In this refuse bin lay religion, ethics and morality, sociopolitical ideologies, and the arts. Nearly everything that human civilization had valued since the earliest symbolic records had been junked by the positivists. What remained—modern science—was a comparative newcomer to the scene. Civilization had done remarkably well without science until as recently as, say, 1610, when Galileo's telescope revealed the moons of Jupiter to be orbiting that planet, thereby providing a theoretical model of the heliocentric planetary system.

Of course, the claim of the positivists could not be sustained outside their tight little Vienna Circle. Common sense told the other 99.44 percent of the inhabitants of Western civilization that the cleansing soap of religious belief was real, bona fide knowledge, and so were ethics and morality. Humans "knew for sure" that God existed, that some things were right and other things were wrong, and that Michelangelo's Sistine Chapel frescoes expressed human knowledge of intrinsic beauty and of great value.

Perhaps the most revolutionary aspect of the dumping of both logical positivism and logical empiricism in favor of the new philosophy of science was the admission to genuine knowledge of all forms of concepts and mental constructs, provided only that they can be expressed as simple propositions (or clusters of propositions) that are not self-contradictory, do not abuse the established meanings of words, and do not switch the domain of discourse in the process of logical analysis. Those basic violations will continue to be punished by outlawing of the guilty propositions, exiling them to the land of nonsense.

Perhaps we can call this updated view of human knowledge the New

Epistemology, and with it the New Ontology. Already, in Part One, the New Epistemology has been assumed in adopting Mario Bunge's outline of the fields of knowledge (see Chapter 1). We did not, however, compare in any great depth the crucial differences between what Bunge calls the research fields (empirical fields) and the belief fields. Examples were given, but not the criteria of demarcation between groups or paired sets of fields. That is the task of Part Two, on which we are now embarking.

Why is this an important task? Establishing firm and logically valid criteria of demarcation is part of the process of accurately and precisely describing each knowledge field and the realm of reality to which it relates. What this amounts to is putting each knowledge field in its own box or compartment, safely isolated from all the other boxes. One of the rules here is that no defined field can occupy more than one box; the corollary being that no box can contain more than one field. When we carry out this process of sorting out of knowledge fields, we quickly come across certain academic subjects that seem to be a mixture of two or more fields. Human (cultural) history is a good example, because history, as usually written by academic historians, consists of two ingredients: (a) what actually happened or was said, and (b) interpretation (whether what happened or was said was right or wrong, good or evil, commendable or despicable). We are forbidden from making a special box to house both (a) and (b) in their mixed state. Instead, we are required to separate the propositions of (a) from those of (b) and put each pile in its own box. This may be difficult to do—it may, in fact, be impossible to accomplish. Unfortunately, having done the mixing two fields of knowledge—empirical and emotive—they may have been irreversibly blended, or homogenized. Once four different foods—say, milk, eggs, chocolate, and vanilla—have been mixed in a highspeed blender a homogenized milkshake is the result. While some of the ingredients can perhaps be identified, others may elude our sense of taste. Our problem is to separate mixed fields into their components, even if this is successful only to the extent that the existence of at least two unlike fields is clearly recognized.

Then why is it important to avoid the homogenization of knowledge fields, besides just being neat and tidy in one's thinking? The reason is that humans are prone to perform deception on other humans in order to promote their ideologies or world views. These deceivers dump out the contents of two or more knowledge boxes, mix up the contents in one pile, then say: "You must believe that everything in the mixed pile is of the same quality—no item is distinguishable from the others." The aim of this deceit (if it is deliberate) is to apply this conclusion to some sociopolitical issue.

Yes, it does matter that knowledge takes a number of distinctly different forms, and that these forms should be definable and recognizable. It matters to the extent that human beings object to one sector of a society being exploited, demeaned, and degraded by another sector.

CHAPTER 9

The Major Classes of Knowledge

We begin with a rather terse outline of the infrastructure of this chapter. My principal source of ideas and concepts lies in the major academic works on epistemology and ontology by philosopher Mario Bunge. To simplify references to his publications, I use a special reference code throughout this and later chapters.[1]

The Two Major Classes of Knowledge

First-order headings (the fundamental dualism or dichotomy):

I. Perceptional Knowledge: Knowledge of Things Perceived or Perceivable (B5: 35, 72)

II. Ideational Knowledge: Knowledge of Things Imaged or Imagined (B5: 35, 73)

Preliminary definitions and formulations:

I. PERCEPTIONAL KNOWLEDGE

In Bunge's system, this epistemic class contains the *fields of inquiry* (B5: 91), or *research fields* (B6: 197-98).

The ontological realm of this class is described as *empirical* or *naturalistic*.

---

[1]Reference code "(B5: 77)" reads "Bunge, vol. 5, p. 77." The reference is to Mario Bunge, *Treatise on Basic Philosophy*, vols. 3, 5, and 6. Copyright © 1977, 1983 by D. Reidel Publishing Company, Dordrecht, Holland. Reprinted by Kluwer Academic Publishers.

The cognitive process (cognition) by which this form of knowledge is initially gained by the brain is *perception*, the process of forming *percepts* (B5: 35-42, 72).

Examples of important knowledge fields in Class I:

- Pure empirical science (natural and social); applied science (technology, engineering); i.e., everything listed in Figure 4.1 under the heading of "empirical science."
- Human history, journalism. Academic philosophy of all knowledge fields.

II. IDEATIONAL KNOWLEDGE

In Bunge's system, this epistemic class contains the *belief systems* (B5: 91).

The ontological realms within this class are described collectively as *transempirical* (nonempirical), or *transnatural*, and include the *supernatural realm* (B5:161).

The cognitive process by which knowledge of the transempirical realm is generated is *ideation* (B5: 35, 72). Although created by the brain through ideation in both epistemic classes, constructs (concepts) in Class II are characterized by being *autonomous*, requiring no perception of or direct reference to things outside the brain. Ideation in Class II includes the creation of *images* (*imagery*) through imagination, which may also be called *mentifacts*.

Examples of important knowledge fields in Class II:

- Pure mathematics, formal logic.
- Religion and theologies, ethical and moral systems, aesthetics, sociopolitical ideologies.

We continue with more descriptions and comparisons of the fields of knowledge.

Epistemic Fields of Perceptional Knowledge

Known variously as *fields of inquiry*, *research fields*, and *empirical fields*, they are characterized by certain ontological principles exemplified by (but not limited to) scientific research. Bunge refers to this collection of principles as "scientific ontology" (B3: 15-18). He offers ten ontological principles, of which the first eight are as follows: (Indented matter is quoted from B3: 16-17.)[1]

1. There is a world external to the cognitive subject.

2. The world is composed of things. (*Things* are also described as "concrete objects.")

3. Forms are properties of things.

4. Things are grouped into systems. ("What there really is, are systems—physical, chemical, living, social.")

5. Every system, except the universe, interacts with other systems in certain respects and is isolated from other systems in other respects.

6. Every thing changes.

7. Nothing comes out of nothing and no thing reduces to nothingness.

8. Every thing abides by laws. ("Whether natural or social, laws are invariant relations among properties, and they are just as objective as properties.")

The above "principles" are assumed (if only tentatively) to permit further analysis; i.e., they are axiomatic, and they may be described as presuppositions or a priori propositions. This raises the question as to whether they may actually be pure products of the imagination. To the contrary, observation-based consensus, including that gained from experimental quantum physics, gives strong independent support that these principles are not mere a priori assumptions. It has been suggested, however, that their acceptance is a form of belief (acceptance on faith alone), and hence a belief system, but we tend to dismiss this suggestion on grounds that the axioms of belief systems are of a quite different ontological class.

Paralleling the above principles (esp. nos. 2 through 6), is a description of perceptional knowledge as quantitative in the sense of requiring the independent existence of time, space, and matter, all of which are measurable in terms of fundamental dimensional units of time (T), length (L), and mass (M), as well as their products and ratios. Is this requirement a necessary and sufficient criterion of distinction between perceptional knowledge and the purely qualitative content of values found in the ideational fields? (More on this later.)

The content of perceptional knowledge can be divided into two fundamentally different categories: timeless knowledge and timebound knowledge (review, if necessary, Chapter 4). *Timeless knowledge* is found in pure physics and chemistry and their basic laws of the behavior of matter and energy. Perhaps one could go further than that and propose that physics and chemistry are composed of nothing but timeless knowledge. Scientific statements in this area describe events or relationships that have an extremely large probability of being repeated in almost exactly the same way. "Timeless" is not intended to imply absence of time as a physical dimension, but only that no absolute time frame is specified. For example, the spontaneous decay of a particular radioisotope proceeds in exactly the same manner every time it happens. Timeless knowledge comprises much of applied science and technology, as, for example, in industrial chemistry or prescriptive medicine. Timeless knowledge is highly amenable to quantification, so that whenever two or more variables are involved, a functional mathematical statement is available. Physicists, especially those of the extreme reductionist school, often express the view that only timeless knowledge is true science.

Focusing on the macrocosmic inorganic sciences (MIS) and the biological sciences (BS) of Figure 4.1, we might be tempted to say that they, too, contain a great deal of timeless knowledge. For example, the principles of theoretical seismology or meteorology might be cited as timeless in content, but I would counter that these are pure physics. If molecular genetics is cited—the DNA code, for example—I would say that insofar as reference is made to DNA as a molecule, then the discussion is one of pure chemistry (biochemistry). If all the timeless physics and chemistry is taken out of the MIS and BS, most of what is left is not timeless. Perhaps there is a fringe area of possible laws within these complex historical sciences, and we have debated this possibility in Chapter 4.

Timeless knowledge may also be a major part of the social sciences—that part which is analytical, i.e., that deals in the search for unifying principles and/or laws. Critical observers would say that such a search is thus far flawed or even futile, but it is nevertheless there in concept and the search for regularities goes on.

Timebound knowledge is basically historical in character or quality and describes complex sequences of changes in form and structure that over long periods of time produce unique structures, as seen in cosmology, historical geology, and evolutionary biology. Included is the study of the effects of recent and current natural phenomena in terms of unique time frames and geographical positions; i.e., the history of currently changing natural forms and structures. The probabilities of such complex phenomena giving rise to an almost identical form at a different time and place are extremely small—approaching zero—and can usually be neglected.

Looking back at Figure 4.1, our schematic diagram of the components of empirical science, we can discern a strategy of two sets of criteria: (a) subjects of inquiry based on the substances, structures, and phenomena being observed and (b) the quality of the knowledge about each, whether timeless or timebound. Think of a two-layer cake, each layer with a very different flavor. The cake can be sliced vertically to separate the subject fields from one another but give the total effect of both flavors. Sliced horizontally, there are only two servings, each of a different flavor.

Human History—Is It Within Perceptional Knowledge?

We now focus on human history. At least some historical knowledge is attached to every form of artifact, so there exists the history of human culture. Human history as an academic discipline is not usually placed within the social sciences. In *Webster's Third New International Dictionary* (1961) human history is listed as one of the humanities, along with "languages, literature, mathematics, and philosophy." What a mélange!

Historical research asks: "What happened? When and how did it happen? Why did it happen?" The logical extension of the past tense is

"What is happening right now in the world of human affairs?" This brings in current events and the news—so the field of journalism, to the extent that it reports on cultural events (vis-à-vis natural phenomena such as earthquakes and volcanic eruptions) is part of human history. Closely allied with journalism (as strictly "objective" news reporting) is the introduction of evidence (testimony) by sworn witnesses in courts of law in the trial of criminal and civil cases.

Human history comes with diverse modifiers: political, social, military, legal, medical, technological, and many others. We also recognize histories of the arts and sciences, of religion, supernaturalism, pseudoscience, language, and literature. Archaeology investigates history through the study of artifacts of past cultures.

Almost entirely, the content of human history is timebound knowledge. In the literature of human history, timebound knowledge is described as *ideographic*, in contrast to timeless knowledge, described as being *nomothetic*. This distinction is attributed by Ernest Nagel to the German philosopher, Wilhelm Windelband, who published it in a 1915 essay (Nagel 1961, pp. 547-48).

Keep clearly in view at all times the distinction we must make between an idea that is genuine ideational knowledge and a statement about an idea. The latter, as a historical document of what someone thought or imagined, becomes in part perceptual knowledge, once it goes outside the originator's mind to take a form capable of entering other minds. Take for example this statement: "Immanuel Kant in 1785 asserted the *existence of categorical imperatives*." The full sentence describes a historical event, whereas the words in italics express belief in an abstract concept generated by Kant's imagination. We could isolate the ideational component as a proposition reading: "Categorical imperatives exist."

Although human history is conventionally set apart from empirical science, it is bound by the same eight ontological principles we listed as essential to empirical science. Check them over one by one for yourself. All deal with a world external to the cognitive subject, with properties and systems of things, and with things that abide by natural law. Historical investigations of artifacts are bound by the same procedural rules and ethics laid out for empirical science research on nature. True, direct experimental replication seen in timeless science cannot be practiced in historical research; but neither can the experimental method be used to repeat complex sequences of past events of cosmology, historical geology, or evolutionary biology. Thus the fundamental distinction between timeless and timebound knowledge remains intact.

We noted in the introduction to Part Two that human history is in practice typically a mixed field, in which emotive expressions get mixed in with statements of fact. The dualism of human history in terms of knowledge content (epistemology) has long been debated. The two components are (1) empirical (what actually happened or was reported to have happened); (2) emotive or *empathic*. This second component consists of *empathic reconstruction*, in which the historian relives the feelings, impulses, and emotions of the agents whose decisions determined events (Durbin 1988, p. 120). Ernest Nagel uses the adjective *value-impregnated* for this component (1961, p. 473). It is the same component we noted in Chapter 4 in commenting on the social sciences. Thus both the social sciences and human history are "tarred by the same brush."

Philosopher Wilhelm Dilthey (1833-1911) considered the empathic component to be essential in the understanding of human history. If that position is accepted, the requirement can serve as a criterion of demarcation between human history and the historical empirical sciences. In that case, we would be obliged to place human history in the ideational class of knowledge. Dilthey's view was endorsed in 1946 by R. G. Collingwood, who went so far as to claim that all human history as told will be subjective and will reflect the world view of the historian's own time (Durbin 1988, p. 120). This extreme subjectivist view is now often described by the adjective *hermeneutic*. Philosopher Carl Hempel, an advocate of logical empiricism, challenged the Dithley/Collingwood requirement that human history must be subjective. Hempel claimed that human history has the same structure as scientific explanation. (For references to Dilthey, Collingwood, and Hempel, see Durbin 1988, pp. 122-23.)

Perhaps the bottom line is simply that human history contains an unmistakable empirical content that belongs in the class of perceptional knowledge, but that we must carefully exclude emotive or moralistic expressions that are often attached to historical accounts. This caveat does not mean that an empirical account of human history cannot go beyond merely stating the "facts." John Ziman states: "Having ascertained the 'facts,' the historian tries to uncover the hidden motives and forces at work, just as the scientist goes beyond the phenomena to the laws of their being" (1980, p. 46). What must be clearly separated as ideational knowledge is any belief statement that carries moral bias or emotive lessons of history; for example: "The Nazis (or Fascists, or Communists, or Americans) were evil persons and did bad things."

Academic philosophy can be claimed as perceptional knowledge on essentially the same grounds as those offered for human history, and with the same caveat. Philosophy examines critically what humans have said or written. Any proposition, whatever its content—perceptional or ideational—is treated as a historical document to be analyzed in terms of its knowledge content. Our placement of philosophy is in close agreement with the principal definitions of "philosophy" given by Peter Angeles (1981, p. 211). Our placement also fits well with what transpires in departments of philosophy in accredited secular colleges and universities—as judged by what they teach and the publications their faculties produce. On the other hand, we would want to rule out the "First Philosophy" of Aristotle, which included description of "that kind of being that is immutable and transcendent" (Angeles 1981, pp. 102-103). Other brands of "philosophy" are to be ruled out because they, too, are systems of ethical and moral values; they go by the modifiers "political," "speculative," and "synoptic" (1981, pp. 215, 272, 286). "Philosophy of religion," in particular, needs to be carefully scrutinized as to its context. Where associated with a seminary of a particular religion or sect, the subject may be partly or largely the uncritical teaching of a theology or dogma, posited to be the truth.

Criteria of Demarcation Separating Empirical Science from Other Perceptional Fields

What, then, distinguishes empirical science from other possible extra-scientific fields within the perceptional knowledge fields—especially from human history. Perhaps the answer is "nothing." Perhaps a better answer is: nothing except (as stated earlier) on purely ontological grounds of the subject areas involved. Then our criterion of demarcation of science is downgraded to something of no really important qualitative significance within the total field of perceptional knowledge.

Following up that conclusion, consider a more fundamental aspect of ontology, namely, the concept of "thing" in terms of "change." Bunge develops this concept in depth:

> If we are to believe science, we must uphold the ontological postulate that all things are in flux. In fact the sciences describe, explain, predict, control or elicit changes of various kinds—such as motion, accretion, division, and evolution. (B3: 215)[1]

Elucidating this statement, Bunge observes that the world encompassed by perceptional knowledge "consists of things that do not remain in the same state forever" (B3: 215). But this observation applies equally well to both empirical science and human history. Thus we cannot find in changing versus changeless things a criterion of demarcation between science and human history. In striking contrast, as we shall see, the element of necessary change is a valid criterion of demarcation between all perceptional knowledge and certain fields within ideational knowledge—especially religion and theology.)

The idea of separating natural things from artifacts, as we have done, can easily be shown to be useless as a criterion of demarcation. True, many subjects of scientific research are natural in the sense of being unrelated to artifacts. Think of cosmology, solar physics, neurophysiology of the human brain, or the evolution of vertebrates; the list is almost endless. Then think about the synthesis of new organic compounds in the field of organic chemistry and of new elements and particles produced by physicists. Are these not artifacts? Much, if not most, of technology consists of the invention and construction of artifacts, but we would not think of eliminating technology from the broad area of empirical science. (For myself, I would not hesitate to segregate and quarantine technology on grounds that it is usually value-driven, but that's another thought.)

To reduce the above criterion of demarcation to absurdity, consider two cases. (a) Anthropologists, looking for the earliest stone tools produced by hominids, collect stream-rounded pebbles, each bearing two intersecting facets made by forced fracturing. The question is, are these natural facets made by impacts with other pebbles in the bed of a turbulent stream, or are they made by pressure-force applied by a human hand holding another hard object? (b) Boys on a hike along a mountain ridge stop to dislodge a precariously perched boulder, sending it hurtling down the mountainside and coming to rest among similar boulders at the mountain base. At other times, similar boulders are set free by a natural process of frost-wedging and follow similar paths. In both cases each kind of event uses the same naturally formed object, applies a similar force, and achieves a similar result. The argument could be made that human intention makes the alternative modes ontologically different, but many artifacts are created unintentionally.

One last effort before we give up! Could it be that the elusive criterion of demarcation lies in the lawfulness of science? The root of the idea is expressed by Bunge's observation that scientific research may be "conceived of as being ultimately the search for objective laws" (B3:

177).[1] He equates lawfulness with "real possibility" and reasons that "then science must be regarded as the study of the really possible."

The late Judge William R. Overton, in his court opinion in *McLean* v. *Arkansas*, listed as the first two of five "essential characteristics" of science "(1) It is guided by natural law and (2) It has to be explanatory by reference to natural law" (Overton 1982, p. 938). The remaining characteristics pertain to testability, falsifiability, and tentativeness. Philosophers Larry Laudan and Michael Ruse have since debated the validity of Overton's list of characterstics (Ruse 1988, pp. 335-66). Laudan vigorously denies the validity of any of the listed essential (necessary) characteristics as viable criteria of demarcation between science and non-science; he regards the very idea of finding a single criterion as outmoded and discredited. Philip Kitcher (1982, Chapter 2) also covers this subject fully, saying much the same thing.

There seems to be no doubt that lawfulness—the reliance on underlying fundamental laws—is a necessary characteristic of empirical science, but is it a sufficient characteristic? Even a quick inspection of human history vis-à-vis empirical science reveals that the answer is no. The interpretations of human history require adherence to the same natural laws demanded by empirical science. Military history, for example, cannot evoke God's command that the sun should stand still as an explanation of the outcome of a battle. Nor do we attribute witch trials to forces of witchcraft itself. We can go on to such other touted characteristics of empirical science as testability, falsifiability, and tentativeness and reach the same conclusion. (The possibility that human history has, besides natural laws, its own set of unique laws—laws of history—is not at issue here.)

If at this point you've become discouraged by not finding any characteristics of empirical science that can serve singly or in groups to define it uniquely as a recognizable, unambiguously separate field of knowledge, you have come a long way indeed. Your discouragement may have turned to apprehension as you realize (as did Larry Laudan) that Judge Overton's dismissal of the secular propositions of creation "science" on grounds that they are really not science is on a very flimsy footing.

What can we safely conclude at this point? Following up the logical sequitur of Bunge's pronouncement that "science must be regarded as the study of the really possible," we must broaden our definition of science to include any and all of the epistemic fields within our Class I, Perceptional Knowledge. We must embrace human history—that which is currently unfolding as well as past—within the broader concept of science and vice versa. This proposal for generalization and unification may elicit groans from the side of science and technology, but it provides a single intellectual power base with which to confront an enemy host based in Class II, Ideational Knowledge.

Confronting our coalition are such powerful hostile groups as the literalist/fundamentalist Christians who promote creation "science," the panderers of pseudoscience and pseudotechnology, and the sellers of anti-science sociopolitical ideologies. But look again! Practitioners of ideational knowledge include the humanists, who uphold science and oppose the repressive pressures of the Christian fundmentalists. There are also many science-supporting intellectuals who exercise value judgments in the field of aesthetics—they are the critics of literature, fine arts, and the performing arts. And don't overlook a large cohort of mainstream scientists and engineers who find no conflict between their theistic religious belief and their science. It behooves us, then, to move on to an investigation of these diverse fields of ideational knowledge.

Epistemic Fields of Ideational Knowledge

Ideational knowledge can be defined in both a negative sense, by what it excludes, and in a positive sense, by what it includes. Of the two alternatives, the exclusional is the simpler: things with which it deals (things which are created by the brain) are not things perceived; i.e., not what actually exists "out there." (Caution: Ideation would perhaps be impossible for a brain totally devoid of perceptional knowledge.) A statement of what ideational knowledge includes is more difficult because it must consist of a list of diverse epistemes, each with its unique ontology. There are, as we shall see, "strange bunkmates" in this class. Perhaps you noticed in our earlier list of examples that we do, indeed, have a mélange to contend with.

First, however, it is necessary that we review the sequence: perception → conception → ideation. All three are assumed to be neurophysiological activities of the animal brain. Bunge covers this entire topic in great detail (B5: 35-42, 72-74, 157-61). Summarizing briefly, the neural process of *perception* starts with detection in a neurosensor, but "the response is a function of both stimulus and the internal state of the organism" (B5: 35).[1] Thus a *percept* is formed: "When perceiving something we construct a percept of it with sensations, memories, and expectations" (B5: 36).[1] A *perceptual system* is thus set up in the brain, and it is essential to all empirical knowledge, but it also functions by spontaneous processes to generate other percepts without external stimuli; examples are hallucinations and dreams (B5: 36-37). Already, you see, we are at work establishing criteria of demarcation between the two great classes of knowledge.

Concepts are formed through the brain's process of *thinking* (conscious mental activity) (B5: 39). *Concept* is almost impossible to define except through tautology. Bunge is not very helpful in this regard, but he gives it a try:

> Sit quietly in a soundproof room, close your eyes, and think of any intellectual problem. You will be thinking up concepts, most likely not isolated concepts but entire systems of them. (B5: 159)[1]

"Concept" has somewhat different meanings in psychology and philosophy. Here, "concept" is equated with "elementary idea," "unit construct," "smallest component of a proposition," and "the building blocks of all other ideas" (B5: 159). What counts is relationships: "Concepts are brain processes (or collections of such) but they do not involve any neurosensors even though some of them do originate in perception" (B5: 160).[1] That last sentence has meat in it. We can now distinguish two ontological kinds of concept: ". . . those which originate in perception and those which do not. We may label them *empirical* and *transempirical* (or transcendental), respectively" (B5: 160).[1] Bunge elaborates:

> The transempirical concepts do not originate in perception, i.e. they cannot be learned from experience but must be acquired by reflection. However, they are not necessarily isolated from all empirical concepts. For one thing some transempirical concepts may be suggested by empirical ones. (B5 161)[1]

The process by which transempirical concepts are formed is *ideation*, the invention of autonomous systems of concepts. Ideation is thus virtually synonymous with imagination, which can be referred to more formally as *imaging*. By "autonomous" we mean "independent of any empirical reality." On the other hand, ideation can be an important tool (perhaps the most important tool) in formulating hypothetical propositions about what is possible in the fields of perceptional knowledge, such as science.

Moving on to kinds of ideological knowledge and their ontologies, consider what specific subfields exist that can be characterized as being both transempirical and autonomous. A partial list of the major categories includes the following:

- *Transempirical fields* of formal logic and pure mathematics.

- Transempirical fields of value-based knowledge including (a) ethics and morality, (b) aesthetics, (c) worldviews (*Weltanschauungen*), and (d) sociopolitical ideologies.

- *Supernatural fields* of religion, theology, spiritualism, magic, the occult. Bunge calls them "unworldly fields" (B6: 228).

- Transempirical fields of pseudoscience and pseudotechnology (false perceptional knowledge pandered as being genuine).

The entire classification system of fields of knowledge presented thus far has both similarities with and differences from Bunge's system, which is the parent system of the two. The similarity lies in the fundamental dichotomy of two major classes. What is here Class I, Perceptional Knowledge, is nearly equivalent to Bunge's *Fields of Inquiry* or *Research Fields* (B5: 91 and B6: 197-8). What is here Class II, Ideational Knowledge, is nearly equivalent to Bunge's *Belief Systems* (elsewhere, *Belief Fields*, 1984, p. 38).

The differences are significant, however, and arise from our substitution of revised class definitions. Substituting the cognitive process of ideation for "belief" has a major impact through forcing logic and pure mathematics, which were typically referred to as "formal science," out of the first class and putting them into the second. This shift will be addressed and justified in Chapter 10. Here, however, we comment on Bunge's selection of the word "belief" as denoting an epistemic field. This seems surprising in view of Bunge's strong assertion that "belief" must not be equated to "knowledge":

> We shall (also) distinguish knowledge from belief, but this distinction is not a mere methodological gimmick. That knowledge is not belief should be clear from reflection on cases such as "*A* knows *B* but does not believe *B*", and "*A* believes *B* although he does not really know *B*". Yet there is a large body of philosophical literature that defines knowledge as a kind of belief; we shall pay hardly any attention to it. (B5: 61-62)
>
> Knowledge, then is not a kind of belief. (B5: 86)[1]

We can straighten out this seeming aberration without difficulty by examining the semantics involved. You need to separate "belief" in Bunge's text into two terms of different meaning. First, there is "believing" (verb, to believe), as a cognitive process—which is not in itself knowledge, but rather a mode of accepting (or rejecting) knowledge. Consider believing as an evaluating process with a range of from +1 (total belief) to –1 (total disbelief). The "thing" that is the object of believing exists independently of the level of intensity of believing shown by a person. Second, there is "belief" (noun), the object of believing; e.g., belief in the Holy Ghost. Beliefs comprise a bona fide form of reality in the supernatural ontological category. That Bunge affirms this proposition is clear in his statement that for a rational person, "his beliefs are special cases of his knowledge not the other way around" (B5: 87). Keep in mind, too, that our definition of knowledge includes that which is asserted (posited, claimed, presumed, believed or disbelieved, imagined) to exist. In short, knowledge can be genuine or bogus—or just about anything you choose to claim about it. Any concept arrived at by ideation qualifies as knowledge in this broad definition, but we cannot restrict it to narrower definitions, such as "knowledge is justified true belief."

Class II, Ideational Knowledge, includes many kinds of beliefs, for example, belief in God, belief in moral rightness versus wrongness in ethics, belief in goodness versus evilness of humanity, beauty versus ugliness in the arts, and so on.

Perhaps at this point, with your mind leaping ahead to criteria of demarcation between various epistemic fields, you notice that the beliefs listed above are entities without substance, and specifically that they cannot be assigned fundamental physical dimensions of time, space, or matter. They are either beliefs in "ghostly things" (Bunge's term), as in the supernatural, or they are abstract value qualities that defy physical measurement. These special properties will emerge in following chapters as we attempt to establish further criteria of demarcation.

The Ultimate Criterion of Scientific Knowledge

We may think we have made an airtight case for separating the epistemic fields of perceptional knowledge (namely, empirical science and human history) from those of ideational knowledge generally and from systems of belief in a supernatural realm in particular. We will be challenged, of course, by those seeking to break down any ontological barriers we try to set up. One group from the field of religious believers in particular will be heard from. Their attack on science runs about like this: A scientific hypothesis, when originally formulated, contains a proposition or set of propositions that is "conjectural" (to use Karl Popper's favored term). After all, the word "hypothesis" means "a thesis that is (at least in part) hypothetical, or conjectural." The word signifies that there exists a gap in scientific knowledge, not yet filled by perceptional information. The attack then focuses on the nature of the conjecture itself. Our adversaries ask: Where does it come from? Their answer: It is generated by the cognitive process of ideation, i.e., it is a product of human imagination. They are on sound ground and we provisionally accept that statement.

What follows is another story. The science-bashers say next: "Science is a belief-laden system, because every attempt it makes to penetrate further into an understanding and explanation of the unknown requires that it generate imagined entities—images, they are—and these can only be held as beliefs because there is on hand no empirical evidence in their support." Their bottom line makes a major jump to an unwarranted conclusion, using relativism as its basis: "Science is a belief system, no different from religion with its supernatural entities." Why is their conclusion false?

Let's set up a simple example. AIDS was recognized to exist as a human disease with a certain collection of symptoms (disorders) leading to death of the victim. Science (prior to the discovery of HIV) proposed Hypothesis A: Aids is caused by a virus, as yet unidentified, that attacks the human immune system. A religious cult proposed Hypothesis B: AIDS is caused by a supernatural force wielded by God as punishment for individuals who sin. The formulation of both hypotheses involves ideation by the same neurophysiological process. Beyond that similarity, they diverge sharply in content and intent.

The first criterion of distinction is that Hypothesis A adheres to axioms and laws of science and is based on prior empirical knowledge; it predicts that observation will confirm the existence of a phenomenon or cause in accordance with those presuppositions. In contrast, Hypothesis B requires, by virtue of the definitions of its component terms, that the supernatural agent can never be identified by perception of physical reality. Pretending to ignore these clear and essential differences in the content and intent of ideational statements as used in two distinctly different knowledge realms is a form of deception practiced by the purveyors of certain systems of religion and pseudoscience, and by science-bashers generally. This blurring of the distinction between two ontological/epistemological fields we identify as the *fallacy of homogenization.*

A second criterion of distinction between the two hypotheses is as follows: Hypothesis A has only tentative status as ideational knowledge; i.e., it is ephemeral, or transitory. It must in time be corroborated to a satisfactory level, or on the other hand judged highly improbable and subsequently dismissed, modified, or replaced. Simply put, the fate of the scientific conjecture is to become (or fail to become) an observed reality. Hypothesis B, in contrast, has a permanent status as absolute truth within its belief system. It is invulnerable by definition of the supernatural knowledge realm in which it resides. Moreover, any number of contradictory hypotheses of a supernatural content can be put forward and all of them will be invulnerable to falsification on any grounds whatsoever; all of them can coexist forever with no necessity that any of them must ultimately be rejected.

Essentially the same statements can be made of ideational propositions in the field of ethics and morality, where contradictory beliefs have equal status, and in the arts, where opposite opinions on quality of artistic expression can be sustained. Similarly, in pure mathematics and formal logic an initial (a priori) axiom can be asserted to be either true or false, and thus lead to two contradictory conclusions that are both logically correct.

A Set Model of the Knowledge Fields

We now present a graphic display of the fields of knowledge, using the elementary set theory of mathematics to define its relationships. We limit ourselves to seven discrete kinds, or *elements,* of knowledge: one is identical with Class I; the other six comprise Class II. The seven are as follows, though we do not number them because they form neither a logical succession nor a hierarchy:

Class I: Perceptional Knowledge (one element)

Science and History

Class II: Ideational Knowledge (six elements):

Mathematics and Logic

Religion

Ethics and Morality

Sociopolitical Ideologies

Aesthetics

Pseudoscience

Limitation to a universe of seven elements is dictated by the graphic display we have chosen (Figure 9.1). It is a figure consisting of seven identical hexagons, six of which surround a central hexagon. All seven fit perfectly together. Each hexagon in the outer ring is contiguous with three of the other six. The central hexagon is contiguous with all the other six, and we assign that position to Class I because science is our first concern and we wish to examine each of its interfaces with the other six fields. (Perhaps you already appreciate that if we used seven nested circles, touching but not overlapping, we would be showing a Venn diagram, and the same description would apply.)

There are two *sets* (the classes I and II). Set I contains as its one element Science and History; Members of Set II contain the other six named fields. Both sets are limited, or finite. Together they comprise the *universal set*, or *universe of discourse*. The relationship of the two sets is that they are *complementary*, relative to that universe. One a priori postulate or presupposition is that neither a *union* nor an *intersection* of the two sets is possible, and thus the two are *disjoint sets*. Set I consists of empirical knowledge; all six members of set II consist of ideational knowledge.

Our arrangement of the six elements of Set II filling the space surrounding Set I is not entirely arbitrary, nor is it mandatory. We have put the three "oughts" together as a subset: Religion, Ethics and Morality, and Sociopolitical Ideologies. All three are telling us what we ought to be thinking and doing, and beyond that, whether what we do is good or evil. Aesthetics stands by itself in that it passes quality judgments of beauty, or of good or poor taste, on the products of the arts and literature, but does not judge them to be morally right or wrong, good or evil. Mathematics and Logic is also a loner, telling us neither what is nor what ought to be, but simply what could logically exist in a rational never-never land; i.e., what "could be." Pseudoscience has no redeeming message; it is simply false science, or "false-is," but invites the prescription that it ought to be trashed.

And now you ask: Where is philosophy in this picture? Philosophy as a form of knowledge can be visualized as a matching seven-hexagonal figure that can be placed as a transparent overlay upon what we have already presented. We need only add the words "Philosophy of . . ." to the labels in each hexagon. In terms of set theory, we now have a *one-one correspondence* of the two layers. Together, they comprise a super-universe of discourse.

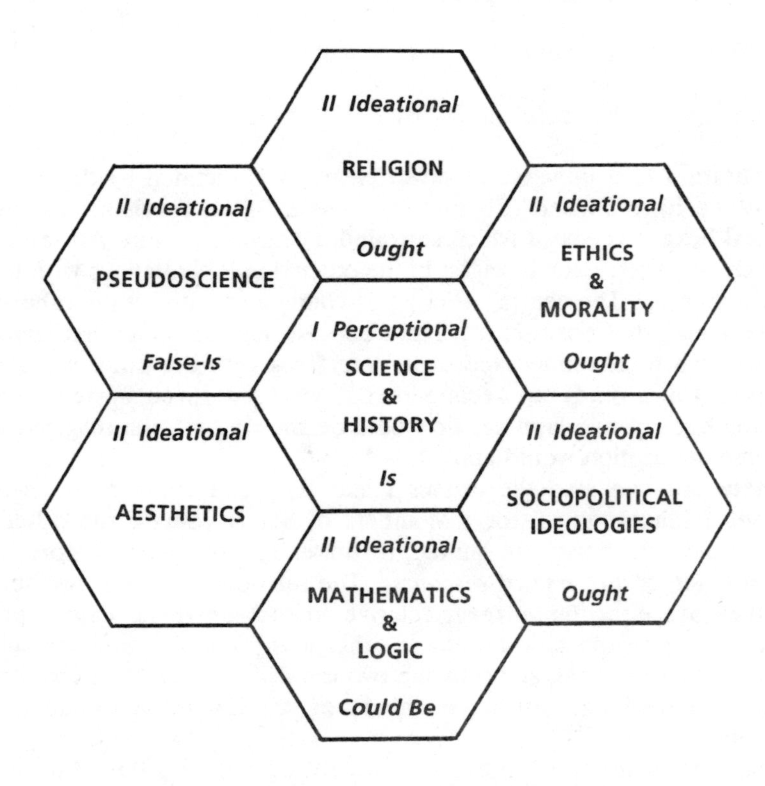

Figure 9.1 A set model of the seven fields of knowledge, each represented as a hexagon. The perceptional field of Science & History occupies the center, surrounded by six ideational fields.

We offer this pictorial model with the caveat that for a full and coherent explanation you must read the full text of this chapter and all those that follow in Part Two. The seven-hexagon model is no more than a simplified pictorial guide—a stylized sketch-map, you might say, of the territory of all knowledge. If each hexagon is like a separate nation on the map, the interfaces of the hexagons can be thought of as international boundaries—some peaceful, others heavily fortified and guarded. If you are a New Ager or a Fundamentalist Christian, perhaps you will want to pin this diagram to the wall and throw darts at it. Science and History makes a nice bullseye!

Credit

1. From Mario Bunge, *Treatise on Basic Philosophy*, vols. 3, 5, and 6. Copyright © 1977, 1983 by D. Reidel Publishing Company, Dordrecht, Holland. Reprinted by Kluwer Academic Publishers. Used by permission of author and publishers.

CHAPTER 10

The Nature and Place of Logic and Mathematics

Here we examine the placement of formal logic and pure mathematics within Class II, Ideational Knowledge. That move may seem a radical departure from Mario Bunge's system, which puts formal science among the research (inquiry) fields. Is this transfer outrageous? Not when we have made the basic distinction between the two classes on grounds of perceptional versus ideational modes of gaining knowledge.

The Placement of Formal Logic and Mathematics

Considering formal logic first, I refer to a half-century-old standard college textbook on logic and the scientific method; it has been on my library shelf nearly that long as a relic of university days. The authors were two Columbia professors of philosophy—Morris Cohen and Ernest Nagel (1934). Their work is scholarly and thorough, and it still pops up occasionally as a cited reference in recent philosophical works. In their preface, these authors comment on the epistemic nature of logic:

> The book has been written with the conviction that logic is the autonomous science of the objective though formal conditions of valid inference. . . . The traditional view of logic as the science of valid inference has been consistently maintained, against all attempts to confuse logic with psychology, where by the latter is meant the systematic study of how the mind works. (1934, pp. iii-iv)[1]

Calling logic "science" is a red herring, but focusing on the adjective "autonomous" keeps us on the trail. Substituting "study" for "science," it reads: logic is the autonomous study of the objective though formal conditions of valid inference. As the second sentence of the quote clearly states, in practice, all terms and relationships in formal logic are entirely free of perceptional or observational content. As in pure mathematics, symbols quickly substitute for real entities in the statement of propositions, which then concern nonexistent objects, and this fact "does not militate in any way against the objectivity of the relation of implication" (1934, p. 20).[1] Comparing logic with pure mathematics, these authors observe that in both fields "we inquire only as to the implication of our initial propositions, without regard to their truth or to whether their subjects are existent or nonexistent, real or imaginary" (p. 20). On pure mathematics they say: "No linguistic fiat or resolution to think differently can change the truths discovered or deduced in such fields as the theory of prime numbers. And this is true of all rigorous logical deduction" (p. 20).[1]

Mario Bunge's assessment of the position of logic and mathematics in their relation to epistemology is expressed as follows:

> These two formal sciences provide the backbone of all the other sciences but they do not investigate cognition and they are not in a position to discriminate factual truth from factual error. In short, they are not members of cognitology. To be sure, logic used to be regarded as the *organon* of inquiry and thus as identical with methodology. But we have learned since the beginning of the modern period that logic may at most discover (logical) error: it is not a guide of inquiry but a constraint on it. We now tend to conceive of logic as the intersection of mathematics and philosophy, as well as a tool of conceptual analysis and criticism in all cognitive fields. (1983a, p. 12)[2]

I find in the words "not members of cognitology" the clue I seek in removing formal logic and pure mathematics from the Class I fields of perceptual knowledge, which depend on cognition of the external world.

Reinforcing the conclusion that logic belongs in the transempirical ontological realm are a number of statements offered by Cohen and Nagel at various points in their treatise:

> The fact that logic is concerned with necessary relations in the field of possibility makes it indifferent to any property of an object other than the function of the latter in a given argument. (P. 12)

> In the calculus of propositions all propositions are considered only with respect to their *truth values* and not for the *specific meaning* of what they assert. (P. 127)

> No proposition [in logic] can be demonstrated by any experimental method. (P. 129)[1]

Turning briefly to pure mathematics, much the same description of its epistemology and ontology applies to it as to logical analysis. Cohen and

Nagel devote a text section to pure mathematics (1935, pp. 129-50). First, they examine the nature of axioms, which in both logic and mathematics are fundamental propositions assumed to be true and used to build theorems. In logic, they express the conditions of valid inference, but they are assumed to be true without the need to be demonstrated (proved) (p. 79). Axioms are "logically prior" to theorems (p. 132). Clearly, axioms and the theorems derived from them are in the transempirical realm. An answer to the question of whether axioms are materially true "cannot be given on grounds of logic alone, and must be determined by the special natural science which empirically investigates the subject matter of such axioms" (p. 132).[1] Continuing, they write:

> But it must also be admitted that the material truth or falsity of the axioms is of no concern to the logician or mathematician, who is interested only in the fact that the theorems are or are not implied by the axioms. It is essential, therefore, to distinguish between *pure mathematics*, which is interested only in the facts of implication, and *applied mathematics*, or natural science, which is interested also in questions of material truth. (Pp. 132-33)[1]

> What we have called pure mathematics is, therefore, a *hypothetico-deductive system*. Its axioms serve as hypotheses or assumptions, which are entertained or considered for the propositions they imply. (P. 133)[1]

Of course, we place applied mathematics, as defined above, within the fold of science and technology as fields of perceptional knowledge. This transfer comes about when the symbol for any variable quantity (x, y, or z) is specificially defined in terms of time, length, mass, or some combination thereof, thus:

Pure mathematics: $y = x \bullet z^2$

Applied mathematics: $E_k = M \bullet V^2$

where E_k is kinetic energy $[\, ML^2T^{-2} \,]$, M is mass,
V is speed $[\, LT^{-1} \,]$, and V^2 is $[\, L^2T^{-2} \,]$

It is perhaps interesting that pure mathematics, while being a trans-empirical field, provides (with logic) the undergirding of natural (empirical) science, and this function is one of the necessary conditions of natural science. Does this mean that science rests, in part at least, on ideational knowledge? That thought should get a rise out of the scientist who eschews anything nonmaterialistic!

That we choose to place formal logic and pure mathematics in the class of ideational knowledge in no way implies that it must buddy-up with other fields in this group, such as religion and the value systems of ethics and sociopolitical ideologies. Indeed, a necessary characteristic of religion and the value systems is that they ignore, defy, or abuse logic. On the other hand, logic and mathematics provide one of the necessary foundations of science, surely equal in importance to the requirement of being undergirded by its ontological first principles and its fundamental natural laws.

Axioms in Logic, Mathematics, and Science

The subject of axioms in logic, mathematics, and science deserves special attention. *Axiom* has several synonyms, or closely synonymous equivalents, among them "presupposition," "primitive proposition," "assumption," and "initial postulate." For science, we have already presented the set of axioms necessary for perceptional knowledge as principles of scientific ontology. These are axioms that describe the realm of external reality. They tell us that such a realm exists and what is to be implied by the word "thing." Sets of axioms are also required in logic and mathematics, but they have an entirely different flavor from the principles of scientific ontology, because they refer to an entirely different ontological class. What is more, the axioms of classical and mathematical logic are also essential to acquiring empirical knowlege. We can go even further and say that the axioms of logic form the necessary foundation of all rational discussion, and that to contravene them is to generate nonsensical statements that must simply be dismissed.

Like mathematics—Euclidean geometry, for example—logic has its axioms or primitive propositions that are accepted a priori as valid. These principles apply in full force to mathematical logic and to all empirical propositions of science, as well. Called the *Laws of Thought* from the times of Plato and Aristotle, they are also called "principles of thought." The word "thought" is, however, misleading in either title, because they have nothing to do with neurophysical processes of thinking. They are primitive propositions, three in number, deemed necessary (and perhaps sufficient, as well) for the formulation of all other propositions and chains of related propositions. The three are as follows (Cohen and Nagel 1934, pp. 181-82; Angeles 1981, p. 153):

1. Principle (Law) of Identity: If anything is *A*, then it is *A*. If proposition *p* is true, then *p* is true.

2. Principle (Law) of Contradiction: Nothing can be both *A* and *not A*. Proposition *p* cannot be both true and false at the same time and in the same respect.

3. Principle (Law) of the Excluded Middle: Anything must be either *A* or *not A*. Either *p* is true or *p* is false.

Philosopher Peter Angeles describes these principles as follows:

> They have been regarded as *ontologically real* (describing the ultimate features of reality); as *cognitively necessary* (no consistent thinking is possible without their use; all coherent thought, and all logical systems, rely upon them for justification; their denial presupposes their use in denying them); as *uninferred knowledge* (the immediate and direct result of a rational examination of the relations of timeless universals). In modern times, they have been regarded as only three among many *principles*, or *rules of inference* that can be invented and used in logic; or as definitionally true (tautologous) and hence irrefutable. (1981, p. 153)

In the calculus of classes, a part of mathematical logic, the same three principles, along with seven more, define explicitly the nature of operations and relations of classes. To restate the principles in this framework, we need to set down the following definitions and relations with respect to classes:

> *Class*: "A group of individuals each having certain properties, in virtue of which they are said to be members of the class." (Example: The class "even numbers" contains the *set*, or collection, of all even integers.) Classes are designated by the letters *a, b, c*, etc. The *universe of discourse* is the entire domain of possible classes, and is symbolized by *1*; it is useful when we wish to relate one particular class to all other classes. We are free to limit the domain of discourse to whichever classes we chose to discuss. A *null class* (or zero class), though logically definable or imaginable, is without any members; it is designated by *0*. (Cohen and Nagel 1934, pp. 122-23)[1]

Certain relations between two classes, *a* and *b*, can be summarized as follows in terms of the operations performed on them. (Note that symbols used here may differ in meaning from those in algebra.)

- Logical multiplication: a × b (or ab); reads as "a and b."
- Logical addition: a + b; reads as "either a or b" (used to designate all members who are individuals of both classes).
- Inclusion: a < b; reads "a is included in b."

- Equality: $a = b$; reads "a is included in b, and b is included in a" (i.e., they have the same members).

The three principles of thought are expressed symbolically as follows, using the above relationships:

1. Identity: for every class, $a < a$. "Every class is included in itself." It follows that $a = a$.

2. Contradiction: $a \times \text{not-}a = 0$. "Nothing is a member of both a and not-a."

3. Excluded middle: $a + \text{not-}a = 1$. "Every individual in the universe is either a member of a or not-a."

How do axioms come to exist? Obviously they are statements generated by the human brain and must come from somewhere—either arising spontaneously by some mysterious process, or derived from other knowledge previously stored in the brain. Axioms are themselves propositions, no different in structure from those propositions that follow from the axioms. The traditional view of axioms, such as those of Pythagoras, until nearly a century ago is summed up in the word "self-evident." All of us have used this term in reference to many notions or propositions that seem to need no explanation to justify their validity. The American Declaration of Independence, enumerating the divinely granted rights of citizens, announces: "We hold these truths to be self-evident . . . ." In philosophy, however, such beliefs are not satisfactory as explanations.

Cohen and Nagel discuss the questions of origin and validity of axioms at some length (1934, pp. 129-33, 141-47). On the proposition that axioms are self-evident they write:

> But this view is a rather complacent way of ignoring real difficulties. In the first place, if by "self-evidence" is meant psychological obviousness, or an irresistible impulse to assert, or the psychological unconceivability of any contrary propositions, the history of human thought has shown how unreliable it is as a criterion of truth. (P. 131)[1]

They note that many propositions once regarded as self-evident are now known to be false. Reliance on such "fundamental intuitions" has led to many failures. "But whether a proposition is obvious or not depends on cultural conditions and individual training, so that a proposition which is 'self-evidently true' to one person or group is not so to another" (p. 131).[1]

Perhaps there is no satisfactory answer to the question: How can the truth of axioms be established? On the other hand, perhaps a better understanding of their nature will reveal an answer. Cohen and Nagel bring some helpful ideas to bear on this problem. They point out that although we place the axioms first in order of statement, logically ahead of the dependent propositions, this is not necessarily the order in which they were first formulated. Perhaps the *temporal* order (succession in time) was the reverse of the logical order. They show that in the case of Euclidian geometry, many of the theorems (propositions) of Euclid were known long before he lived and were thought of as independent propositions. "The axioms were thus in fact *discovered later* than the theorems, although the former are *logically prior* to the latter" (p. 132).[1] The process of axiom formation seems to be one of searching through a large accumulation of existing initial propositions and picking out those bare statements both common and necessary to all the others. Having completed this selection process, we have a collection of propositions (axioms) that carry two important qualities: fertility and independence. "Consequently, *fertility* is one property which axioms should possess; this means that they should imply many theorems" (p. 143).[1] As to independence, "A set of assumptions [axioms] is *independent* if it is impossible to deduce any one of the axioms from the others" (p. 143).[1]

Another ray of hope for establishing the validity of axioms lies in the observation that "a proposition which is an axiom in one system may be a theorem in another" (p. 141).[1] Where in one system, we find a set of propositions (theorems) logically demonstrated, and see that they are identical with a set of axioms in another system, the latter set stands validated "by the same token," as it were. Using this principle, despite its appearance of circularity (i.e., being tautologous), Cohen and Nagel claim that "there are no *intrinsically undemonstrable propositions*" (p. 142).[1] They point out that what is a logically prior proposition in a given system can always be found in the position of the logical successor of a prior proposition in another equivalent (but not identical) system; and vice versa.

Why dwell at such length on trying to establish the logical validity of axioms? If it cannot be done, all of science is threatened by a charge that it rests on axioms that must be accepted solely on faith. This means that while the house of science is solidly built of naturalistic parts, its foundation is constructed of transnatural materials incapable of supporting the overlying structure. Christian fundamentalist creationists are quick to seize on this conclusion, because it gives power to their two-pronged attack on science, claiming that (a) both science and religion are based on blind faith and thus science is really a form of religion; therefore (b) both science and their brand of religion should be taught on an equal-time basis in public schools. Western theistic religion makes itself immune to counterattack on the same lines by requiring belief (blind faith) in a supernatural realm, the contents of which are unknowable in any empirical sense.

An Example of Formal Logic in Scientific Reasoning

We offer an example of how formal logic underlies the scientific method, and more broadly, all investigations falling within the fields of perceptional knowledge. The subject under which the example falls is the syllogism and false uses thereof, as treated by Cohen and Nagel (1934, pp. 96-100).

One valid form of the syllogism reads: If A is B, then C is D; A is B; therefore C is D. In conventional symbols: $p \supset q; p; \therefore q$. This reads: "If p is true, then q is true; p is true; therefore q is true." As an example, if naturalistic organic evolution is true, then fossils should be found in such-and-such sequence in strata; evolution is true; therefore, fossils will be found as predicted. The antecedent ("A is B") in the hypothetical major proposition is the scientific hypothesis being put forward; the finding of fossils in a certain sequence (e.g., according to homology) is a deduced consequence ("C is D"), but finding them in the sequence predicted will at best serve only as passive corroboration of the hypothesis.

We consider next an invalid form of the syllogism in which the affirmation of the consequent in the minor proposition is wrongly used to infer the truth of the antecedent of the major proposition. The form of this invalid syllogism would be: $p \supset q; q; \therefore p$ reads. "If evolution is true then fossils should be found; fossils are found; therefore, evolution is true." Concerning this invalid argument, Cohen and Nagel state:

> It is therefore a fallacy to affirm the consequent and infer the truth of the antecedent. . . . It is sometimes committed by eminent men of science who fail to distinguish between necessary and probable inferences, or who disregard the distinction between demonstrating a proposition and verifying it. For example, if the theory of organic evolution is true, we should find fossil remains of extinct animal forms; but the discovery of such remains is not a proof, is not conclusive evidence, for the theory. (1934, pp. 98-99)[1]

The point being made is that as long as one or more alternate explanations to evolution can be put forward for the origin of fossils in the sequence in which they are found, the fossils constitute no proof of the theory of evolution. Modern fundamentalist creationists, for example, call on divine recent and sudden creation of all organisms followed by the hydrodynamics of a one-year flood to explain the fossil succession as it is observed. Others of religious leanings, but relatively more liberal in thought, favor continual divine intervention to guide organic evolution (continuous creation); others of a nontheistic view postulate a teleology in which a mysterious supernatural force (entelechy, élan vital) acts to move evolution toward its supposed goal. As long as such alternatives can be put forth as possible explanations, evolutionists cannot make the fiat claim "evolution is a fact," as have distinguished scientists such as Norman D. Newell and Stephen J. Gould (see Strahler 1987, p. 329). To do so is to engage in a logical fallacy.

A second fallacy in the construction of syllogisms is to deny the antecedent and infer the falsity of the consequent (Cohen and Nagel, p. 100). In symbolic form: $p \supset q$; not-p; ∴ not-q. In words: "If evolution is true, then fossils occur . . . ; but evolution is false (contrary to Genesis); therefore, fossils do not occur (as evolution requires)." This fallacy is used by the fundamentalist creationists.

On the other hand, our syllogism about the scientific hypothesis can be recast in a logically valid form as a test of its validity. We use the form that reads: $p \supset q$; not-q; ∴ not-p. We are denying the truth of the consequent, which must therefore deny the truth of the antecedent. In our example, we would be saying "but fossils are not found in the predicted order; therefore the hypothesis of organic evolution stands falsified." This is the well-known falsification criterion of Sir Karl Popper, for decades the standard fare for scientists under logical empiricism. Creationists use this mode of the syllogism when they declare the established stratigraphic succession with its fossil sequence to be grossly in error, and hence that evolution stands falsified. Their argument is logically correct, and can be refuted only on observational (empirical) grounds.

This exercise in logic raises an interesting question: Should creationists be granted equal time in the science classroom to present a logically valid argument that could negate evolution if relevant field evidence could be marshalled? In tit-for-tat, should evolutionists be allowed to teach the logically invalid argument that the fossil record proves evolution to be true, which thereby commits the logical fallacy of affirming the truth of the antecedent from the truth of the consequent? Perhaps we would need to establish as a rule of debate that only logically valid arguments are to be allowed in the science classroom.

Such musings aside, it seems clear enough that the epistemic fields of perceptional knowledge are undergirded by the principles of formal logic, and this must be a necessary condition of operating in those fields.

The Origins of Numbers and Number systems

The nature of an epistemic field can sometimes best be revealed by the history of its cultural evolution. So we turn now to a review of the historical development of both logic and mathematics. It seems best to reverse that order, starting with mathematics. We will find that logic underwent a gradual evolution from the purely language-based Aristotelian logic to the highly abstract symbol-based modern mathematical logic, which is to say that logic became increasingly mathematicized. The entanglement of logic and mathematics throughout the development of both is so nearly total as to defy separation, but we will try to trace two threads separately, if that is possible.

Western mathematics in its earliest forms as developed by the Babylonians and Greeks consisted of two distinct areas. One was strictly empirical, or observation-based, as well as being highly practical: the counting (enumeration) of things and measurement of things in terms of their absolute magnitude (length, distance, volume, weight). This activity of quantitative evaluation, which was essential in commerce, surveying, and navigation, required an integer number system plus some arbitrary measuring sticks (or pots, or weights) to serve as things that were to be counted. The enumerative sequence or series began with the numeral 1. Arithmetic made possible under this enumerative system included addition, subtraction, multiplication, and division. Subtraction of two sets of equal numbers yields "nothing" or "zero," (e.g., from a set of 5 marbles take away 5 marbles, and you are left with "no marbles"). The problem of zero becomes very important in applied arithmetic, and we will refer to it shortly.

The other form of mathematics in this early period was what is now known as Euclidean geometry developed by Euclid, a Greek mathematician at Alexandria, Egypt, ca 300 B.C. It dealt with relationships between and among points, lines, and areas. This system of knowledge was logic-based; it required a set of axioms, from which numerous propositions were derived and tested. Although Euclid's *Elements* included what is often described as a "theory of numbers," it was in the sense of geometry, or geometry-based. Points, lines, areas, and volumes were ascribed no finite widths or lengths in the quantitative/scale sense required for the practical system. Such enumerative aspects of this geometry as were included dealt not with absolute measurement, but with relative lengths, areas, or volumes.

Two centuries earlier than Euclid, however, was the Pythagorean School, a Hellenic secret society organized by Pythagoras about 500 B.C. Their mathematical studies included the logical, philosophical, and mystical aspects of a theory of numbers; it used the integer series starting with 1 but did not recognize "zero." From that school we recognize today "Pythagoras's principle": "Empirical observations cannot furnish pure mathematical proof, which must be provided instead by deductive chains of reasoning" (Kramer 1970, p. 19).

The Pythagoreans had also been deeply interested in geometry, and they sought to relate numbers to geometrical forms. Numbers were represented by dots, grouped into geometrical patterns, such as triangles and squares. The case of the ratios of sides of triangles, an example of early number theory, is illustrated in Figure 10.1, showing a right triangle with sides in the relative lengths of 3, 4, and 5. Both dots and areas are shown. You can enlarge or reduce this diagram to any degree you wish, but the information remains the same, as expressed in the theorem it illustrates: the sum of the squares of the two legs of a right triangle equals the square of the hypotenuse.

How did "zero" get introduced into our practical number system and into pure mathematics? Take the practical system first. Mathematician Tobias Dantzig, in his book *Number: The Language of Science*, reviews the various repeating integer sequences that have been used in what he calls "number language" (1954, pp. 11-17).[3] Although the binary system (base two) may have been the most ancient of them all, and was extolled by Leibniz, it never caught on—until the age of the electronic computer, that is. Instead, the five human fingers, one set on each hand, seems firmly established as the ancient source of our decimal system (base ten). For our amusement, Dantzig asks what our history of culture would have become if humans had evolved with only "inarticulate" fingerless stumps for hands. He sees the decimal system, as a "physiological accident" (p. 15). If Providence is responsible, he adds, we "will have to admit that Providence is a poor mathematician."[3] Almost any other number except a prime would have been better in terms of its divisibility. The duodecimal system (base 12) was promoted by Buffon (1707-1788) for universal adoption, as it has 4 divisors, whereas 10 has only two. Note that our time system of hours, minutes, and seconds is duodecimal. Our compass bearing system of 360 degrees of arc uses a sexagesimal system (base 60), divisible by 2, 4, 6, 10, 12, 15, 20, and 30. Lagrange (1736-1813), on the other hand, opted for using a prime number, such as eleven, but that idea fizzled despite its mathematical elegance. Concludes Dantzig: "So may the decimal system stand as a living monument to the proposition: Man is the measure of all things" (p. 17).[3]

Positional numeration is a process of "lining things up" in an order, as contrasted with having them grouped in random nonlinear aggregations. When placed in line, the things can then be more easily counted (enumerated) and the count tallied. Tallying was first done by making a series of notches on a stick or scratches on a rock, or by placing objects such as nuts or pebbles in a row. In these records, there is no such thing as zero, and it would not occur to the recorder to put down a "zero" notch before making the notch designating the "first thing." The Greeks

---

[3] Reprinted with the permission of Macmillan Publishing Company from NUMBER: The Language of Science, Fourth Edition by Tobias Dantzig. Copyright 1930, 1933, 1939, 1954 by The Macmillan Company. Copyrights renewed © 1958, 1961, by Anna C. Dantzig, 1966 by Henry P. Dantzig and George B. Dantzig, and 1982 by Mildred B. Dantzig.

had no number "zero." The Latin number system (Roman numerals) doesn't contain zero; ten is X, twenty XX, fifty L, one hundred, C.

How and why, then, did "zero" come to be introduced as a necessary number? A clue lies in Dantzig's chapter title "The Empty Column." Tally devices, such as the counting board and the abacus provide the explanation. Almost everyone has played with an abacus, or at least seen one being used. The basic idea behind the abacus is shown in Figure 10.2A; a set of parallel rods held in a frame carry sliding beads, ten of them for each rod. The rod is longer than the beads it carries, so the beads on one rod can be divided into two groups with an empty space between. At the start, all beads are placed in the "nothing" position, against the top of the frame. As counting or adding takes place, beads on the rod farthest right are transferred down to the opposite end of the rod. What do you do when the number of beads reaches 10? Slide all beads back up to the start

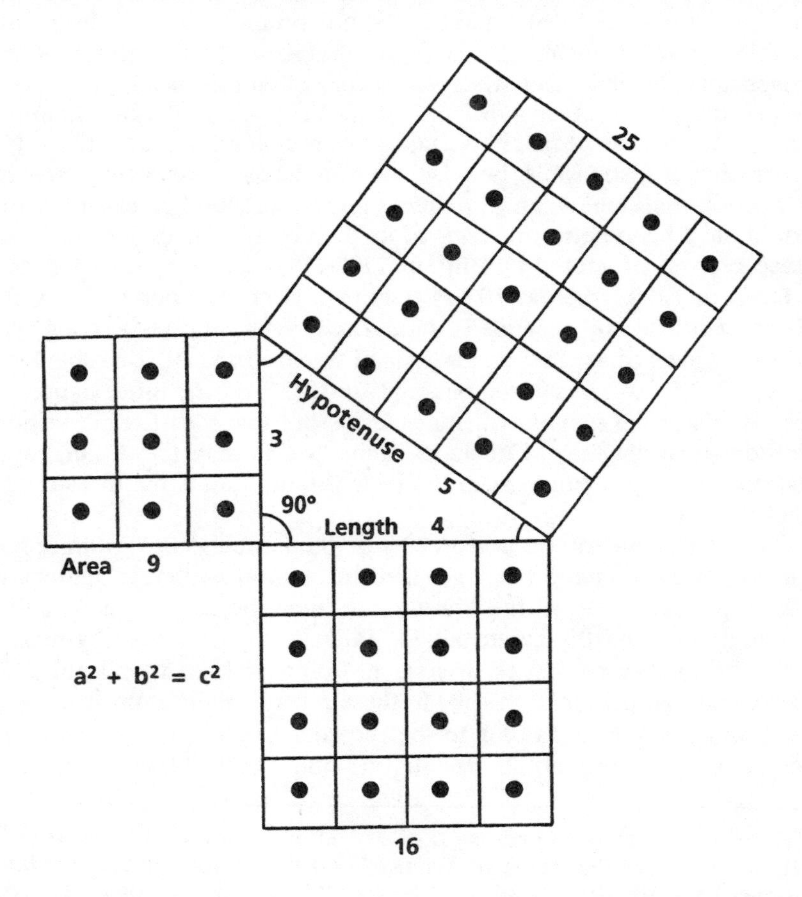

Figure 10.1 Squares of the sides of a right triangle with lengths of 3, 4, and 5, shown by both dots and squares.

position and bring down the lowermost bead in the column next left, i.e., in the column of 10s. This is the same thing our pocket calculators do. A symbol is then needed to *record* that the empty first column is indeed empty. It could be any symbol you would like to select—a black square or dot, or a picture of a bird or fish. Zero (0) has become the symbol of choice, and is has also become a number in its own right. To understand why this should be so, suppose that your abacus shows, from right to left, no beads on rods *a* and *b,* three beads on rod *c*, and four beads on rod *d*, as in Figure 10.2B. How would you write that total on a slate or on paper?

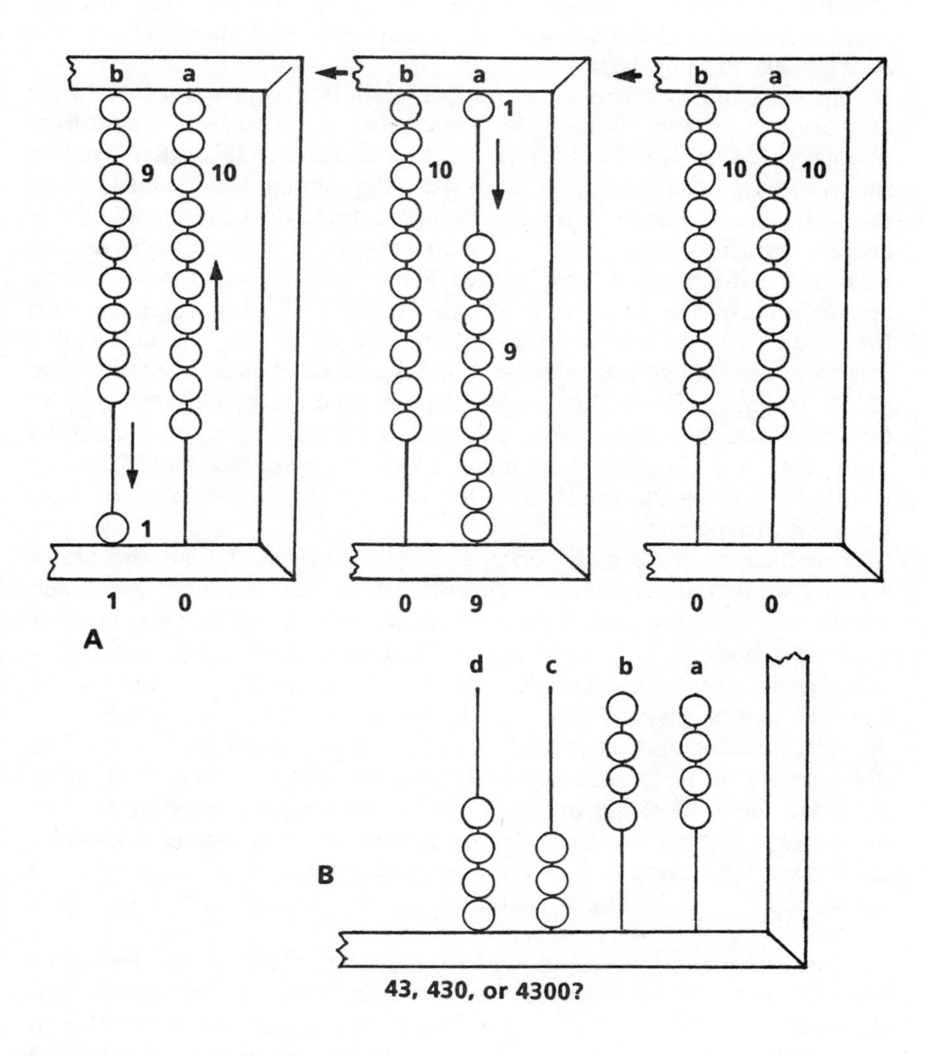

Figure 10.2 Principle of the abacus.

With no zero number available you have no choice but to record 43. But by introducing the symbol for zero, you can correctly record it as 4,300. Each zero, then, symbolizes a power of ten applied to the number or numbers to the left: 4,300 means 43×10^2.

Note that your entire available number sequence (class) still begins with the number one (1). We can say that the number 1 is the ancestor of all the following successive integer numbers. What comes before "1," you ask? The answer, "Nothing," or "Absence of any member of the class of the universe of numbers." So then you ask: Why does modern mathematics designate the same symbol "0" as the start of the number sequence, serving as the ancestor of 1 and all following numbers? We will address this question later.

The meaning of zero as a number, denoting a power of ten first appeared in Europe during the Thirteenth and Fourteenth centuries (Dantzig 1954, p. 32). The idea of zero as a numeral to fill a blank column seems to have come to Europe from India through the Hindu word *sunya.* Actually, our modern Western numerals 1 through 9 originated in the ancient *Brahmi* numerals, where zero appears as a small circle or dot following 9 (Menninger 1969, p. 418). We have reproduced them here in approximately the original form, as Figure 10.3. According to Karl Menninger, the oldest known written zero is in the Hindu Gvalior inscription of A.D. 870, in which it appears as a small circle in the number "270" (p. 397, Figure 226). The Brahmi numerals were adopted in modified form into East Arabic, still used in Turkey with zero indicated by a dot (pp. 417-19). European numerals of the Fifteenth and Sixteenth centuries, bearing the small circle for zero appear to have evolved from the West Arabic numerals.

The history of zero is perhaps more complicated than the above account would have it. Menninger also traces the use of zero to the Greek astronomer Ptolemy (Second century A.D.), noting that he "was familiar with the symbol *0*—an abbreviation of the Greek word *oudén*, 'nothing'—as a sign indicating a missing place . . . ." This use was in the context of the Babylonian sexagesimal system (base 60) as applied to the notation for degrees, minutes, and seconds of arc, where the symbol served to indicate the absence of an integer degree (Menninger 1969, p. 399). This usage preceded the earliest use of zero in India, cited above, and suggests that the concept was carried from Greece to India by followers of Alexander the Great, who advanced as far east as northern India in his invasion about 320 B.C., allowing a spread of the Hellenistic culture into that country.

The Indian word *sunya*, meaning "empty" or "blank," was translated into *sifr* in the Arabic language, and was later introduced into intellectual circles in Europe as the word *cifra*. The mathematician Gauss, writing in Latin (1801), used *cifra* in the same sense. In English we have changed the word into *cipher* with the same meaning (Dantzig 1954, p. 32).

In Europe, after the arabic numerals and mathematics were introduced there arose an intellectual struggle between two factions over the two meanings of "zero." Those who upheld the traditional definition of zero as a null class and not a number became known as the *Abacists*, whereas those who used it as a number, following the Babylonian-Indian-Arabic practice, were called the *Algorists*. The struggle continued from the Twelfth to the Fifteenth centuries with the Algorists clearly winning out by the beginning of the Sixteenth century (Dantzig 1954, pp. 33-34). The zero of the Algorists, says Menninger, "liberated the digits from the counting board and enabled them to stand alone" (p. 398). Dantzig goes even further, saying: "In the history of culture the discovery of zero will always stand out as one of the greatest single achievements of the human race" (1954, p. 35).[3] This development enabled all mathematics to emerge, hence, all science.

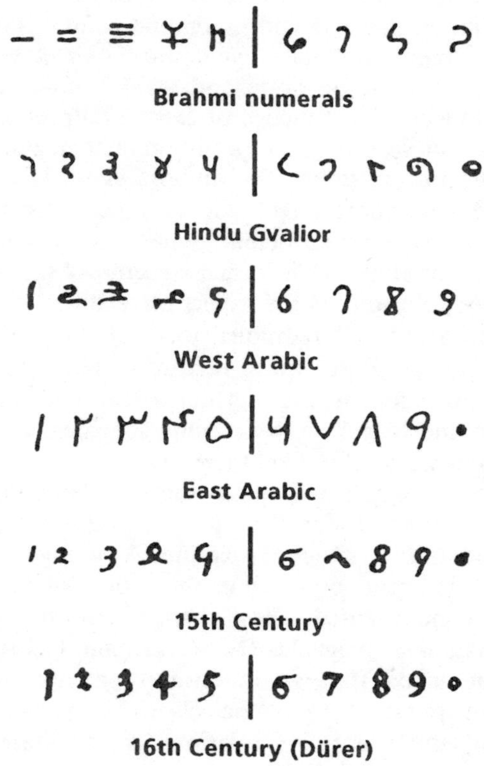

Brahmi numerals

Hindu Gvalior

West Arabic

East Arabic

15th Century

16th Century (Dürer)

Figure 10.3 The evolution of our modern western numerals 1 through 9. (From Karl Menninger, Paul Broneer, trans., *Number Words and Number Symbols*, p. 418, Figure 239, the MIT Press, Publisher. English translation copyright © 1969 by the Massachusetts Institute of Technology. Used by permission.)

While the use of zero in the number 10 and subsequent multiples thereof is fully explained and justified, there remains the problem of why the same symbol is placed before 1 as the first of the integer series where it represents the "ancestor" of a line of logical "successors." We will return to this problem later.

The Rise of Modern Mathematical Logic

The formal logic of Aristotle and those before him contains laws of reasoning that remain valid today, but by the mid-1800s their inadequacy for application to mathematics became obvious. The inadequacy lay not in what traditional logic successfully covered, but with what it did not discuss. This deficiency was felt in fields of both formal logic and pure mathematics.

Cohen and Nagel have criticized the traditional Aristotelian logic system for its limited purpose and its limited content (1934, pp. 110-11). The purposeful direction of that logic system, they state, was normative; i.e., toward what should be done and what should not be done, for example, in composing the sequence of existential propositions that make up the syllogism. In so doing, the traditionalists neglected the essential properties of the propositions. The authors state: "Because traditional logic has stressed this [normative] side of logical forms, it has failed to consider such forms with sufficient generality and has neglected to undertake a study of all possible formal structures" (1934, p. 110).[1] What was missing, they point out, is the aspect of relations between things and between propositions: ". . . traditional logic has long been remiss in not studying systematically these logical relations which are the basis for the complicated inferences in the mathematical and natural sciences" (p. 111).[1] Cohen and Nagel set down quite specifically the deficiences of the old system, but we need not list them here.

A renaissance of logic set in about the middle of the 1800s, and the remarkable feature of it was that logic and pure mathematics were virtually merged into a single discipline. You might say that simultaneously logic was becoming more mathematical while mathematics was becoming more logic oriented. These changes are usually credited to two English mathematicians, Augustus De Morgan and George Boole, both of whom were convinced that an indefinitely greater number of valid inferences were possible than the classical system could provide. Development of the theory of relations is considered De Morgan's greatest contribution, while Boole concentrated on applying symbolic conventions to all phases of the logical process. Describing Boole's epoch-making 1854 book, *An Investigation of the Laws of Thought*, Cohen and Nagel state: "It showed, with undeniable power and success, that the methods of mathematics are applicable not only to the study of quantities, but to any ordered realm whatsoever, and in particular to the relations between classes and between propositions" (p. 112).[1]

Thus the era of symbolic, or mathematical, logic was ushered in. Among philosophers who participated was the American pragmatist, Charles S. Peirce (1839-1914). On the subject of symbols in modern logic, Peirce wrote: "The warp and woof of all thought and research is symbols, and the life of thought and science is the life inherent in symbols" (Cohen and Nagel 1934, p. 117). Perhaps the most influential and productive persons engaged in transforming the study of logic to its new form were the English philosophers Bertrand Russell (1872-1970) and Alfred North Whitehead (1861-1947). Jointly they published a monumental three-volume work titled *Principia Mathematica* (1910-1913)—a work so formidable, indeed, that it is said that only two persons have ever read it from cover to cover, and often with the added quip that "the authors themselves are included in that estimate" (Dantzig 1954, p. 97). In 1907, however, Russell had published his *Principles of Mathematics* and he is justifiably regarded as the prime mover of the new school of mathematical logic, building on some necessary axioms from the Italian mathematician Giuseppe Peano (1858-1932). Russell also drew heavily on an obscure 1889 work of the German philosopher Gottlob Frege (1848-1925), whom he credits with "first logicising mathematics, i.e., in reducing to logic the arithmetical notations" (Russell 1920, p. 7).

The Theory of Natural Numbers

In his popular book, *Introduction to Mathematical Philosophy*, Bertrand Russell tells us that "all pure mathematics, in so far as it is deducible from the theory of natural numbers, is only a prolongation of logic" (1920, p. 25). Enlarging upon this concept, he states:

> All traditional pure mathematics, including analytical geometry, may be regarded as consisting wholly of propositions about the natural numbers. That is to say, the terms which occur can be defined by means of the natural numbers, and the propositions can be deduced from the properties of the natural numbers—with the addition, in each case, of the ideas and propositions of logic. (p. 4)

What are the "natural numbers"? We have only to look as far as Russell's third page to find our elusive zero firmly implanted as the first symbol in a series of whole numbers, thus: "0, 1, 2, 3, . . . n, n + 1, . . . , and it is this series that we shall mean when we speak of the 'series of natural numbers' " (p. 3).

On the preceding page, Russell has told us: "To an average educated person of the present day, the obvious starting point of mathematics would be the series of whole numbers, 1, 2, 3, . . . etc." He adds: "Probably only a person with some mathematical knowledge would think of beginning with 0 instead of with 1, but we will presume this degree of knowledge. It is only at a high stage of civilisation that we could take this series as our starting point." So much for those poor, uncivilized Greeks and Romans who had no zero in their series of natural numbers! Of course, we have all used zero in this manner from the start of our study of arithmetic, but does that familiarity justify its inclusion?

We need to challenge Russell on his smugness. A good case has already been made for the necessity of zero in our decimal system, but it's there to symbolize an empty column. Nowhere beyond 1 does zero occur by itself as a digit. The best thing we can do at this point to avoid confusion is to change over to a new symbol for the "empty column"; i.e., one that signifies a power of ten. The symbol # is available on all standard keyboards and will do just fine. The sequence of powers of ten then appears as follows:

1# 1## 1,### 1#,### . . . 1,###,### . . . etc.

So much for that faithful workhorse, the cipher, now divested of any possibility of standing alone as a number beyond 1.

Russell places the entire logical validity of the system of natural numbers on a set of three *primitive ideas* and five *primitive propositions* set down in 1889 by Peano (see Peano 1973, p. 113). Everything else in mathematics (and therefore science) hinges on these a priori propositions. They are, as Russell states "hostages for the whole of traditional pure mathematics" (1920, p. 5). That is indeed a tremendous burden! According to Russell (p. 5), Peano's primitive "ideas," or "fundamental notions," are:

0, number, successor

We should check the meanings of these words, which are in themselves terms, rather than propositions. The meaning of *successor* is quite clear as "that which follows in sequence" and refers here to the next number of a natural order. *Number* means here the entire class of natural numbers, as laid out above. In the second edition of his *Principles of Mathematics*, Russell states that "number" is synonymous with *finite integer* (1937, p. 125). The meaning of *0* remains to be defined by Peano's first five primitive propositions, which, according to Russell, are as follows (1920, pp. 5-6. See also Pickert and Görke 1974, p. 72.):

(1) 0 is a natural number.

(2) The successor of any natural number is a number.
(3) No two natural numbers have the same successor.

(4) 0 is not the successor of any natural number.

(5) Any property which belongs to 0, and also to the successor of every natural number which has the property, belongs to all natural numbers.

These five propositions are assumed, of course, but they are mutually independent, and no one of the five can be deduced from the other four (Russell 1937, p. 125). Only the first four are basic to the system; the fifth follows by mathematical induction from them. Although Peano's propositions are not logically necessary, they must be free of contradictions, if what follows is to be logically flawless (Dantzig 1965, pp. 66, 75).

Upon cross-checking Russell's version of Peano's first five primitive axioms (propositions) in other publications, I found a disturbing discrepancy. Mathematicians Edna E. Kramer (1970, p. 59) and G. Pickert and L. Görke (1974, p. 94) have used "one" (1) for Russell's "zero" (0) throughout their lists, also historically attributed to Peano. An English translation by H. H. Kennedy of several of Peano's works became available in 1973. In it is Peano's 1899 paper, "The Principles of Arithmetic, Presented by a New Method" (Peano 1973, pp. 101-134). The paper is introduced by his translator as "the first statement of Peano's best known achievement, the postulates for the natural numbers" (p. 101). Much to my surprise, I found "one" (1) used throughout, with an explanation that "1 means unity," but no mention of zero (p. 113). The only conclusion open to us at this point is that Russell took it upon himself to substitute "zero" for "one." Without zero in place, he could not have begun to build his system of symbolic mathematical logic.

I suggest that we should not accept the first proposition as Russell states it, because it is contradictory with respect to the other four. This can be done on grounds of class affinities. Russell has assumed that 0 is in the same class as all the numbers beginning with 1; i.e., that all are finite integers. This makes no sense if we consider the domain of the universe. The set of numbers starting with 1 constitutes a homogeneous class of things—finite integers that can be counted (enumerated), and this is a single class. Zero, by definition, means "empty," a concept that has held from the origin in the Hindu word *sunya*, through the Arabic *cifra*, to the modern English *cipher*. "Zero" unambiguously lies in a separate null class (a class with no members) in relation to all members of the class of numbers beginning with 1.

Russell was fully aware of our complaint, for he introduces a proposition with these words: "Take, for example the natural numbers (excluding 0) . . ." (1920, p. 17). This stated exclusion may have led him to say in the next paragraph: "We want to make one bundle containing the class that has no members: this will be for the number 0. Then we want a bundle of all the classes that have one member: this will be for the number

1." Yet, knowing full well this distinction between 0 and 1, he leaves his set of propositions in place, as given to him by Peano.

In the second edition of *Principles*, Russell redefines his "logical theory of cardinals" to include the following:

> (1) 0 is the class of classes whose only member is the null-class. (2) A number is the class of all classes similar to any one of themselves. (3) 1 is the class of all classes which are not null and are such that, if x belongs to the class, the class without x is the null class. . . . (5) Finite numbers are those belonging to every class s to which belongs 0, and to which n + 1 belongs if n belongs. (1937, p. 128)

Whereas (1), (2), and (3) make the correct distinction between 0 and the finite integers, (5) seems to obliterate that distinction by lumping 0 with the rest; it is thus contradictory.

A second kind of objection to the first proposition relates to the established meanings of words. It may be all very well for Russell to counter: "No matter what you say or think or argue, or what is the established meaning of words, I am inserting 0 before 1 as the ancestor to 1." That is a fiat he throws down upon us, and perhaps we must live with it. On the other hand, does mathematical logic allow a proposition to include clear violations of established meanings of words? Symbols have no empirical meanings, unless in applied mathematics real meanings are specified, so the answer is yes. If such violations are permitted, leading to the formulation of a bogus proposition, it must be dismissed on grounds of being nonsensical. The violation here is in setting up a false definition of "zero" as meaning "something" rather than "nothing." This kind of nonsense would not have arisen under the traditional word-logic.

Let us follow up these objections with the following demonstration: We line up on a table several discrete objects easily recognized as such and present the layout to an observer with instructions to count (enumerate) them, at the same time touching each object in turn. Our observer begins to count, naming the first object as "1" and ending with "5" as the last one. "Wrong!" calls Russell, who has been watching intently. "You should have begun your count with zero, which is the true ancestor of the set, so the last object will be 4, and not 5." This demand is tantamount to requiring that the definition of all the words signifying numbers be redefined. When this is done, you will need a revised dictionary, and upon looking up the meaning of "one" you will discover that it means "being more than one in number." Then, each normal human being can be correctly described as having one eye and on each hand four fingers. What we have exposed here is a failure to use the words "zero" and "one" in their established meanings.

To pursue this argument, let us go Russell one better. If a zero placed in front of 1 is acceptable as the ancestor number, what is to stop us from placing in front of it yet another symbol (let it be &), which we shall declare to be the ancestor of zero? Now, the new number "ampersand" is our ancestor, and our natural numbers consist of the following series:

&, 0, 1, 2, 3, . . . n, n + 1, . . .

There can be no limit to the number of new symbols thus added.

Furthermore, suppose that in the demonstration described above, we ask our observer to watch closely as an attendant picks up and removes all the several objects that were on the table. We now ask the observer: "How many objects, if any, remain on the table?" The observer answers: "None remain." We then ask: "Please record that answer by an appropriate symbol on this blank piece of paper." The observer inscribes "0." Russell objects: "That's incorrect; zero is reserved for the first of the things removed." Observer: "Then what do you suggest I write?" Russell: "Be patient and I'll think of something."

Mathematics, starting with arithmetic, has a necessary place and function for zero as an independent digit in the number system, and we should not try to get rid of it, but rather to accommodate it through a rational explanation. We have suggested above the concept of zero being a null class within the same domain of discourse as the number class; the two classes together constituting the universe. We need only accept Peano's primitive propositions as he wrote them in 1889. We need only add that zero is a symbol designating a null class that contains no members of the number class. No matter how many members may occupy the number class, we can always wipe them out by introducing and subtracting an identical number set, yielding the null class with the symbol "0."

Mathematical Induction Versus Logical Deduction

> A certain body of indefinable entities and indemonstrable propositions must form the starting-point for any mathematical reasoning; and it is this starting-point that concerns the philosopher. (Bertrand Russell, 1937, p. 129)

The analytical process used by Peano and Russell consists of first formulating a set of arbitrary entities and propositions serving as axioms or presuppositions. The mental process of deriving from those axioms any number of propositions that follow by rules of logic is described as *deductive*. Deduction can take place free and clear of empirical knowledge and produces what we have described as ideational knowledge.

A contrasting method of deriving knowledge is *inductive*, usually described as "proceeding from the particular to the general" (Dantzig 1954, pp. 66-67).[3] As explained in Chapter 1, the inductive method is used in science and requires observation that yields perceptional knowledge. An important idea here is that the validity of a scientific statement (probability of its being correct) improves (or diminishes) with repetition of the observation, whether by the same observer or by many observers. At some level of successful repetition with no failures, the proposition is extended in value to the status of a law. You might put this idea in other words as follows: "Because this outcome has been the same every time the experiment or phenomenon has been observed, we are reasonably safe in taking action as if it always will be so." This inductive process cannot, however, establish truth in any absolute sense. It is always a gamble, because at some future time an observation to the contrary may appear, weakening or even falsifying the supposed law. Let us now attempt to apply this kind of scientific induction to the Peano-Russell theory of numbers.

Induction in mathematics, when viewed as analogous to the scientific method of repeated observations, consists of actually carrying out the succession of numbers or of products of numbers to a finite length, hoping that no failure will be encountered. This may explain why *mathematical induction* has been called "reasoning by recurrence" and to express the "principle of recurrence." With the integer number succession, we do this by counting ever higher. Children are intrigued with the question: How high can you count without stopping? To one thousand? one million? one billion? Exhaustion finally sets in or time runs out. Never can any human prove the validity of the number theory by actually naming all the individuals of that population. In practice, then, the method rests on the idea that some "reasonable" number of correct solutions of a theorem, no incorrect solutions having been encountered, allows us to claim that the theorem is correct for all possible values of the variable quantity or quantities it contains.

Dantzig gives an example of an attempt to apply mathematical induction, defined as reasoning by recurrence:

> Consider the quadratic expression ($n^2 - n + 41$). . . . We set in this expression $n = 1, 2, 3, \ldots$ up to $n = 40$: in each of these cases we get a prime number as the result. Shall we conclude that this expression represents prime numbers for all values of n? Even the least mathematically trained reader will recognize the fallacy of such a conclusion: yet many a physical law has been held valid on less evidence. (1954, pp. 67-68)[3]

Dantzig follows with an example of a failed attempt at this form of mathematical induction (p. 69). Pierre de Fermat's first theorem states that (2 raised to the power 2^n) + 1 is a prime number for all values of n. Fermat showed that the theorem holds for $n = 0$ and $n = 1, 2, 3,$ and 4. Leonhard Euler, however, showed that it fails for $n = 5$, thereby disproving the theorem (see also Kramer 1970, p. 501). Dantzig summarizes:

> This process of induction, which is basic to all experimental sciences, is for ever banned from rigorous mathematics. Not only would such a proof of a mathematical proposition be considered ridiculous, but even as a verification of an established truth it would be unacceptable. For, in order to prove a mathematical proposition, the evidence of any number of cases would be insufficient, whereas to disprove a statement one example will suffice. A mathematical proposition is true, if it leads to no logical contradiction, false otherwise. The method of deduction is based on the principle of contradiction and on nothing else. (1954, p. 67)[3]

Unfortunately for students and other victims of mathematics, there is yet a third definition of "induction" besides (1) scientific induction based on observation and (2) mathematical induction based on the principle of recurrence. The third meaning is more obscure than the first two.

Russell's version of Peano's theory requires a crucial step, which is to certify the placement of the number 0 in the same class as all the other numbers. The first four of the Peano/Russell propositions can be used by logical deduction to conclude that 0 comes first, preceding 1; but here only *relations* (spatial or temporal) are stipulated. Nothing in the first four propositions says anything about the *properties* of any of the numbers. This void allows us to suppose that zero could be an apple, whereas all the other numbers are oranges, and we have already exploited that opportunity. Proposition (5) is introduced to specify properties (qualities) independent of relations. It tells us that all the numbers, including 0, are oranges; (5) is thus a necessary fiat, coming out of nowhere, and it is not logically derived from (1) through (4). This step has been called the *induction step*, and the process of applying it is called *mathematical induction* by Russell (1920, p. 6, 21). Great caution is urged at this point, because we have here a third kind of induction. Philosopher Peter Angeles defines it as follows:

> 3. A form of nondeductive inference in which the conclusion expresses something that goes beyond what is said in the premises; the conclusion does not follow with logical necessity from the premises. (1981, p. 132)

To be consistent, however, not only the fifth proposition but also the the first four of the Peano/Russell propositions are nothing else but fiats that are pure fabrications out of his imagination. Our opening quotation from Russell sums up the situation nicely. Peano's "indefinable entities and indemonstrable propositions" are not tied to any empirical knowledge, and this does concern us here. Should we not be free to confess quite frankly that number theory accepts as its base the existing number sequence (including 0) used for centuries in enumerating and calculating? If so, we can show how that counting system embodies the ancestor/successor principle. From that point on, number theory moves ahead on schedule to accommodate all branches of mathematics.

This dependence of science on a transempirical logic/mathematics may seem strange or incongruous at best, until we realize that the transfer from ideational to empirical realms it requires is accompanied by a transformation of the former systems as follows. Formal logic assumes that the universe (domain) of discourse deals with things and propositions that are only logically possible; i.e., that are transempirical. When transferred to science, the universe of discourse becomes one of physical reality, requiring capability of being observed. Pure mathematics, when transferred to science, undergoes a transformation in which the abstract variables of a transempirical realm are given physical definitions in terms of time, space, and matter. But with this ontological transformation, the same rules of pure mathemtics apply with equal force.

Beyond question the integer number system is derived from an artifact of long standing that has worked well in arithmetic. This thought leads us to wonder if nature presents us with some common phenomena that could directly inspire such a number system. Water drops of uniform size falling in regular time succession from the tip of a leaf might be one, for there is always a first drop followed by a second, etc. The growth rings of a tree, exposed in a cross section of its trunk might be an other. In both examples, the number of integers is finite. For both, a null class precedes in time the first integer event.

Mathematicians of the formalist school, to which Peano, Russell, and Hilbert belonged, would have none of this nonsense about external reality being a source of inspiration for pure mathematics. Russell did, however, seem to question that pure mathematics could keep its "purity," for he ventured that "entities which are perceived by the so-called senses . . . are not commonly regarded as coming within the scope of philosophy; but it seems highly doubtful whether any such exclusion can be maintained" (1937, p. 129).

A Number System Based on Continuously Variable Quantities

Lacking in the Peano-Russell number theory system is a concept of continuous variation of quantity. Instead it may be described as a "quantal" system composed entirely of quanta, that are discrete particles (things) each one clearly separated from the others. Each particle is imagined to be a completely bounded entity with respect to all the others. A "quantal domain" of knowledge (i.e., its universe) consists of two classes: a population class and a null class. All individuals in the population class are defined as being identical in quality definition. The null class contains no individuals of that definition. The population class contains precisely the number system that the average educated person of the present day would choose—one consisting of numbers starting with 1, and reserving 0 for the absence of any numbers. Using only these integers we can, moreover, combine them in various ways to form other discrete "quanta" or "particles" that can be formed into sequences, for example, ratios of two numbers such as 1/1, 1/2, 1/3, 1/4, ... 1/n. The Pythagoreans did just that.

To bring zero into a number class is, however, logically possible in a system consisting only of a continuously variable quantity extending toward the high end to infinity and at the low end approaching zero (nonexistence) as its limit. This system can be converted (transformed) into a quantal system such as that of Peano and Russell by creating from the continuum a class of discrete individuals. The continuum is imagined to be cut into segments, much as one might chop a worm into segments to use as bait for fish. A nylon thread known as a monofilament better suits our model, because it has no natural segmentation.

We will make use of the concepts of continuum, continuous functions, limits, infinity, and of infinite divisibility and infinitesimals. Particularly important is the infinitesimal, a variable quantity that approaches zero as a limit. We may want to call this the "continuum-infinitesimal system" of numbers. Dantzig has written:

> The importance of infinite processes for the practical exigences of technical life can hardly be overemphasized. Practically all applications of arithmetic to geometry, mechanics, physics and even statistics involve these processes directly or indirectly. . . . Banish the infinite process, and mathematics is reduced to the state in which it was known to the pre-Pythagoreans. (1954, pp. 136-37.)[3]

"Continuity" is a key word here; what does it mean? What is a *continuum*? We must reconcile this concept with that of the Peano-Russell quantal system of numbers that treats only aggregates of discrete numbers. Think first of continuity as an essential attribute of a line. In geometry we are taught that a line is generated by a moving point. Where do numbers fit on that line? Dantzig suggests that we think of a point "as a *limiting position* in an *infinite process* applied to a segment on a line" (1954, p. 141).[3]

Dantzig attempts to derive the continuum from the series of natural numbers (p. 142). The attempt is logically based on the successor axiom; i.e., repetition without limit; it is the *serial process*:

Given the sequence a, b, c, d, e, f, g, etc.,

by successive addition, a, a + b, a + b + c, a + b + c + d, etc.,

we derive the natural sequence of numbers, 1, 2, 3, 4, 5, etc.,

which can be carried out to approach infinity.

In contrast, an *evanescent sequence* approaches zero as a limit, for example, the diminishing series: $1/1$, $1/2$, $1/3$, $1/4$, . . . $1/n$. We would say: "The limit of $1/n$, as n approaches infinity, is zero." On the other hand, the increasing sequence $1/2$, $2/3$, $3/4$, $4/5$, $5/6$, . . . $n/_{n+1}$ approaches 1 as the limit: "The limit of $n/_{n+1}$ as n approaches infinity is 1." Thus, any whole rational number can be generated by sequences that converge on the number as a limit, approaching it either from lower-to-higher or from higher-to-lower; for example:

1.9, 1.99, 1.999, 1.9999, . . . approaches limit 2 from below

2.1 2.01, 2.001, 2.0001, . . . approaches limit 2 from above

Thus all integer numbers, as well as zero, are capable of being described in the same manner, namely, by a limit approached but never reached. And this also means that the limits themselves can have no existence in the empirical realm; i.e., they are abstractions and lie in the ideational realm of pure mathematical logic.

We next introduce the concept of *direction* in number theory. Mathematician Edna Kramer in her monumental work, *The Nature and Growth of Mathematics* (1970), tells us that for all his sophistication it seems strange that Pythagoras did not grasp the idea that opposite directions of travel on the same line between two fixed points could be distinguished by positive and negative signs thus: eastward from A to B as +1000 km; westward from B to A as –1000 km. Kramer observes: "One wonders whether, in this respect, he once again rejected the worldly practicality of mathematicians he met in the Middle East" (1970, p. 29). Had Pythagoras grasped the notion that the starting milestone in his journey must bear the inscription "zero km" he could have set up two sequences: Eastward, 0, +1, +2, +3, . . . +1000; westward 0, –1, –2, –3, . . . –1000. The final step from journey to theory would have been to rearrange these two sequences with respect to a shared zero point, thus:

–1000, . . . –3, –2, –1, 0, +1, +2, +3, . . . +1000

Next, we recognize this continuous series from negative to positive numbers as a *number line* that equates the notion of number to the notion of length of line (Kramer 1970, p. 30). In the figure below, the number line is shown as a *coordinate system* with the origin of the system dividing the line into two rays: *OX* (positive) and *OX′* (negative):

X′				0				X
X′	–3	–2	–1	0	+1	+2	+3	X

Time →

Integers can now be marked at equal intervals and labeled accordingly. Furthermore, other kinds of numbers that can be approached (but never reached) can be inserted; for example the irrational square-root of 2 (1.41421 . . .), or the transcendental *pi* (3.1416. . .), and negative values of the same. All conceivable ordered numbers are accommodated by the number line. These include rational fractions, powers and radicals, irrational numbers, and complex and imaginary numbers (Kramer 1970, p. 30). Each of these kinds of numbers is defined as a limit, as for the integer whole numbers.

To change metaphors, zero appears here as a divide between two "streams" that "flow" in opposite directions. This arrangement conforms with Peano's fourth proposition, so that –1 is the successor of 0. Unfortunately, science needs a continuously through-flowing stream from left to right, as the arrow below the diagram indicates. Now, –1 becomes the successor of –2, 0 the successor of –1. Switching to the empirical realm, this one-directional version accommodates the sliding of arbitrary measurement scales with respect to one another, as for example, converting the Celsius scale to the Kelvin scale, in which the latter is obtained by adding 273 Celsius degrees. For time measurements the unidirectional model is obviously superior because an arbitrary zero-time ("time-zero") can be placed wherever needed to yield an all-positive sequence.

The quantal number system also provides for certain of the above kinds of numbers, but it does so by using ratios of the integer whole numbers, there being no concept of continuum. These number ratios tend to "fill the space," so to speak, between the whole integers. In the continuum/infinitesimal system, "space" between integers is filled by an infinite number of infinitesimals; the gaps between the rational numbers are filled with irrational numbers; and the gaps between them are filled with transcendentals. So we arrive at a measure of the "distance" between successive integers through integration of all the dx-es between the limits n and n + 1. This "distance" x is not dimensionally defined, although length would be a suitable dimension, but the important conceptual point is that the "distance," Δx (delta-x), between any two integers, n and n + 1 is always the same.

How does the continuum/infinitesimal system relate to the real world?

Does it provide better models than does the quantal system? For time and space it is perfectly suited for measurement purposes, as the fundamental dimensions of time and length are imagined as true continua. The number system of limits and infinitesimals is a pure artifact readily superimposed on nature. For matter, it is a different story. A direct electric current seems to be a continuum of flow, but consists of individual electrons. Fluid flow of a river may appear as a continuum, but it consists of discrete water molecules. Quantum mechanics describes all matter as consisting of particles, or treatable as particles. This applies to the four forces, weak and strong, that are explained as being carried by particles. The mathematician David Hilbert, taking note of the quantum theory, commented as follows:

> But even energy, it was found, does not admit of a simple and unlimited divisibility. Planck discovered the energy-quanta. And the verdict is that nowhere in reality does there exist a homogeneous continuum in which unlimited divisibility is possible, in which the infinitely small can be realized. The infinite divisibility of a continuum is an operation which exists in thought only, is just an idea which is refuted by our obervations of nature, as well as by physical and chemical experiments. (As quoted in Dantzig, 1954, p. 238.)[3]

In transferring any axiom-based logical system of numbers to its normative form in the empirical world of science, the former loses none of its integrity as a knowledge category within the ideational field. What we have been driving at in this chapter on the epistemological place of pure mathematics is admirably summed up in a rather long paragraph by Dantzig, well worth quoting in full:

> The mathematician is only too willing to admit that he is dealing exclusively with acts of the mind. To be sure, he is aware that the ingenious artifices which form his stock in trade had their genesis in the sense impressions which he identifies with crude reality, and he is not surprised to find that at times these artifices fit quite neatly the reality in which they were born. But this neatness the mathematician refuses to recognize as a criterion of his achievement: the value of the beings which spring from his creative imagination shall not be measured by the scope of their application to physical reality. No! Mathematical achievement shall be measured by standards which are peculiar to mathematics. These standards are independent of the crude reality of our senses. They are: freedom from logical contradiction, the generality of the laws governing the created form, the kinship which exists between this new form and those that have preceded it. (1954, p. 231)[3]

Credits

CHAPTER 11

The Supernatural Realm: Religion and the God Concept

We turn now to other knowledge fields that can fairly be described as belief fields because they require that one believe in some concept or concept system. Recall from Chapter 1 that I criticized the news media and science writers, as well as many scientists, for using the expression "scientists believe that . . ." in introducing empirical propositions that are hypothetical with varying degrees of support from observation. That usage of "belief" cannot easily be eradicated. My practice is to define "to believe" as "acceptance of a proposition on faith alone, as being true without being supported by any empirical evidence." The definition is perhaps best illustrated by a belief in a God, in the style of the Western Judeo-Christian faiths. Theologians of that belief agree that God exists in a supernatural realm of reality, so that by definition, that belief cannot rest on observation of nature by humans. One of the difficulties that news writers create by using the expression "scientists believe that . . ." is that anti-science factions can accuse scientists of holding unsupported beliefs in their empirical propositions. That makes science appear to be scientism, and easily leads to the relativistic charge that science and religion are, after all, much the same kind of knowledge system. This charge is typical of the expressed opinions of the New Agers as well as the fundamentalist Christians—two disparate groups, neither of which has much use for the other. In this chapter and in Chapter 12 we look for criteria of distinction between the group of fields that includes religion, ethics/morality, and sociopolitical ideologies, on the one hand, and perceptional knowledge (science and human history) on the other. Of these several interfaces, the interface of religion with science is perhaps the most controversial of all, generating extremes of emotion on both sides.

"Religion" and "Religious" Defined

Before attempting to describe and delineate the place of religion among the belief fields, we need to establish the meanings of the words "religion" and "religious" and then to apply them to the discussion that follows. For guidance I turned first to philosopher Vergilius Ferm, in whose college classes I studied general philosophy as well as the philosophy of religion. He had authored our textbooks in both areas, but nowhere in those books or in his lectures did he give even the faintest hint of where he personally stood on religion and how it figured in his world view (Ferm 1936, 1937).

Professor Ferm makes clear enough that both "religion," as a generic or collective noun, and "a religion," as a concrete noun, are almost impossible to define in any acceptable manner because of the enormous diversity of belief phenomena they encompass:

> "Religion" in general, has reference to *all conceivable meanings* given at one time or other by religious people to their religious experience. It is the collective term of all conceivable meanings. (1936, p. 54)

A way out of this morass of meanings, Ferm suggests, is to focus upon the adjective "religious" as the broader term, "covering any conceivable behavior, act, situation, object, or experience involving a certain kind of participative reference, actually or potentially, on the part of an individual or group of individuals" (p. 54). He then attempts a definition of what it means to be *religious*:

> For one to be religious, we would say, is to effect in some way and in some measure, individually or socially, a vital adjustment to w(W)hatever is reacted to or regarded implicitly as worthy of serious and ulterior concern. (P. 54)

Ferm holds his definition "to be fundamental, a *sine qua non* of the genuinely religious spirit." The words "vital adjustment" he sees as "inclusive of a total response, the whole nature, reacting, affecting cognition, feeling, and will" (p. 55). The final two words might be taken to hold the key to our problem of distinguishing religion from science, namely, that the concern is *ulterior* to (beyond, or outside of) the empirical realm; i.e., that it is transempirical. On the other hand, the same definition could in that respect apply just as well to "ethical." Therefore, I would want to specify that "ulterior concern" identifies with a supernatural realm and the nonreal entities it is claimed to contain. My specification seems justified in view of Ferm's insertion of the capital "W" for optional use in "whatever," for this would suggest the possible presence of a higher authority or force as the object of reference; i.e., a god-object.

While perusing the literature on religion another philosopher came to my attention, taking up where Vergilius Ferm left off and greatly strengthening my reluctant conclusion that overwhelming diversity is the outstanding attribute of religion. My rescuer proved to be Archie J. Bahm, then professor emeritus of philosophy at the University of New Mexico. A specialist in the philosophy of religion, he had published a book titled *The World's Living Religions* (1971). His essay on this subject in the humanist journal *Free Inquiry* bears the controversial title "Humanism Is a Religion" (Bahm 1984, p. 44).

Bahm points out that the word "religion" has several legitimate meanings, a situation that makes for confusion and disagreement among humanists. Bahm tries to make things simple by reducing the meanings to two: theistic and scientific. He writes:

> In its theistic sense, "Religion is belief in god." The many variations on conceptions of god include polytheism, pantheism, duotheism, belief in the supernatural, etc. This is the most popular meaning of the term, at least in Europe and America where ethical monotheistic religions prevail. To use the term religion in the scientific sense, one must take into account that a factual study of all the world's religions will reveal that some major religions are not ethical monotheisms and that some are explicitly atheistic. (1984, p. 44)

Bahm goes on to divide the world's religions into two broad classes: (1) self-help religions, of which most are Asian, and (2) other-help religions, to which the ethical monotheistic religions belong. In the self-help religions each person must work his (or her) way to the goal of life. Many of the self-help religions involve a "law of karma" saying that one's personal intentions and actions can retard or promote progress to the goal. In the other-help religions a personal deity is held responsible for bringing an individual into the world and for helping that person to achieve the goal of life. The self-help religions are largely nontheistic, Bahm states. He gives two outstanding examples:

> Two major Asian religions—Jainism and Theravada Buddhism—are explicitly atheistic. Jains believe that eternal souls are perpetually reincarnated in accordance with the law of karma. Since one must work one's way to the goal by one's own efforts, no savior is needed. Since both souls and bodies are eternal, no creator is needed. . . . Jainism is explicitly atheistic, not only because no god is needed to create or save, but also because any appeal to a god for help has evil consequences. Theravada Buddhists believe that there

> are no souls because they believe that everything is impermanent, and souls are supposedly permanent. . . .Theravada Buddhism is explicitly atheistic, both because its basic doctrine of impermanence excludes the existence of any permanent god and because any appeal to such a god would be cheating and result in evil consequences. (P. 44)

It may take some effort on the part of Western monotheists, whether they are fundamentalist creationists or theistic scientists, to accept Jainism and Theravada Buddhism as religions. The former group sees such atheist religions as evil and its adherents as deserving of eternal damnation. But whether a specific religion is viewed as good or evil should not in any way detract from its being a genuine religion—or at least that seems to be what Bahm is saying, and it is also my position. We both look at religion with the scientific view, i.e., religion as a real phenomenon, capable of being described empirically and without emotive responses as to its being good or evil.

And now, will Professor Bahm support me on adopting Ferm's broad definition of "to be religious" as "a vital adjustment" to something regarded "as worthy of serious and ulterior concern"? Indeed, he does in this paragraph:

> My own conclusion, shaped and reshaped several times following such extensive searchings, is: Religion is concern for the ultimate values of life as a whole. Generalizations about the nature of values and of how concern about them is expressed through beliefs and practices extend the definition. But the key words are "concern for ultimate values," not "belief in god." (P. 44)

Using this broad definition of "religious" arrived at by both Ferm and Bahm easily leads us to the decision that religious knowledge is clearly ideational, and in that case nothing further would need to be said on the subject. But in so doing we would only be dismissing as irrelevant the knowledge claims of those who promote the theistic versions of religion and their theologies in direct opposition to modern science.

So from this point on we will narrow our view to the Western monotheistic religions of Judaism and Christianity, for these represent the power structure within which modern science arose. Sociopolitical dominance lay with the Roman Catholic Church, then in power, and its spinoffs in the dissident Protestant sects. For all of these versions, religion requires a supernatural realm containing an omnipotent and omniscient God who created and sustains the universe. Initially, science was dedicated to describing God's magnificent work, i.e., to natural theology. But instead, this highly successful endeavor had the effect of eroding God's creative role in a series of devastating steps. By the late 1800s, following the arrival of Darwin's theory of evolution by natural selection, God was viewed by many intellectuals as having joined the company of the unemployed of society. Since then, the task of Christian theology has become a campaign to reinstate God's position as a useful and productive agent of human society, and particularly to reinstate the teleology that the evolutionists had eliminated along with God.

Philosopher of science Tad S. Clements (State University of New York at Brockport) faced the same problem of finding a working definition of religion in its relationship with modern science. He has used much the same solution as ours in limiting the field of view:

> I propose to limit *religion*, for the purposes of the present investigation to supernaturalistic religions, that is, to religions in which belief in some reality allegedly above or beyond nature, together with special, privileged modes of apprehending and adapting to such supposed realities, predominates. (1985, pp. 548-49)

Perhaps we could look at our solution as identifying a "family matter," with religion the parent and science its offspring. We can perhaps best understand the waywardness of this precocious child and its rejection of the domineering parent through their evolutionary history. Set in its ways, the parent would not change, but neither would the child conform, so the two parted company. Philosopher Mario Bunge put his finger on the root cause of the separation when he pointed out that whereas science constantly changes in response to new information, religion (along with all pseudoscience and other belief fields) changes little or not at all, and that when change does occur "it is forced by controversy and outside pressure rather than by the results of research" (1984, pp. 39-40).

Ontological Models Relating Science to Religion

Before continuing our examination of the science/religion interface in the western European culture we need to establish a broad philosophical view of belief systems that specifically recognize or deny a supernatural (transnatural) realm. The philosophical field of ontology, which examines the nature of reality, provides for a number of different models that can be helpful here.[1] Six such models are pictured in Figure 11.1. Each model must concern itself with the origin of the universe (cosmology) and, more specifically, with the event of initiation of life on Earth (biopoesis), and with the diversification of that life into numerous forms, including the human species. Each model must deal with the nature of reality and the role of time. The role of purpose or design may also be included, but only where required by the model.

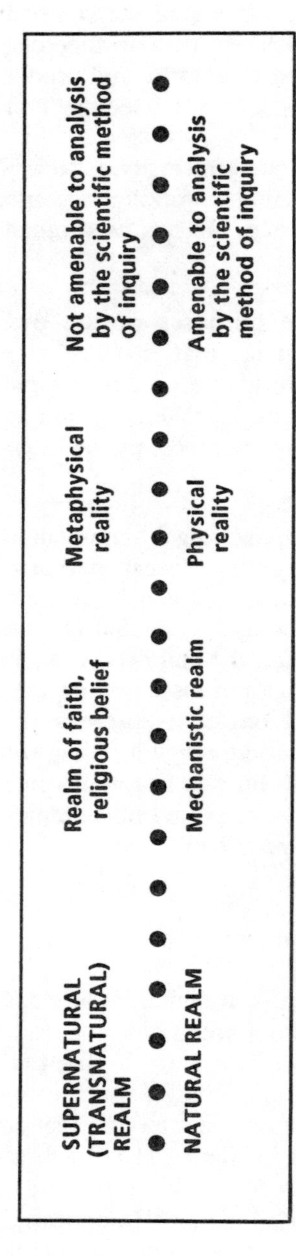

(Above: Key to Diagrams A through F)

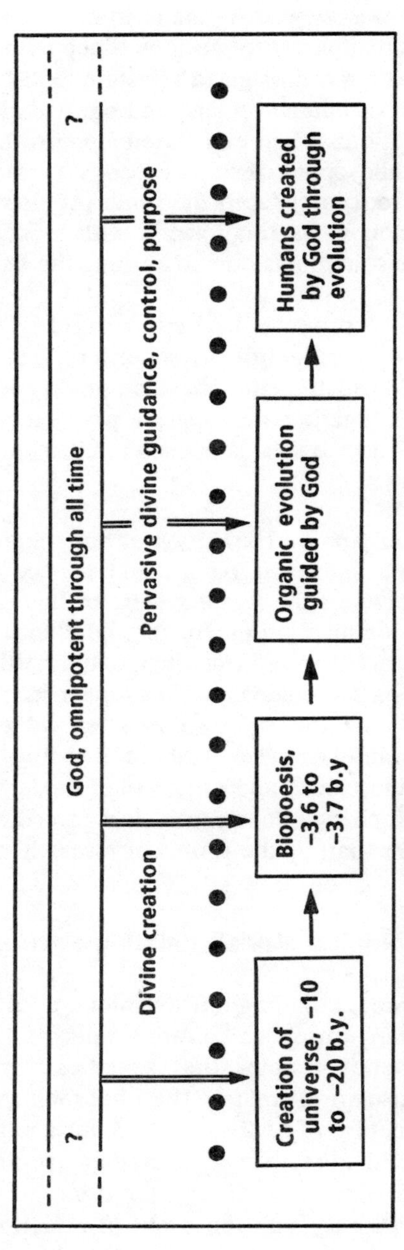

A. Theistic-Teleological Dualism ["b.y." = billion years]

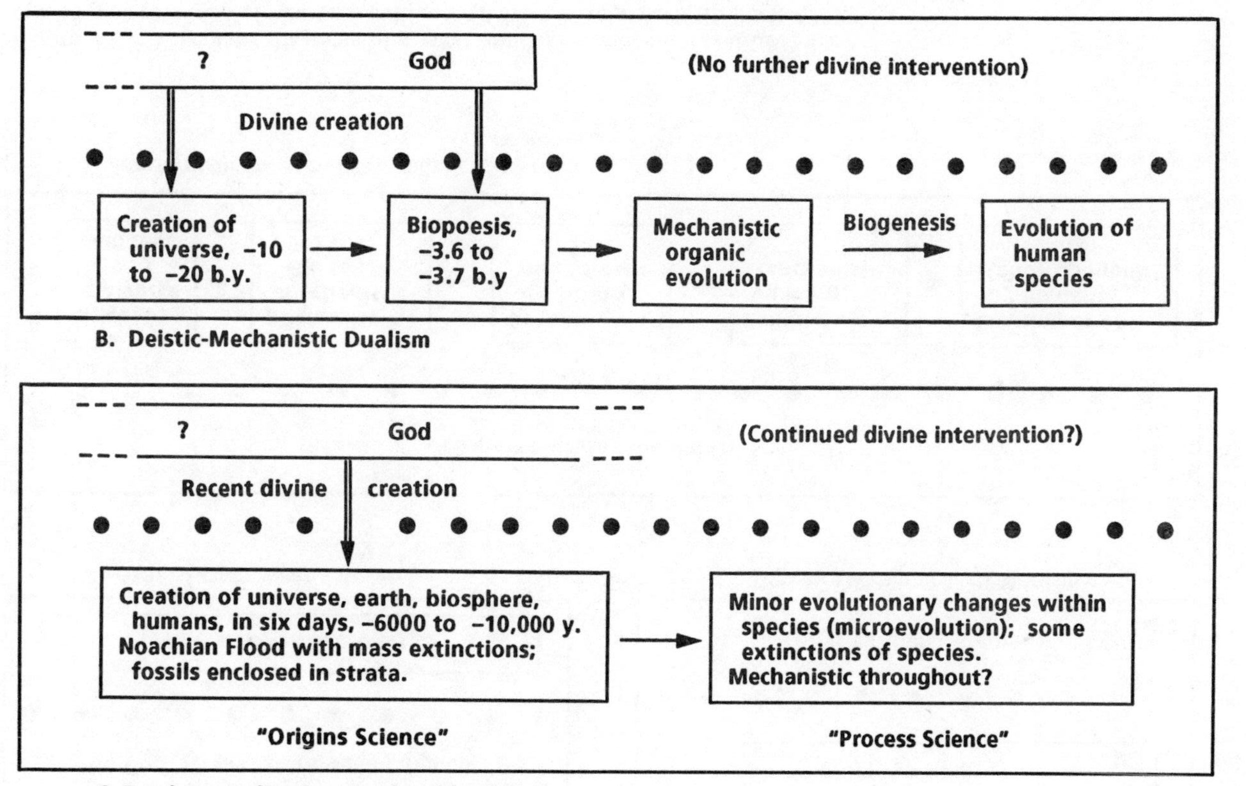

B. Deistic-Mechanistic Dualism

C. Fundamentalist Creationism (dualistic?)

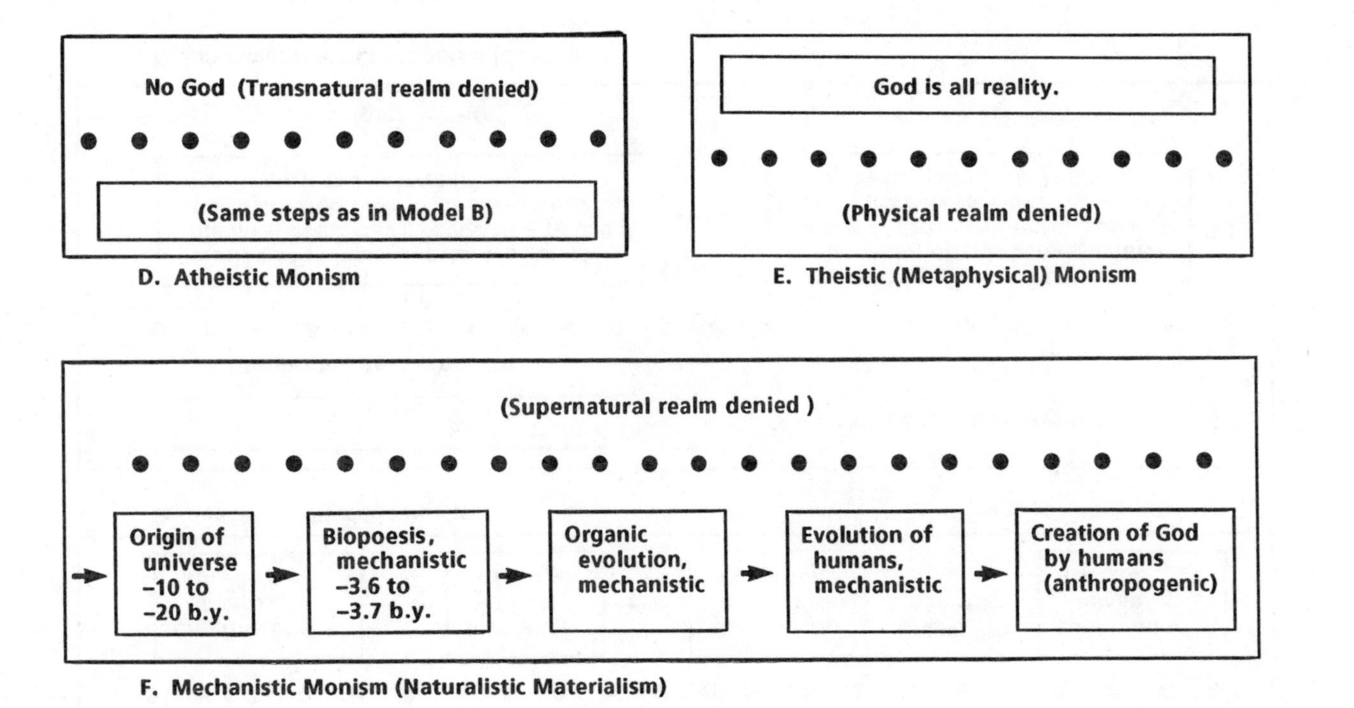

Figure 11.1 Schematic diagrams of several ontological models extant in philosophies and religions of Western cultures. (Adapted from A.N. Strahler, 1983, *Journal of Geological Education*, vol. 31, pp. 88-89. Reproduced by permission.)

The ontological models are presented as schematic diagrams in which certain definitions and rules are imposed for the sake of clarity and uniformity. All models are based on two possible realms within which reality is presumed to reside. In each diagram a horizontal line of dots separates a *natural realm* from a *transnatural (supernatural) realm.* A given hypothesis or model may require both realms or only one. Where both realms are postulated to exist, their spatial relationship is not specified; they may occupy the same space.

The natural realm includes all physical reality that is amenable to analysis by empirical science. This ontological realm pairs with science as a field within the class of perceptional knowledge. The supernatural ontological realm pairs with one of the belief fields in the class of ideational knowledge; here it specifically deals with Western theistic religion.

First, consider the classification of these models as monistic or dualistic. *Ontological monism* asserts that there is only one kind of reality. Here, ontological monism is taken to mean acceptance of either the supernatural realm or the natural realm, but not both. *Ontological dualism* asserts the reality of both realms. Usually, ontological dualism postulates some interaction or influence between realms, and usually the direction of flow of influence is from the supernatural to the natural. The case of a totally separate and independent existence of the two realms can simply be dismissed as uninteresting.

Model A, representing *theistic-teleological dualism*, recognizes the existence of God in the supernatural realm. In the philosophy of religion, the concept of God as a continuously active agent influencing the natural realm is known as *theism*. In the natural realm, we show a succession of physical constructs produced through cosmic and geologic time. They follow the current, or conventional, scientific opinion or consensus, which we may call "mainstream science," to distinguish it from creation "science." First is the origin of the present universe, some 10 to 20 billion years ago (-10 to -20 b.y.), possibly by the Big Bang mechanism. The existence of God and/or universal matter prior to the cosmic explosion is not specified here. The Christian Church seems to hold to the belief that whereas God is eternal and has always existed, the universe was created by God out of nothing (*ex nihilo*). On the other hand, the model allows us to suppose that matter and energy have always existed; i.e., that they are eternal. This alternative follows historically the Greek atomists, who postulated that the universe consists of a finite number of atoms that are permanent and eternal, without a beginning or end (Angeles 1980, p. 32).

The mainstream science view is that about -3.6 to -3.7 b.y. at the earliest, biopoesis occurred on our planet. *Biopoesis* is the formation of living matter from nonliving matter. The word "biopoesis" derives from the Greek *poiesis* and Latin *poesis*, meaning "creation." It was introduced by N. W. Pirie (1951) and adopted by Preston E. Cloud (1980, p. 386). An alternative term, *abiogenesis*, should perhaps be limited to its historical context with *biogenesis* in the nineteenth-century argument over ubiquitous spontaneous generation of life from nonliving matter under present environmental conditions. We are careful here to limit "biogenesis" to its long-established meaning as the principle that all life comes from preexisting life. The principle of biogenesis thus comes into effect following biopoesis in time sequence. You should know, however, that "biogenesis" is used for "biopoesis" in at least one comprehensive textbook on evolution: Dobzhansky, et al., 1977, p. 349.

An important feature of the theistic-teleological model is the belief that God caused biopoesis; i.e., God created life from nonlife. In this model, divine creation provided the starting point of organic evolution, which then proceeded under pervasive divine guidance toward a higher goal. For some persons, this goal is believed to have been the rise of the human genus, *Homo*. (Other versions of divine intervention include numerous events of creation of life forms.)

Notice that in the theistic-teleological model, the flow of influence across the boundary of the two realms is in one direction only. The same is true for models B and C. If we subscribe to the principle that the sum total of energy and matter in the universe cannot be increased or decreased (law of conservation of matter and energy), the control exerted by God on the natural realm can carry no flux of either energy or matter. This conclusion is consistent with our characterization of the supernatural realm as that which is not amenable to analysis by the scientific method. Just how God controls the natural realm is a mystery that rests totally in faith or religious belief.

While the theistic-teleological model is probably satisfactory for large numbers of educated and devoutly religious persons of the Judeo-Christian tradition (excepting the fundamentalists), it grates rather harshly on the sensibilities of the mechanistic (nontheistic) scientists. Under the theistic model, science cannot explain all of nature, because divine intervention is unfathomable. In ways unknown and unknowable the stochastic process in nature is being stacked to achieve an unknown purpose.

Model B, *deistic-mechanistic dualism*, follows the same program of physical constructs in the natural realm as in Model A. The essential difference is that the deistic model shows God's creative work to have ceased with biopoesis so, from that point to the present, organic evolution has proceeded on its own in a completely mechanistic manner. The cessation of divine causation and guidance following creation is referred to in philosophy of religion as *deism* (Angeles 1980, p. 52). The concept is seen in Cartesian philosophy and may have been adopted by none other than Charles Darwin, whose final sentence in *The Origin of Species* (1859) reads:

> There is grandeur in this view of life, . . . having been originally breathed by the creator into a few forms or into one; and that, whilst this planet has gone cycling on according to the fixed law of gravity, from so simple a beginning endless forms most beautiful and most wonderful have been, and are being evolved.

One has to be cautious here in reading Darwin's private mind. In a statement by Ashley Montagu including the above quote, I came across a footnote explaining that the words "by the Creator" were not in Darwin's first edition, but were inserted in later editions "as a concession, probably, to his wife's religious views" (Montagu 1984, p. 14).

Once life was started by God, biopoesis never again occurred, either by divine creation or by natural (mechanistic) processes. Thus a totally mechanistic system has prevailed over approximately the past 3.6 b.y. This deistic model obviously has some major advantages for the mechanistic scientists, since it keeps God "out of their hair" while they are investigating evolution and geologic history; they are also free from being "bugged" by teleological suggestions. Any evolutionary changes that might look like the long-range strivings of organisms toward a distant goal must be accounted for by physical circumstances alone. What's more, attributing biopoesis to God gives the geologist and biologist an aura of respectability with the religious community; any charge of atheism can be protested. Perhaps for some scientists, this belief assuages guilt for having excluded God from performing any useful role in evolution. At the same time the supreme mystery of God's breath of life can be enjoyed and even savored as a numinous experience. A special form of deism is found in the theological hypothesis that God is dead.

An important variation of the deistic-mechanistic model limits divine activity to the single act of creation of the universe. Biopoesis then becomes just one of many purely mechanistic developments taking place billions of years after the universe was formed. This variation of the model places scientific astronomy on a purely mechanistic basis, free of any supernatural influences, while at the same time allowing astronomers to express a belief in God and His/Her great act of special creation. Thus the astronomers and cosmologists also escape being branded as atheists.

Model C, *fundamentalist creationism*, is another dualistic model. It represents (more or less) the doctrine of recent creation held by Christian fundamentalists. It is unique in dividing the content of knowledge into two divisions, one succeeding the other in time. The first body of knowledge, dealing with recent creation and the Flood of Noah—the question of "origins," that is—properly falls into the belief field as religion; the second, into conventional (mainstream) science. Laws of science, although created by God, continue in force to the present. For a believing scientist, this temporal dichotomy removes most of the stress of having to cope with possible divine intervention in the present world of nature. The creationist tenets do, however, allow room for miracles performed by God in post-Flood time.

Model D, *atheistic monism*, eliminates the transnatural realm entirely. This is pure ontological monism, for only the natural realm exists. Atheistic monism is abhorrent to the theistic religious establishment consisting of Judaism, Christianity, and Islam, but is considered intellectually respectable by philosophers and is supported in the writings of a long list of distinguished scholars. (See essays by Ernest Nagel, Sidney Hook, Sigmund Freud, Antony Flew, Bertrand Russell, and others in Angeles, 1976.)

Model E, *theistic monism*, (metaphysical monism), holds that only the transnatural realm exists. (Peter Angeles has pointed out to me that the use of the words "transnatural" or "supernatural" in the context of monism is unfortunate because the prefixes "trans" and "super" imply a dualism. The trouble seems to lie in the selection of words. Perhaps for "transnatural" in this case we should substitute "spiritual.")

In philosophy, theistic monism is seen as a type of *idealism*. An extreme type, it is perhaps best illustrated by Christian Science. The metaphysical doctrine of Christian Science, as formulated by its founder, Mary Baker Eddy, goes as follows: All is infinite mind and its infinite manifestations, for God is All-in-all; Spirit is immortal truth; matter is mortal error; Spirit is real and eternal; matter is unreal and temporal. Illness and sin are considered to be illusions. All Christian Science teachings derive from the acts and sayings of Jesus. Obviously, these biblical sources do not concern themselves with questions of cosmology and evolution.

Model F, *mechanistic monism* or *naturalistic materialism*, like atheistic monism, eliminates the supernatural realm entirely and only the natural realm exists. In this sense, it appears atheistic, or at least nontheistic. It does, however, incorporate God into the natural realm under a mechanistic concept: God is anthropogenic; God is created by humans. This position is fully presented in Chapter 13 under the heading of "A Naturalistic Origin of Religion."

Mechanistic materialism in modern philosophy may have originated in the writings of Ludwig Feuerbach (1804-1872), who commented: "God is nothing other than the deified . . . mind or spirit of man." Dale Riepe, describing Feuerbach's philosophy, states:

> God was the affirmation of our wishes. He was a projection (wish fulfillment) from our sensuous experience. The supernatural came into being in man's efforts to understand and control the world. What man wished to be, he projected "onto a perfect being—God." (Riepe 1985, p. 223)

Martin E. Marty paraphrases Feuerbach's postulate thus: "Yes, God is nothing but a product of faith, a projection" (Marty 1985, p. 103). Marty sees this ontological transfer of theism into the naturalistic realm as a great advance in philosophical reductionism.

With this brief description of several ontological models as a framework for viewing Western religion in reference to science, we return to our search for criteria of demarcation between science and religion.

The Place of Western Religions and Their Theologies

We have designated religion and theology as supernatural fields along with spiritualism, magic, and the occult. The position of the last three is so clear that we need not refer to them further. In contrast, religion/theology is thoroughly interwoven with ethics/morality, worldviews, and sociopolitical ideologies, and therefore requires close scrutiny. Religion/theology needs to be compared separately with science and with ethics/morality in terms of criteria of demarcation. Here, however, our task is to examine possible criteria of demarcation between science and religion. "Science" as used here, is equated with the content of Class I, Perceptional Knowledge, as per our earlier conclusion in Chapter 9 as to what constitutes science.

We consider criteria of demarcation in two categories: The criterion from logic and the criterion from ontology and epistemology. Alternative propositions from science and religion are compared with regard to the assumptions of each and corollaries thereof.

The Criterion from Logic: A. For science, a proposition (known as the hypothesis) stands in ideal terms of formal logic as materially either true or false (correct or incorrect, perfect or imperfect). B. For religion the proposition (a statement of dogma, as in a creed) stands only and absolutely as true (correct, perfect, final).

Corollaries of each of these two primitive propositions are as follows: A. For science the proposition (hypothesis) faces one or more of the following fates: (1) tentative acceptance (through corroboration) or rejection (by falsification), and if rejection, then replacement; (2) modification or revision leading to its rejuvenation; (3) shelving for future consideration. B. For religion the proposition, as revelation (revealed knowledge), faces only one fate, i.e., to stand forever as ultimate truth, unquestioned and immune to critical examination, modification, replacement, or rejection. To declare subsequently that the proposition was in the first instance in error would place in jeopardy any substituted proposition declared to be the new truth. Thus the inerrancy must remain firm.

We all know that in practice, religious dogmas are revised from time to time, contrary to this requirement. Changes in Roman Catholic dogma announced by the pope have been met with mixed responses from the constituent membership—not always favorable. Perhaps the question is whether it is the pope who is infallible, but not God; i.e., that God changes his mind on occasion, but the pope is infallible in reporting those changes. The Mormon Church in 1890 withdrew its sanction of polygamy, and quite recently adjusted its dogma to allow blacks to participate in the ministry. Such admissions of prior errancy, though made under pressure from the membership, cannot help but erode confidence within that body. The "creation scientists" of the Institute for Creation Research under the leaderhip of Dr. Henry M. Morris have held firmly to the inerrancy of the Bible as it describes recent, rapid Creation and the Flood of Noah. (In the vernacular, this position is called "Bible thumping.") The ICR has, however, on occasion abandoned certain specific points of evidence previously put forward in support of their interpretation of the Genesis account. These reversals are relevant to the empirical realm of science, and not to religious doctrine, which remains unchanged.

The Criterion from Ontology and Epistemology: A. For science, the content of the proposition (hypothesis) is in the ontological realm of the empirical (naturalistic) and in the epistemic field of empirical knowledge, derived from perception of that which is external to the observer. B. For religion, the content of the proposition is in the ontological realm of the supernatural (immaterial agency or force) and in the epistemic realm of belief knowledge generated by the mental process of imagination (although typically charged to a process of revelation).

Essential Features of Christian Theology

Because modern, or "western," science arose in western Europe, where Christianity was in the position to guide and influence all political and ethical systems, we need to focus on the essential features of Christian theology, comparing and contrasting them with those of the emerging new science.

For Christianity, we need to inquire into certain groups of concepts that are necessary to it; i.e., the *dogma* of Christian theology. First is the Christian description of God in regard to knowability, functions, capabilities, powers, and limitations (if any). Next, we need to examine the postulated mechanisms of transfer of God's knowledge and power to the material world, which involves one-way transgression of the supernatural/natural interface. Revelation transfers information while miracles transfer force, creating material change. Also in the category is the transfer of God, or part of God, across the interface in the form of a Messiah, God's son Jesus. Another important function is transfer in the opposite direction, namely, from the material world into the supernatural realm. Individuals or their souls make this transit after death to assume immortality in heaven or hell. Jesus made a return trip back to the supernatural, and is expected to repeat this round-trip journey in the Second Coming. The accompanying illustration, Figure 11.2, attempts to graph these goings-on. One question that emerges is whether these transfers or transmigrations must be accompanied by a change from the immaterial state to the material state and vice versa.

First, however, we need to elaborate on the characteristics of the supernatural realm, as distinct from the natural realm. The essence of the difference lies in the distinction between *unknowability* and *knowability*. Mark Francis, writing on Herbert Spencer (1820-1903), observes that Spencer's generally materialistic philosophy included provision in the

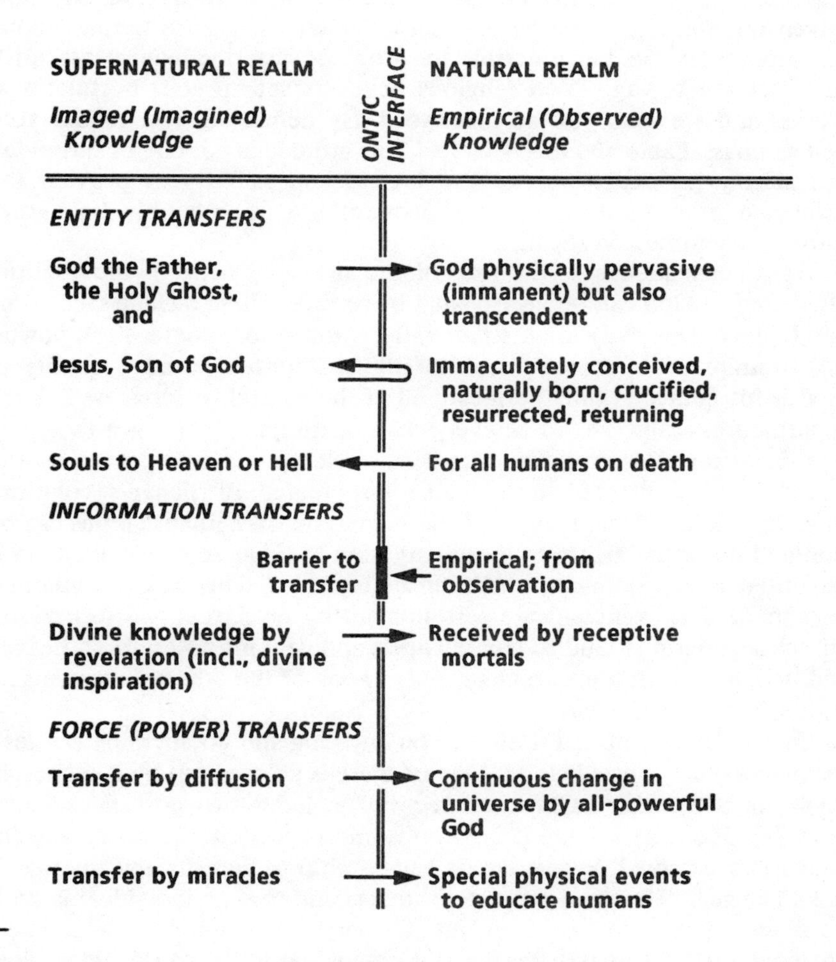

Figure 11.2 A crude attempt to show traffic flow between the two ontological realms.

"unknown" for the "unexplainable"; that there is "an Absolute" that transcends "human knowledge and even human conception" (Francis 1985, p. 651). In his earlier years Thomas Huxley (1825-1895), a friend of Spencer, held a place for the Unknown as a spiritual realm, thought to be unknowable in contrast to the knowability of the material realm. Later, Huxley abandoned this position in favor of straightforward materialism (Francis 1985, p. 336).

In contrasting the Western theistic religions with science, the most important criterion of distinction is that the supernatural or spiritual realm is unknowable in response to human attempts to gain knowledge of it in the same manner that humans gain knowledge of the natural realm (by experience); i.e., that the interface cannot be pierced in that direction. Given this fiat laid down by the theistic believers, science simply ignores the supernatural as being outside the scope of scientific inquiry. Scientists in effect are saying: "You religious believers set up your postulates as truths, and we take you at your word. By definition, you render your beliefs unassailable and unavailable." This attitude is not one of surrender, but simply an expression of the logical impossibility of proving the existence of something about which nothing can possibly be known through scientific investigation.

As a consequence of the foregoing conclusion, none of the qualities attributed to God can be investigated by science. These include God being good (devoid of evil), omniscient (all-knowing), omnipotent (all-powerful), omnipresent (eternal), and infallible. Another putative quality of God is his ability to fully permeate all of the natural realm as well as the supernatural realm; i.e., to be everywhere at all times. (This is a theory of *immanentism*.) Also following from omniscience and omnipotence is the belief that God created the universe, has created all change (evolution) since the initial creation, and will continue to do so. Perhaps all this can be summed up in the assertion that in addition to being unknowable, God is unlimited in any parameter that can be imagined. This lack of limitation permits God to communicate with humans by revelation and to perform miracles; it permits God to create angels and demons, to establish heaven and hell, and to transport human souls to one or the other of those places where they exist forever.

Giving God unlimited license to be anything and do anything can lead to some logical contradictions. One of these is pointed out by Kai Nielsen (1984, p. 543). Judeo-Christian doctrine includes two postulates about God: (a) God is an infinite being who is transcendent to the world, and (b) God has a personal, loving, caring relationship to humans and the world. Asks Nielsen: "How can we give to understand that an individual is both transcendent to the world and at the very same time stands in some personal relation to it?" My answer comes easily. In certain ideational fields of knowledge, among them religion, morality, and aesthetics, contradictory propositions can be sustained with impunity, because logic does not apply. Where logic is not a requirement of the knowledge system, it is futile to fault statements because they commit logical fallacies.

Using the above description of the Christian monotheistic model of the supernatural realm and its contents can lead to the discovery of a criterion of distinction between that religion and science that may prove to be sufficient in itself: Whereas science is necessarily a monistic ontology, Christianity is of necessity a dualistic ontology. Science is monistic in recognizing only a material universe, containing only those things that are capable of being observed. In contrast, the Christian postulate that "God in the supernatural realm is unlimited in power" is unthinkable without some external object or objects on which that power can be exerted. Can you imagine a God existing alone with absolutely nothing else in the universe but himself? The natural, or empirical, realm in the form of the physical universe is therefore absolutely necessary to form the receiving object of God's power. To see how this works, let us focus on the phenomenon of the miracle.

Belief in Miracles and Revelation

Philosopher Antony Flew has analyzed in great detail the phenomenon of religious belief in miracles (Flew 1985, pp. 452-58). He states that miracles can be defined in terms of two essential elements, carrying two paradoxical implications:

> On the one hand there has to be a strong natural order, an order that cannot be broken or upset by any person or power (or combination of persons or powers) within the universe. On the other hand, we have to have an occasional and essentially exceptional overriding of that strong natural order, by a supernatural power conceived as lying beyond or behind that universe. Such interventions from outside would either constitute or endorse revelations of that supernatural power(s). . . . Miracles must be impossible to both man and nature, though not, of course by the same token to a God or any other putative supernatural power. (1985, p. 453)

The scientist, when operating entirely within the subject of science, takes a nontheistic position, and therefore accepts only the strong natural order, but not the supernatural power; for science this is a monistic ontology. The theist, however, must accept both the natural order and the supernatural power; i.e., a dualistic ontology. This is obvious from the definition of a miracle as a manifestation of the overriding of the natural order by the supernatural power. There is, then, no escape from the conclusion that Christian theism requires belief in both the supernatural realm and the natural (empirical) realm. The monistic-versus-dualistic criterion set forth here fits as a derivative into the more general criterion stated at the outset of this section, namely, the criterion from ontology and epistemology.

Some theologians have attempted to take the starch out of the above argument by saying that in performing a miracle, God conscientiously follows all natural laws, and therefore does not commit "irregularities." Flew strongly objects to such an interpretation. After all, if God follows all the laws, we don't have a "miracle" in the first place! Flew notes that Thomas Aquinas gave a clear and unequivocal definition of miracles as "those things that are properly called miracles which are done by divine agency beyond the order commonly observed in nature." Flew quotes from an article by Eric Mascall in *Chamber's Encyclopedia* insisting that miracle "signifies in Christian theology a striking interposition of divine power by which operations of the ordinary course of nature are overruled, suspended, or modified" (Flew 1985, p. 453).

Divine revelation is another phenomenon that requires the ontological dualism we have described. There must be a receiver in the natural realm who translates the revealed message from its supernatural context into its natural context. Revelation is regarded by theologians as one of the ways in which God sends proof of his existence to humans. Revelation has been analyzed in some detail by Tad S. Clements, who calls attention to a fly in the ointment: Some theologians hold that to receive genuine revelations (including what are called divine inspirations) "an attitude of trust, openness, uncritical acceptance (or at least a willingness to relax our critical defenses) is, or may be, a necessary condition for divine knowledge" (1985, p. 552). Clements rightly points out that such a mental state is not conducive to presenting a scientific attitude toward incoming information. In other words, the observer is actually situated on God's side of the ontological barrier that is supposed to have been penetrated. I would add that the revelation claimed to have been received is limited to a claim of an event producing no externally observable physical phenomenon to be independently observed or recorded. On the other hand, events touted as miracles are typically observable as physical phenomena, and on further scientific investigation may end up being fully explained by natural processes.

Evaluation of miracles and revelations brings us to consider the question raised earlier: Does the transfer through the interface between the supernatural and the natural in either direction require that the entity making that transfer also change its ontological realm? Take God, for instance, under the dogma of immanentism. If God can diffuse himself everywhere throughout the natural realm, does he also remain in the supernatural state? If so, how can he cause physical change to occur in the natural realm? The case of Christ may be different, for he is considered to have changed state when ascending into heaven after his crucifixion. Upon arrival there, Christ must have assumed the power of immanentism, as befitting a Son of God. When humans die, they become souls in the supernatural realm, hence undergo a one-way change of state. But if they are to be able to communicate with living humans on Earth, they must possess immanentism. So it seems that for all players in the heavenly realm, immanentism is an endowment. That means that the question of how energy or matter in the empirical realm can be manipulated by the heavenly individuals is a very serious one. We await advice from the theologians on this question.

The Religious Arguments from Design and Improbability

Christian theologists seem to be of the opinion that science desperately needs God's help to explain fully the universe, our planet and its biota, and most importantly, its humans. These amazing features, the theologians claim, couldn't just have happened by chance and without a supreme intelligence to engineer them. Hundreds and perhaps even thousands of journal articles have written to support that belief, and there are even journals devoted largely to that purpose. Needless to say, neither those articles nor their journals have the status of legitimate scientific literature. Two arguments are needed by the theologians to support their claim; both attempt to establish the existence of God in the supernatural realm. These are the argument from design and the argument from improbability; they are closely linked. Unless the existence of God can be first be demonstrated, no case can be made for God's creative role. We will find that the arguments are circular (tautological), because they use God's postulated product to prove his existence, while at the same time, his existence is required to create the product.

The *argument from design* can be traced back to a set of arguments for the existence of God put forth by Thomas Aquinas (1225-1274) in his most important work, titled *Summa Theologica*. These arguments have been clearly set forth and analyzed in great detail by Peter A. Angeles (1980, pp. 17-43). Saint Thomas offered "Five Ways" by means of which he believed he could prove the existence of God. Angeles notes that "all of the Five Ways are elaborations of concepts taken directly from Aristotle and indirectly from Plato" (p. 20). The common theme of these Five Ways is that each describes an object or condition that can be identified in the universe and that each requires a motivating agent. For example (in my words), observed motion requires a "mover," observed effect requires a "causer," necessity of existence requires a "needer," scales of value require an "evaluator," and, lastly, design requires a "designer." It is the last "Way" that concerns us here. Professor Angeles quotes St. Thomas's description of the fifth way as follows:

> The fifth way is taken from the governance of the world. We see that things which lack knowledge, such as natural bodies, act for an end, and this is evident from their acting always, or nearly always, in the same way, so as to obtain the best result. Hence it is plain that they achieve their end, not fortuitiously, but designedly. Now whatever lacks knowledge cannot move towards an end, unless it be directed by some being endowed with knowledge and intelligence; as the arrow is directed by the archer. Therefore some intelligent being exists by whom all natural things are directed to their end; and this being we call God. (Pp. 19-20)[2]

The argument from design closely follows the argument of the other four ways in requiring an agent (mover, causer, needer, or evaluator); it is this requirement that gets the whole argument into a logical difficulty known as *infinite regress*. If we require God as the causative or creative agent, then we must ask: "Who caused or created God in the first place?" The answer calls for another godlike agent to create God, and before that yet another such agent, etc., *ad infinitum*, and so we end up with an infinite regress.

The logic of the infinite regress is unassailable as the arguments are laid out. The only escape is to rule out infinite regress by postulating that there is only one agent—God—in control, and that he is eternal. In describing this option, Angeles explains: "God is Eternal. He is not caused by anything else. There never was a time at which God was not. He always has been. Thus if God is Eternal, the question 'What caused God?' is meaningless" (1980, p. 31).[2]

The eternal nature of God also rules out any possibility that God is self-caused, in itself a logical impossibility (Angeles 1980, p. 31). The premise of an eternal God cannot be challenged by science. The message here seems to be that it is futile for nontheists to invoke the infinite regress as an argument against divine causation and design. The creationists have in the eternal God a "waterproof" argument; like a duck's back, its impervious shield sheds all attempts to penetrate it.

The argument from design resurfaced prominently in England in the popular writings of theologian William Paley (1743-1785). It can be found in his popular books arguing for the existence of God. One of these, titled *Natural Theology; or, Evidences of the Existence and Attributes of the Deity*, was published in 1802 and is often quoted today, perhaps because Paley used what is now called the *watchmaker analogy*, a favorite of fundamentalist creationist debaters in modern times. Perhaps in no other topic in the creationist lecture circuit are the gullible cultists more easily deceived than by this distorted version of the argument from design and its companion piece, the argument from improbability.

Paley's watchmaker ploy allows the creationist orator to pull from his vest a large gold watch. (No women, I trust, have ever taken part in this nonsense.) Paley's orator first asks (rhetorically) if anyone in the audience could tell how the watch originated merely by examining the intricate parts inside it; then quickly follows by asking if anyone would go so far as to assert that the watch formed spontaneously from a collection of inanimate single atoms brought together in one place. This possibility being absurdly remote, the orator confidently concludes: "Every watch must have a watchmaker." This is, of course, the argument from design, backed up by the argument from improbability.

Before going any further, we should be aware that the argument from design, taken by itself, is automatically invalidated on grounds of being a tautology. The word "design" is defined in our dictionaries as a purposeful creation of an intelligent mind, which has to be either a human or a superhuman being. No other meaning is implicit in the definition. So "design" is already linked unambiguously to "designer" by definition—neither could exist in concept without the other. To argue that the discovery of design is proof of the existence of a designer is therefore specious.

If there is to be an effective argument favoring the existence of a designer, it must stem from the nature of the thing identified as the design; i.e., something that is so unique and so complex in its structure or organization as to seem virtually impossible of arising through spontaneous and natural random processes. So we must turn next to the argument from improbability.

The Christian theologians' attack on a wholly naturalistic origin of the universe uses the *argument from improbability*. To illustrate how the argument runs, we will use the scientific problem of biopoesis, the origin of life on Earth. To understand the argument we can use a familiar illustration—a calculation of the probability of a person being dealt a perfect bridge hand of thirteen spades. The odds can be calculated as one in 635,013,559,600 (Doolittle 1983, p. 94). In contrast to a deck of 52 playing cards, a living cell consists of millions of atoms. In the writings of mainstream scientists there are many probability statements that can be lifted out of context to show the extremely minute probability that the atoms of a single cell will spontaneously come together to form a typical living cell. One of these can be found in a book titled *Energy Flow in Biology*, by biologist H. J. Morowitz (1968, p. 7). There we read that the probability is less than 1 in 10 raised to the tenth power raised to the eleventh power. That would expand to one followed by a hundred billion zeros. The calculation is based on the assumption that perfect randomness exists in the derivation of a population of combinations and that the individual variates (atoms, in this case) are entirely independent of each other. Those conditions apply to molecules of a monatomic gas in the closed-container experiment used to illustrate the second law of thermodynamics (see Chapter 6). The argument from improbability is simply that odds of that order of magnitude effectively rule out random chance and can lead us to only one conclusion: the cell must reflect the work of a supernatural creator.

Applied to the origin of a cell of living matter from nonliving matter, however, an entirely different set of conditions must be stipulated. First, the basic components of living matter are atoms of various elements and groupings of elements into ions and simple molecules. These units of matter do not necessarily behave independently of one another when in close proximity. Because of positive and negative charges, chemically similar or unlike units may attract one another. They may have a strong tendency to join together by chemical bonds that are difficult to break. Take the example of the snowflake, or any crystalline solid forming from a solution or melt as energy is lost from the system. Crystals of marvelous orderliness arise from a prior disordered state—witness the perfect diamond or sapphire. Everyone will agree that mineral crystallization is a natural process in conformity with laws of physics and chemistry, carried out without the aid of a supernatural agent. The calculated probability that the carbon atoms of a perfect 10-karat diamond could have taken their places by pure chance, in the absence of interatomic forces in exact order in the crystal lattice from an initial amorphous state as free carbon atoms, is so small that we can only conclude (using the theologians' argument) that it was impossible and never happened.

Second, the theological argument based on extremely small probabilities of chance of assembly of a cell or other unit of living matter fails completely to include the possibility that the end result would come about by a succession of small steps, each one a relatively simple increase in complexity of form or function, and each new step being capable of maintaining itself with the available supplies of matter and energy.

Third, the fundamentalist Christians fail to take into account the enormous spans of time available for the steps to be accomplished. They may be ruling out the possibility that there could have been, say, two hundred million years available for nonliving molecules to advance by small steps to the level of complexity of a simple cell capable of replicating itself.

Is Naturalistic Monism the Answer?

Finally, we return to those scientists who espouse naturalistic monism as their grand hypothesis of the nature of reality and the workings of everything in the universe. Unlike simple atheism, which summarily denies the existence of God and all other supernatural phenomena, naturalistic monism has a place for the God-phenomenon within the natural realm. God is incorporated into the natural realm under a purely materialistic concept: God is *anthropogenic*, meaning that God is created by humans. "So Man created God in his own image, in the image of Man created he God" (Gen. 1:27, rephrased). God exists in ideas, and ideas are physical realities produced in the brain by neurophysiological processes capable of being investigated and explained by scientific study.

Naturalistic monism recognizes that those of the human species have become differentiated into many cultures, evolved over thousands of years. God takes many forms and displays a wide variation in function. For some cultures, God is multiple—a collection of deities, as in Hinduism. For other religions God is one entity, but the God of each religion differs in specifications from that of other religions.

The anthropogenic hypothesis of God (including with God any other entities that are assigned to the supernatural realm) has the strong logical position of accounting for all forms of God and accepting them all as reality. It is important to make clear that physical reality attaches to the ideas and images of God and that they exist in the natural realm (i.e., the model-image is real). In contrast, the imagined entity is nonreal (nonexistent). Naturalistic monism is not atheistic; it does not deny the existence of God. Quite to the contrary, it has a rational position for all God-models within the total mechanistic system.

Naturalistic monism, which we labeled earlier in this chapter as *mechanistic monism*, Model F of the ontological models, maximizes the opportunity for humans to encompass under a single general hypothesis or theory all physical and biological forms and processes of the cosmos. This is possible because it has a rational accommodation for all phenomena capable of identification and description. Under naturalistic monism, such diverse disciplines as psychology, sociology, ethics, religious philosophy, and political science can be covered by the same general systems theory that applies in the natural and physical sciences. This is surely the ultimate in reductionism. In this context, religion becomes amenable to analysis by science and the supposed conflict between two particular religions or between a given religion and a given naturalistic "philosophy" (such as secular humanism) is restructured into a case of two diverse phenomena capable of coexistence, each in its own habitat—a human mind.

In Chapter 13 we shall attempt to trace the origin of religion and the God concept from the earliest times of development of social behavior patterns within groups of humans. Science has a major role in that investigation.

Credits

1. Portions of this section of text are taken from the author's earlier work: Arthur N. Strahler, 1983, "Toward a Broader Perspective in the Evolutionism-Creationism Debate," *Journal of Geological Education*, vol. 31, pp. 87-94. Used by permission.

2. From Peter A. Angeles, *The Problem of God: A Short Introduction,* Prometheus Books, Buffalo, N.Y. Copyright © 1980 by Peter A. Angeles. Used by permission.

CHAPTER 12

The Places of Ethics, Ideologies, and Aesthetics

We turn now to knowledge fields that, while not religious in the sense of dealing with the supernatural realm, can fairly be described as belief fields because they require that one believe in the intrinsic value or worth of some concept or concept system. Here, we look for a criterion or criteria of distinction between the group of *value fields* that includes ethics/morality, sociopolitical ideologies, and aesthetics, on the one hand, and perceptional knowledge on the other. For ethics and morality "value" refers to whether a particular action or result is right or wrong, good or evil. In the sociopolitical field, very similar kinds of values are paramount considerations, ethics and political ideologies being closely interwoven. In both areas, normative, or prescriptive, knowledge is generated with the objective of determining what ought to be done, or what ought to be the objective of the system. Aesthetics, operates in quite a different value mode that involves sets of arbitrary standards of excellence of performance or execution.

The Humanist View of Ethical and Moral Systems

The history of philosophical views on ethics is rather fully and systematically reviewed by philosopher Antony Flew at a level suitable for mature, thoughtful persons outside the circle of professional academic philosophers (Flew 1980, Chapter 2). Those of us outside that circle rely on such philosophers who make their subjects accessible to us, even though our use of it is often faulty or inadequate. In this case, there is something special to be searched for in Flew's presentation because he is an active secular humanist—a contributing editor to a leading humanist journal, *Free Inquiry*—and a member of the secretariat of the Academy of Humanism. (Other academic philosophers who are listed as members of that Academy or its secretariat are Paul Kurtz, Sidney Hook, Alfred J. Ayer, Kurt Baier, Mario Bunge, Sir Karl Popper, and W. V. Quine.) This is interesting because secular humanism is (as I see it) clearly a belief system resting on asserted human values and strongly flavored by their own set of "oughts." The secular humanists often claim their ethical system to be founded on applied logic, or on science, or on both. They have tried to substitute these empirical sources for morals that religion has always held to be revealed by entities in the supernatural realm. Can the humanistic philosophers keep their commitments to academic philosophy free of entanglement in their emotive devotion to a value system? Can they cope with double-think? These are questions to keep in mind as we go along.

Humanists as a group are divided from within on the question of whether a science of ethics is possible. Archie J. Bahm comments on this question and names individuals on both sides (Bahm, p. 218, in Storer, 1980). He quotes Morris B. Storer as having asserted that "we will never have a science of ethics"; Corliss Lamont as saying "the Humanist contends that a true science of ethics is possible and will yet be established"; Roy Wood Sellars as having not only regarded "ethics as an empirical science" but also stating that "ethics is a very old science and has had a varied and distinctly controversial career very much as have epistemology and metaphysics [ontology]"; John Dewey, American humanism's most influential philosopher, as early asserting that "ethics is the science that deals with conduct, in so far as it is considered right or wrong, good or bad."

What can we glean from Flew's review of the history of inquiry into the nature of ethics and morality? Skipping entanglements with Plato and Socrates, let us begin with the Scottish philosopher, David Hume (1711-ß1776). His *Treatise of Human Nature* (1739-1740) appeared at a time when natural theology was in vogue and Newtonian mechanics had become well established as an insight into what God had wrought. Hume's more mature work of 1748 and 1751 dealing with "the principles of morals" preceded the English/Scottish school of uniformitarian geology that lay a few years ahead for James Hutton, John Playfair, and James Hall; they were soon to begin the serious challenge to recent divine creation. In this milieu of God-still-in-the-saddle, Hume chose to emphasize humanity as the focus of attention and importance, rather than emphasize nature, as the Copernican Revolution had tended to imply.

The Naturalistic Fallacy and Sociobiology

Hume is generally credited with identifying in his *Treatise* the *naturalistic fallacy*. This appears in a passage in which he notes that in every system of morality the author invariably slips from the formally correct use of the copula "is" (or "is not") in stating propositions concerning human affairs to the use of "ought" or "ought not" in propositions on morality (Flew 1980, p. 27). The fallacy is described by Flew as follows: "From the premise that all or most of us are thus or thus inclined, and hence that it is natural to us so to behave; it is immediately, but invalidly, inferred that such action is at best morally licit, if not obligatory" (p. 28).

The crux of the naturalistic fallacy lies in the ontological jump from the empirical (naturalistic) realm to the transempirical realm, which is a no-no. The fallacy gets a workout in our modern day and age as a device to break down the traditional (Mosaic) religion-derived morals; i.e., to justify permissiveness. The distinguished paleontologist, George Gaylord Simpson, gave this example: The late Alfred Kinsey, a biologist whose later researches turned to human sexual behavior (well known through the "Kinsey Report"), claimed to have discovered that homosexuality was rather more prevalent than previously thought. On the grounds that the phenomenon exists in nature, he recommended that laws prohibiting the practice be modified to reflect the conclusion that it cannot be morally wrong (Simpson 1969, pp. 130-48, in Strahler 1987, p. 502). Simpson was addressing himself to a subject dear to the secular humanists, namely, "the search for naturalistic ethics, that is, an ethical or moral system rationally related to the nature of things or to the material universe" (1969, p. 130). Flew, perhaps reflecting the basic problem facing secular humanism, seems somewhat ambivalent about the status of the naturalistic fallacy. Can it be ignored? Can it be declared a non-fallacy? But he seems to support the validity of the fallacy, for he concludes that we cannot fault Albert Einstein's pronouncement: "As long as we remain within the realm of science proper, we can never meet with a sentence of this type: 'thou shalt not kill'. . . . Scientific statements of facts and relations . . . cannot produce ethical directives" (Einstein 1935, p. 114, in Flew, p. 28).

A possible bail-out from entanglement with the naturalistic fallacy may perhaps lie in the recent resurgence and burgeoning of sociobiology, which suggests that certain forms of human behavior which appear to be expressions of morality are a genetic endowment arising from biological evolution by natural selection (as distinct from cultural evolution). When in 1975 Edward O. Wilson published his book titled *Sociobiology: The New Synthesis*, he not only delivered a buzzword of monumental proportions, but also caught the fancy of the secular humanists, who quickly adopted him as a favorite son. Although Konrad Lorenz and other ethologists had previously formulated the idea of certain forms of social behavior being genetic endowments, the coining of a new word for it seems to have set off the burst of attention. What sociobiology is claimed to have done is to have transferred ethical and moral principles from the transempirical to the empirical realm, getting rid of the onus of the "ought." It is important to understand that sociobiology deals in extremely general propensities of human behavior, such as aggressiveness or altruism, and these are of little help in debating the moral issues of, say, abortion, pornography, substance abuse, and promiscuous or premarital sexual activity.

As one would expect, the theme of "man-centeredness" stressed by

Hume is emphasized by Flew and other leading humanists. "Hume provides the first inspiration in modern times for all those who believe that value cannot be, and is not, embedded in the structure of things, but instead is, and must be, some sort of projection of human desires and human needs" (Flew, p. 37). The theme is strongly developed by Paul Kurtz, today's leader of secular humanism, in his 1980 *Secular Humanist Declaration* (Kurtz, 1983, p. 17).

Is a Science of Ethics Possible?

The modern secular humanists strongly endorse the "greatest happiness principle" introduced by Jeremy Bentham (1748-1832) and developed by later utilitarian spokespersons John Stuart Mill and Henry Sidgwick (Flew, p. 55). But it is worth noting that these utilitarians did not include the concept of teleology in their position. For them, life is simply something that exists as a biological process and must be accepted, but it can be manipulated for the greatest good and pleasure for the greatest number. Paul Kurtz expresses this aquiescence to the product of a purely naturalistic organic evolution:

> Living beings tend instinctively to maintain themselves and to reproduce their own kind. This is the primordial fact of life; it is precognitive and prerational and is beyond ultimate justification. It is a brute fact of our contingent natures; it is an instinctive desire to live. (1983, p. 159)

Of course, the same statements can be made about all earthly organisms generated through organic evolution, or at least the metazoans. Existence without purpose is demanded for any clearly defined material system within its own boundary. Purpose in the utilitarian sense can only enter when the system to which purpose is ascribed is a subsystem of a larger system which is the beneficiary of the purpose. Thus Kurtz's statement simply states the existence of the "is" and lends nothing useful to the problem of the status of the "ought."

Existence of organisms without ultimate purpose is an empirical concept fully in line with Darwinian evolution by natural selection, but squarely opposed to the Judeo-Christian doctrine of divine purpose. The latter position requires that humanity, the earth, and the physical universe be nested hierarchal material subsystems within an all-encompassing supernatural system. Secular humanism denies the supernatural system, and with that denial, teleology becomes a null issue—it simply evaporates.

If something—a material system such as humanity—has no purpose within or for itself, can it have value in the sense of intrinsic worth? The humanists say it can. For example, philosopher Archie J. Bahm, a founding member of the American Humanist Association, states:

> Humanists agree that persons are or have intrinsic value. Persons are ends-in-themselves. There is something ultimate about human nature and human values that cannot be reduced to anything else. (1980, p. 210.)
>
> Despite diversities in kind and origin, the ultimate bases of morality are the intrinsic values of persons. Humanists, at least, can trace the values of all rights and duties back to the intrinsic values of human beings. (P. 217.)

As another example, Paul Kurtz notes that "human life has meaning in its own terms" (1983, p. 152). Statements such as these are clearly transempirical and an expression of belief. They embody the fundamental value assumption that must underlie all systems of morality and ethics claimed to be purely secular in nature. ("Secular" here means only "nonreligious"; it does not necessarily imply a materialistic ontology, but may happen to also fall into that category.) It is of no avail to argue that, within the total system of humanity, individuals and subgroups of humans can be of value (i.e., be useful or helpful) to other individuals or to other subgroups. The essential point is that the premise "human life has meaning in its own terms" has equal value status with the corresponding contrary proposition "human life has no meaning in its own terms."

Thus humanistic ethics—or any other secular form of ethics—cannot be asserted to be founded in empirical science on the premise or presumption that the human race as an entity in itself has "intrinsic value." Any moral principle that follows from the premise that "human life has meaning in its own terms" must, of logical necessity, be transempirical, and thus form the first link of a chain of value judgments. For every proposition that "*x* (a course of action) is good (or better, or best)," there is the counter-proposition that "*x* is bad (or worse, or worst)." No empirical foundation exists in any part of either chain. Kurtz makes this clear: ". . . that life is worthwhile is not amenable to a descriptive confirmation; it is not capable of being tested as other hypotheses are. Rather it is a normative postulate, on the basis of which I live" (1983, p. 158). That is why there can be no such thing as "scientific ethics;" that would be oxymoronic. Ethics and morality securely hold their position in the class of ideational knowledge, in the subclass of transempirical knowledge dealing with value judgments.

Perhaps a distinction should be made between two different meanings in the humanists' use of the term "scientific ethics." The first is presumed above, namely, that the substance of ethics can be derived directly from the available body of scientific information; i.e., that a specific scientific statement dictates a specific course of moral action. The second meaning is that humanists can or should adopt the scientific method as the operational model or framework for deriving morals. Perhaps this meaning can be discerned in a series of statements by Bahm, referring to the many kinds of choices humans must make in matters of their physical needs in the light of differing environments and cultural conditions (1980, pp. 210-11). He writes:

> To the extent that we have generalizations, we have bases for an inductive science of ethics. Observed likenesses provide data for hypotheses about how people do behave, about what kinds of behavior beget better results than others, and about how people ought to choose in order to obtain the better results. Such hypotheses, stated as principles, can be tested repeatedly when the same problems recur. (P. 211)

In this system description there can be seen a parallel to the induction/deduction feedback circuit described in Chapter 1. The important difference lies in the presuppositions of the two programs. The set of presuppositions of science establishes the fundamental properties of an external reality; that of humanistic ethics is the nonempirical postulate of "intrinsic value of humanity." Given the latter, it may be possible to formulate hypotheses of desirable behavior, make deductions and test them against observation, and perhaps even derive the equivalent of moral "laws." In the above sense we can perhaps tolerate Bahm's case for "an inductive science of ethics," but with the proviso that his term be changed to read "a science-like method."

In Chapter 13 we take a closer look at the role scientific knowledge may play in formulation of an ethical system, bringing to bear knowledge gained in recent years through sociobiological research.

The Criteria of Demarcation of the Ethics/Morality Epistemic Class

What, then, is the criterion of demarcation of the ethics/morality epistemic class? One crucial characteristic of systems of this class is that for any ethical proposition asserting positive value there exists a corresponding counterproposition of equal logical status asserting negative value. (This is an expression of ethical relativism, described in Chapter 5.) The criterion is not sufficient to demarcate such systems from other value systems in the transempirical area (aesthetics, for example) but is sufficent as demarcation *vis-à-vis* science (along with all other members of the empirical

Class I, Perceptional Knowledge). Science allows counterpropositions to be held tentatively only as competing hypotheses, only one of which can ultimately prove to be correct while the other will ultimately be shown as incorrect, or both may prove to be incorrect. Whether this same criterion also demarcates ethics/morality from religion is a good question and we treat it in the section on religion.

A second criterion of demarcation (already alluded to above) is that moral values possess quality but not quantity. For perceptual knowledge generally and for science and applied mathematics specifically we defined "quantity" in terms of time, space, and mass. The criterion rests on the premise that quality of moral values, as defined here, has no physical dimensions, which is to say that moral qualities have no place in time or space, and possess no mass (and no energy).

At this point we can apply the second criterion to principles of moral philosophy. One of these is the principle of the greater good, which is that a group judgment—a consensus—overrides an individual judgment that is contrary. Is this a rational principle? Henry Margenau, discussing this problem in his 1964 book, *Ethics and Science*, uses the term *hortative power* (or "decisive force") in reference to the grading of various moral imperatives (the oughts) in a hierarchy of priorities (pp. 122-23). He refers to "lower planes" and "higher planes" of value (p. 123). Like most others who discuss moral imperatives, Margenau accepts the principle that group judgments carry greater hortative power than individual judgments on paired alternative oughts. He observes that "collective happiness . . . can be defined and recognized as an objective principle," whereas "individual happiness is subjective" (p. 122). He seems to think this distinction is a logical one. It is precisely this position that I challenge here.

It can be argued that the strength of a moral judgment or imperative is not a function of the frequency with which an individual value assessment is replicated. But this argument does not apply in science when assessing what is scientifically equivalent to the hortative power of a moral ought; i.e., an empirical proposition. When a physical experiment is performed repeatedly and independently to test such a proposition and always produces the same outcome, it gains power in proportion to the number of repetitions, that power being expressed in the confidence which can be placed in the possible correctness of the descriptions, relationships, or changes observed. At some point, and with no negative results reported, the power becomes sufficiently great to allow the proposition to become designated as a law. (Nothing here contravenes the correct logic of the positivists that no amount of testing that a proposition successfully passes can prove its correctness.) In contrast, in ethics, the hortative power of an imperative moral proposition is not increased through its endorsement by a larger number of individuals, because the moral proposition is purely qualitative and devoid of physical properties; its value is not measurable in any empirical sense. It is meaningless to sum individual values or to extract a mean or median value from that sum. Put this proposition into mathematical form as follows:

$$V = v \times n$$

where V is total value (group value), v is one individual's valuation, and n is the number of individuals involved in making the group judgment.

This equation is meaningless in application to moral imperatives, which are ideas that have no measurable properties. Mathematically, the value, v, is dimensionally zero, and multiplication by zero yields zero. A single imperative is pure essence that cannot even be subjected to enumeration. A caution sign is posted at this point. Consider the ought "You shall not kill another human," usually ranked as the most important of all moral imperatives. You may want to point out that the act of killing is a real physical act in the empirical realm, and that is a correct statement. But we are examining only the moral value it carries but does not state, which is that killing is wrong, bad, or evil. It is the *wrongness*, the *badness*, and the *evilness* that is the moral judgment at stake. After all, carnivorous animals kill as a matter of survival, and we don't say to them "don't kill," nor do we conclude that it is wrong for them to kill. (Animal-rights advocates, presumably carnivores themselves, do say "don't kill animals.")

Continuing then with the argument, we point out that only the individuals themselves who take part in the group judgment can be enumerated as empirically real things. So what we are actually dealing with is the principle of one person/one vote, which is no less than a principle that a group of many individuals acting in concert is physically stronger than one; it is the law of rule by force of numbers of force-bearing objects. The poll is not a vote on the intrinsic value of the moral imperative, but only on the number of persons who support it. In contrast, in science a physical experiment itself constitutes the individual variate in a sample population of many trials. Because the experiment uses a physical process or event, the empirical probability of the proposition being tested then increases or decreases with the number of trials performed.

Ideologies and Worldviews

Another of the transempirical fields involving value judgments consists of the *ideologies,* which are belief systems, at least in large part. Mario Bunge distinguishes "two genera of ideologies: worldly and unworldly (religious), and two species of the former—worldviews and sociopolitical ideologies" (1983b, p. 228).[1] Here were are concerned with the two species of the worldly genus. Bunge states that everybody holds some worldview (in

German, *Weltanschauung*), but it need not be either a religion or a sociopolitical ideology. *Worldview* is described by Bunge as "a body of extremely general beliefs about nature and man, is usually a mixture of deep thoughts and platitudes, well-confirmed generalizations and superstitions. In its popular version it often comes in the form of maxims, stories, and parables" (p. 229). He states further that "the value judgments and moral norms included in a worldview are noncognitive, even through they can be analyzed and even justified in the light of knowledge" (p. 229). To the extent that a worldview includes knowledge of science and human history it is perceptional knowledge. The worldview is thus a mixture of all classes and kinds of knowledge. As a popular belief system tailored by experience to be different and unique for every individual, worldview defies systematic analysis, and we shall pass it by.

Sociopolitical Ideologies

Sociopolitical ideologies are concerned particularly with human societies on questions of how they should be structured, what their aims should be, and how they should perform to achieve their aims. Bunge recognizes four distinct areas of concern that may be involved in a particular sociopolitical ideology: (a) "ontological theses concerning the nature of human individuals, groups, and societies"; (b) "theses concerning the *economic, political and cultural problems* faced by societies of a given type at a given time"; (c) "*value judgments* about persons and their activities, races, classes, institutions, etc.; in sum, what is good for the individual, the group, or the community"; (d) *action (or inaction) programs* for the solution (or conservation) of social problems and the attainment (or thwarting) of individual or societal goals" (1983b, p. 229).[1]

I see the first two sets of theses (a and b) as being investigations of real phenomena and hence, if rigorously treated and not consisting of false or bogus pronouncements, as falling into our Class I, Perceptional Knowledge. The first (a) would fall under sociology as a science; the second (b) under human history.

The last two sets of theses (c and d) are founded in value judgments, which we have identified as transempirical. Bunge argues that the problems they address can and ought to be studied in the light of scientific knowledge (p. 230). This is the position of secular humanism on ethical systems. What sociopolitical ideologies prescribe for society is, however, a set of value principles comprising an ethical system. The ideologies are prescriptive (normative) and are concerned with the real world, and in that sense have empirical concerns. Perhaps sociopolitical ideologies can be described as systems of applied ethics. In all cases, the ideology is based on the fundamental initial premise that human society has intrinsic value arising from the supposed existence of a purpose or goal.

In this connection it is interesting to find that in shaping communist ideology, Marx and Engels adopted from Hegel's thesis/antithesis dialectical process a strange kind of reasoning known as *dialectical logic*, which claims that contradiction can exist in reality. If so, the "Laws of Thought" basic to Aristotelian logic (explained in Chapter 10) can be ignored or perhaps be considered invalid. The Law of Contradiction, in particular, has been implicated here: No proposition can be both true and false. The theory of *dialectical materialism* underlying communism holds that everything in the universe belongs within the material realm; i.e., what we have already described as a scientific ontology characterizing the epistemic fields of perceptional knowledge. The theory then posits that everything in this realm contains elements that are simultaneously in opposition, as if forces were exerting continuous tension on some given social structure or concept. Visualize, as a model, two teams engaged in a tug-of-war, always applying tensional forces that may, if equal, produce stasis and, if unequal, motion in either direction.

Inequality in the opposed forces leads to social change that may take the form of growth and development of societal institutions. It is the supposed permanence and immanence of the opposed forces that leads to the notion that contradiction can exist for any proposition, such that it can be both true and false at one and the same time. Peter Angeles points out the flaw in this reasoning; it is semantic. The opposing forces are not correctly described as "contradictory" in the logical sense, but rather as "in contrast," "continuing," or "in opposition." He adds: "The concept of noncontradiction, together with the other two Laws of Thought, is definitionally and irrevocably true" (1981 p. 156).

In our system of analysis, for the transempirical value-judgment based sociopolitical ideologies we say only that two or more mutually exclusive hypotheses (of what "ought" to be) are of equal status and can coexist indefinitely, neither one being required to be eliminated or replaced by the other. Neither has any truth value other than logical possibility, because any product of the imagination is admitted to the ideational class. You might think of this relationship as "live and let live." No propositions of this category are empirical, so that the notions of corroborating and falsifying we apply in science are simply irrelevant and immaterial.

Perhaps the manipulation of social problems can be inspired and facilitated by scientific and technological information, as Bunge suggests. He then notes: "But in fact many sociopolitical ideologies have been openly hostile to science or to technology, whereas others have produced grave distortions in the latter" (1983b, p. 230-31).[1] Examples he cites are the cases of Nazism versus modern physics and Stalinism versus modern biology. Creationists who are fundamentalist Christians strongly repudiate the fascist and communist ideologies, which they say are atheistic expressions of social Darwinism spawned by the godless theory of evolution by natural selection.

The Aesthetic Fields

Much the same analysis we have offered above can be applied equally well to the epistemic status of aesthetics in the arts, and including criticism of artistic works. The fundamentalist Christians have mounted a full-blown attack on the arts. This is manifested in their negative reaction to certain motion pictures, branded as blasphemous, and they have not hesitated to attack the graphic and performing arts on grounds of indecency and pornography. Diametrically opposed opinions as to the merits or demerits of a particular artistic offering, such as a painting, dance, symphony, or poem, have equal status and can coexist indefinitely. Consensus is ineffective as a decider because, as in ethics, the entities of the domain are purely qualitative. The same can be said of fictional literature and historical literature containing quality value assessments, and the literary criticism that follows.

Science and the and arts get along just fine—no conflicts of interest and no nasty territorial contests, such as those that plague the science/religion interface. Artistic endeavor need not be inspired by supernatural agents, but if an artist chooses to claim God's inspiration there will be no repercussions. I doubt that there would be a big flap if Handel's *Messiah* were performed in the auditorium of the local public high school, but I could be wrong. A great requiem mass—whether composed by a Verdi or a Bernstein—could raise some questions of classification: Is it religion or art, or both, and should that make a difference?

Many scientists enjoy participating in the arts either as passive recipients or active performers. If anything, their science research stands to be improved by the rest and relaxation provided by an afternoon's landscape painting or an evening's session of the Beethoven late quartets. If there is any merit in the claims that "discovery" of a great science theory arrives by sheer inspiration, why not indulge in the arts as a stimulant to that creative imagination?

With the close of this rather brief chapter we have completed our analysis of the six hexagons of Class II, Ideational Knowledge, that surround the hexagon of Class I, Perceptional Knowledge. (This is a good point at which to reexamine Figure 9.1.) Their six interfaces show a variety of qualities of relationships. Three of them—Religion, Ethics and Morality, and Sociopolitical Ideologies—are antithetical and hostile; theirs are the boundaries of warring contiguous states, and a besieged Science and History must construct and maintain its ontological Berlin Wall to keep them out. Two of them are friendly, unguarded boundaries and their occupants make good neighbors: Mathematics and Logic are in symbiotic relationship as business partners with Science and History; Aesthetics is a highly-esteemed friend with whom Science and History enjoys relaxing parties and gossip about the other neighbors. Pseudoscience is an abhorrent neighbor, to be ostracized and met only with a cold stare; that boundary needs a tall and well-maintained chain-link fence topped by barbed wire.

Those ongoing wars with the aggressively hostile neighbors need more analysis. Our next chapter is historical in content, reviewing a strongly debated scenario in which science and culture are interacting forces of change—the "nature/nurture" debate come full circle. Although this debate lies within the empirical domain of science, it spins off a historical explanation of Religion, Ethics and Morality, and the Sociopolitical Ideologies. The final chapter provides details of the conflicts along that battered wall between Science and Religion.

Credit

CHAPTER 13

How Science Impacts Ethics and Religion

Although we have examined the belief field of religion and the value field of ethics/morality, we have not inquired into how these fields developed through human history. How did religious and ethical systems originate? Before speculating on this problem, we need to find out what biologists and anthropologists can tell us about the evolution of the human mind and culture, including their guesses as to the early origins of ethics and religion.

I turned first to the writings of Theodosius Dobzhansky, George Gaylord Simpson, and Bernard G. Campbell. In the period 1940-1962, Dobzhansky served as professor of biology at Columbia University. His office was only a few doors away from mine, and I saw him occasionally. He was then at the height of his achievements as a geneticist and was particularly known for his experimental studies with fruit flies. In later years, his interests broadened to deal with the evolution of humans and he became involved with theories of human cultural evolution and the future of humankind. He dealt with these subjects in his 1962 work, *Mankind Evolving* (Dobzhansky, 1962), and later in *The Biology of Ultimate Concern* (1967). George G. Simpson, paleontologist and zoologist, was long a curator of the American Museum of Natural History and also a professor at Columbia University (1945-1959) and at Harvard University (1959-1970). It was at Columbia, where we were faculty colleagues, that I had an opportunity to meet him. He, too, in later years, turned to problems of human evolution, discussed in *This View of Life* (Simpson, 1964) and *Biology and Man* (1969). Bernard G. Campbell, formerly professor of anthropology at the University of California, Los Angeles, specialized in hominid evolution; he received his higher education at Cambridge University. I found most informative his major work, *Human Evolution: An Introduction to Man's Adaptations* (Campbell, 1985); it has a particularly valuable chapter on culture and society, with a section on the evolution of ethics.

Intellect is defined as "the capacity for knowledge; the capacity for rational or intelligent thought, esp. when highly developed; the power of knowing as distinguished from the power to feel and to will" (*Webster's*

Ninth New Collegiate Dictionary, 1985). As such, intellect is the power of the human brain to generate the fields of knowledge we have been analyzing in the previous four chapters. In this chapter, we look back in time for clues as to how this intellect originated and evolved.

The Human Mental Faculty

Ernst Haeckel, a German biologist and philosopher who earned the reputation of being "Darwin's German bulldog," wrote in his 1874 work *The Evolution of Man* to the effect that there is no fundamental difference between humans and the higher animals in their mental faculties. Can such a statement be sustained today?

The genus *Homo* may have entered the scene as far back as two million years ago, for a short time contemporaneous with *Australopithecus africanus*, from which *Homo habilis* is thought to have evolved. From *H. habilis*, the genus may have evolved into *H. erectus*. Then, about a half-million years ago, an archaic form of *H. sapiens* appeared. There followed Neanderthal Man, now considered a subspecies of *H. sapiens*. Modern *H. sapiens* appeared in Europe some two hundred thousand years ago.

From our standpoint, the most interesting evolutionary development in this sequence was the increase in the size of the brain, in the proportion of the brain that is cerebral cortex, and in the surface area of the brain. From an average brain volume of about 450 cubic centimeters (cc) for *A. africanus*, there was an increase to an average value of 1350 cc (range 1000–2000) in *H. sapiens* (Campbell 1985, pp. 231-33)[1]. Important in this great increase in brain volume was the disproportionately great increase in volume of the cerebral cortex, and in increased surface area accommodated by infolding of the surface layers of the cerebral hemispheres.

The above statistics on brain size might be used to support Haeckel's contention that human mental makeup differs from that of the nearest related mammals only in degree, the difference being quantitative but not qualitative. What, if anything, does increase in brain size have to do with possible qualitative differences? Is there associated with increase in brain size a fundamental new dimension in the mental faculties of *H. sapiens*, as compared with hominid predecessors and other higher mammals?

To find an answer to such questions, we need to review the consensus held by modern anthropologists about important changes in the mental activities and capabilities of the brain as *H. habilis* evolved from *A. africanus*, and subsequently into *H. erectus* and *H. sapiens*. Campbell

---

[1]From: Bernard Campbell, *HUMAN EVOLUTION*, Third Edition. Copyright © 1985 by Bernard Campbell. (Aldine Publishing Company, New York.) Adapted by permission.

writes: "In our consideration of human evolution and human culture it is necessary to refer to what has been described as a unique human mental characteristic: conceptual thought" (1985, p. 330).[1] (A footnote on his page reads: "Mental is the adjective of mind, the functional and subjective aspect of the living brain.")

Just what is *conceptual thought*? To get an answer we need to analyze the path of information entering the mind, and the storage of that information. This is a topic discussed in our investigation of science and the scientific method. Recall from Chapter 1 that a percept, the mental image of the external environment, consists of two types of information (1) input through the senses, and (2) memory of previous experience (Campbell 1985, p. 330). Primate perception is very special, as compared with that of other organisms, and particularly with that of other higher mammals. Campbell states:

> Primates, more than all other mammals (except bats), live in a three-dimensional world; their eyes are stereoscopic, and their movements are in all three planes of space. They must have precise ideas of spatial relationships, for arboreal locomotion involves a knowlege of space far greater than that which may be necessary to ground-living forms. . . . The integration of spatial data from the senses to form a composite perception of the environment has clearly gone further among primates than among other groups of animals. (Pp. 330-31)[1]

Campbell also concludes that the memory component of perception in primates inevitably comes to include, in addition to experiential record of events, some generalizations about spatial relationships. He states that by manipulation, "The higher primate can extract an object from the environment, free it, as it were, from spatial implication, and build up a perception of it as a distinct object, not merely as part of a pattern. In time, the primate will come to perceive the environment not only as a three-dimensional pattern but also as an assemblage of objects" (p. 331).[1] Campbell speculates on the effects of these special perceptive faculties:

> When humans began to hunt, human perception evolved accordingly. Using our primary sense, vision, humans evolved the ability to identify objects on the move without reference to their relationship to the fixed part of the environment; they saw them as totally separate from their environment. Here was a fundamental improvement in perception and something novel among land animals: a carnivore that hunted by sight.
>
> It is clear that manipulation and hunting came in turn to make the human perceptual world different from that of all other animals.

> Human analytic perception, more than any other factor, opened the door to the development of conceptual thought and eventually to symbolic culture. (P. 331)[1]

Although what Campbell is describing is hypothesis, it shows a rational approach to the linkage between biological evolution of the unique human perceptive faculty and the human culture that is unique in the biological realm.

Campbell next turns to concepts. In discussing science and the scientific method, we defined a concept as "an abstraction from the particular to the class" (Campbell 1985, p. 332).[1] He finds in the evolution of perception the need to imply "some degree of abstraction from experience, from the particular to the class" (p. 332).[1] Although, like perception, such abstraction is not a conscious activity, conceptual thought (which is a conscious activity) may have had its origin in the advanced perceptive process described above. Campbell considers that conceptual thought is an activity limited to humans; they alone can have conceptual thoughts about objects that are not concurrently visible to the thinker. In other words, humans are uniquely capable of imagination. He elucidates further on imagination: "Imagination is the consciousness of sets of concepts, which are the classification of experience" (p. 332).[1] Imagination permits humans to escape mentally from the present into the future, in order to plan or devise some act or artifact for future use. And now we come to what Campbell considers another unique mental faculty of humans:

> It has been said that what distinguishes humans from animals is the length of time through which human consciousness extends. In animals, this dimension is small, stretching a little way into past and future; in humans, it grows both qualitatively and quantitatively. The evolution of conceptual thought gives humans greater power to live in the past and in the future by abstraction from the past. (P. 333)[1]

Campbell finds in mammals evidence for a classification of experience that can be called "unconscious conceptualization," but in contrast, the conceptual thought engaged in by humans is conscious. He goes on:

> Human conceptual thought appears to be characterized particularly by its conscious nature, but it is no doubt the result of a steady process of evolution from less conscious and indeed unconscious concepts in primates. The human achievement was the fully conscious concept of things not possessed but needed; the recognition of game, weapons, women, or children as classes brought with it the classification of more and more of the environment and the possibility of foresight of future needs. Leaving behind the narrow

> limits of present time experienced, humans entered the broad expanse of past memory and future concepts. (P. 333)[1]

Dobzhansky recognizes the conscious nature described above as *self-awareness*. For humans, he states:

> In point of fact, self-awareness is the most immediate and incontrovertible of all realities. . . . Only by analogy can I infer that other humans have self-awareness—they usually act as I do in situations in which I know that my self-awareness is involved. . . . Human self-awareness obviously differs greatly from any rudiments of mind that may be present in nonhuman animals. The magnitude of the difference makes it a difference in kind, and not one of degree. Owing primarily to this difference, mankind became an extraordinary and unique product of biological evolution. (1977, p. 453)[2]

Note that this statement does not exclude the presence of some level of self-awareness among the other primates and other mammals. The point is that self-awareness is considered to be much more acute and wide-ranging in time and space than for other members of the broader taxa. Dobzhansky adds another special form of human awareness that is unique: *death awareness*. "There is no indication that individuals of any species other than man know that they will inevitably die." He goes on:

> Foreseeing the remote future and planning for future contingencies require capabilities that we know exist only in the human mind. Self-awareness and death-awareness are probably causally related and appeared together in evolution. They appeared because they enhanced the adaptedness of their possessors. The adaptive role of self-awareness is sufficiently obvious, no matter how elusive self-awareness may be. It is an integral part of the complex of adaptations that include the use of symbols, language, and hence acquisition and transmission of culture. (P. 454)[2]

Language, Memory, and Intelligence

Campbell continues with his exposition of the unique qualities of the human mind (1985). In the field of language, he points out that the language of humans is based on the use of symbols. While the symbols must be learned (a cultural phenomenon), the conceptual thought that is required for the formulation of language is a product of evolution. Thus, animals do not communicate with language that uses word symbols; their communication is based on signs (or signals). In animals, "communication is a means by which one organism can bring about change in one or more others; . . . it is the means by which one animal can trigger a response in another. . . . In all animals communication is nonverbal; only among humans is language evolved to supplement nonverbal signals" (p. 342).[1]

Campbell states that all evidence points to the conclusion that the ability to use language is a genetically determined character of modern humans. The language itself is part of culture—an artifact: "This remarkable, priceless, and uniquely human kind of behavior rests on our peculiar endowment of a large, slowly-maturing brain and the peculiar controlled musculature of our vocal tract" (p. 355).[1] Summing up the importance and uniqueness of language, Campbell states: "Language has made possible the vast development of human culture, and brought us the unique human consciousness of ourselves and others" (p. 354).[1]

On the subject of memory, Campbell distinguishes three levels of memory. Experiential memory is a continuous record of the stream of experience. This form of memory operates at the subconscious level and is found in animals as well as humans. Conceptual memory is "the accumulation of abstract concepts about experience" (Campbell, p. 335).[1] While conceptual memory is found in a simple form in animals, it is highly complex in humans. Humans have the unique ability to experience emotions without reference to directly attached objects. Humans have the ability to store in memory highly abstract concepts based on symbols, as, for example, in mathematics. A third class is word memory, closely linked with but distinct from conceptual memory. Only humans can possess word memory.

On the subject of intelligence, Campbell states:

> Intelligence has been defined as the relating activity of mind—the ability to realize the connection between discrete objects and events. It clearly involves imagination, which may be considered to be the presentation from the experiential memory of memory traces (knowledge) not obviously directly connected with the immediate experience. It may involve conscious recall of memory, but the significant feature of imaginative thought is that a much broader range of memory traces is brought to the interpretation of day-to-day experience than in the process of simple unconscious learning, for only thus could the imaginative connection between events be realized. (P. 338)[1]

Intelligence cannot exist alone, Campbell states, "it is a capacity that interacts with knowledge, the accumulation of experiential and conceptual memory" (p. 338).[1]

Campbell's complete picture of the evolution of human mental faculties is carefully reasoned and is based on generally acceptable psychological theory of mental activity. It makes the point that, by the process of biological evolution, the human mind became unique in quality from that of all other primates and all other animals. This hypothesis of mental evolution requires us to question Haeckel's statement to the effect that human mental activity does not differ qualitatively from that of other animals. The rise of human culture accompanied the evolution of the human brain functions and indeed would not have been possible without the emergence of the unique human ability for conscious conceptualization and the specific phenomenon of imagination extending far into the future, but making use of experiential memory extending far into the past. Ethics as we know it among humans could not have developed prior to the development of those unique human mental faculties and the ability to communicate by use of language.

Although I felt satisfied with Campbell's and Dobzhansky's analyses of the evolution of the human brain and its unique mental capacities, I decided to seek a third interpretation, which I found in the writings of G. Ledyard Stebbins, an eminent authority on genetics and evolution. He received the Ph.D. degree from Harvard University in 1931 and has fulfilled a long career of teaching and research in the University of California, first at Berkeley and later at Davis. His book, titled *Darwin to DNA, Molecules to Humanity*, appeared in 1982 and is intended for the general reader (Stebbins, 1982). I found his style extremely lucid, and I strongly recommend the volume to those wishing to fill out a background on modern views of the evolutionary process. For the most part, his views on evolution of human mental functioning parallel those I have just reviewed, but with some variations in terms and some new insights.

How Distinctive Is Humanity?

In answer to the question "How distinctive is humanity?" Stebbins notes the many observed similarities in behavior between apes and humans. Despite seeming likenesses in many respects, Stebbins finds fundamental differences that far overshadow the apparent similarities:

> Given this great degree of similarity, must we conclude that humans are only sophisticated apes endowed with enormous technical prowess but otherwise only quantitatively different from animals? I think not.
>
> Human society must be regarded as containing many novel traits. The family life of humans, including our attitudes toward relatives and friends, could be reasonably well predicted on the basis of a thorough knowledge of chimpanzee and gorilla societies. Nevertheless, these apes have not developed any form of organization that resembles even remotely the human division of labor, disciplined

> armies, religions, nations, and international organizations. Modern human societies are qualitatively as well as quantitatively different from all existing animal societies. (1982, p. 363)[3]

Stebbins names three distinctively novel human characteristics: artisanship, conscious time-binding, and imaginal thinking. *Artisanship* is a specific capability not mentioned by Campbell or Dobzhansky. Stebbins considers artisanship to be "the transcendent outcome of evolution in tool-making" (p. 363).[3] He notes that it is present in rudimentary form in chimpanzees, but he adds: "One could not predict the Parthenon, the pyramids, or the Empire State Building from watching pre-Olduvian hominids 'fishing' for termites with sticks" (p. 363).[3]

Conscious time-binding is essentially the capacity for conceptual memory we have already noted. It refers to the ability of humans to plan far into the future on the basis of experiential memory going far into the past. Stebbins suggests that this ability developed in connection with cooperative hunting and the taming of fire. He calls attention to carved animal bones, interpreted as lunar calendars, as evidence of conscious time-binding (p. 364). *Imaginal thinking* we have covered in some detail under the heading of human imagination and its uniqueness. Stebbins stresses the importance of symbolic language in promoting both conscious time-binding and imaginal thinking.

Stebbins attributes the unique combination of the three human characteristics just cited to biological evolution of the human brain: "From the biological viewpoint, these characteristics are based on nothing but a quantitative increase in the number of cells that form the surface layers of a single organ, the neocortex of the brain. This genetically determined change, interacting with the rudiments of culture that was already present when it began, triggered a direction of cultural evolution that brought novel and transcendent social behavior" (p. 365).[3]

I should insert a note of clarification here as to Stebbins's use of the word "quantitative." The individual brain cells to which he refers are viewed as essentially similar (qualitatively the same) as those in the brains of the apes and the hominid ancestors of the present human species. This should not be interpreted to mean that the difference between the human brain and the ape brain is therefore merely a quantitative difference. The vast increase in the number of cells in the human neocortex, as compared with that of the apes and early man, gives to the whole brain as a functioning organ a qualitative uniqueness. Stebbins clarifies this concept in an earlier chapter of his book, where he explains how *quality* differences arise as a result of complex *quantitative* differences (1982, pp. 147-50).

Stebbins summarizes the enormous consequences of the evolutionary development of the human brain in three categories to its extraordinary level of uniqueness:

> These three qualities are the foundations of the three principal realms of human knowledge. Natural science and engineering are the outgrowth of prescientific artisanship. History, political science, and other branches of the social sciences are fundamentally ways of directing social behavior to avoid disaster and to improve the material state of mankind. The humanities—literature, the arts, and philosophy—are extensions of imaginal thinking. The fact that artisanship, conscious time binding, and imaginal thinking evolved gradually by means of quantitative physical changes that affected a complex of behavioral characteristics does not detract in any way from the novel or transcendent quality of the human way of life. (Pp. 365-66)[3]

What kind of evolutionary pressure induced this remarkable development of the human brain? Science writer Roger Lewin, reporting in *Science* on a meeting of paleoanthropologists in Cambridge, England, in 1987, reviewed the "group-against-group" hypothesis (Lewin 1987). It identifies an environmental force judged to have been more effective in human brain evolution than other forces traditionally listed under Darwin's "hostile forces of nature" (Darwin 1859, Chapter 3).

The group-against-group hypothesis, attributed to Nicholas Humphrey of the University of Cambridge, was expanded upon by Richard Alexander of the University of Michigan. Alexander is quoted as saying: "The only plausible way to account for the striking departure of humans from their predecessors and all other species with respect to mental and social attributes is to assume that humans uniquely became their own principal hostile forces of nature" (Lewin 1987, p. 669). Lewin explains:

> The key engine in this evolutionary drive, suggests Alexander, is a positive feedback resulting from the close match between the competitors—human versus human—in the battle for survival. "In social-intellectual-physical competition [members of the same species] are likely to be—as no other competitors or hostile forces can—inevitably no more than a step behind or ahead in any evolving system of strategies and capabilities. Evolutionary races* are thus set in motion that have a severity and centrality as in no other circumstances." The result is a "runaway" evolutionary trajectory that is analogous to the mechanism favored by some biologists for the phenomenon of sexual selection that, among other things, produced exaggerated features such as the peacock's tail.
>
> [*Here, "race" means a contest, not an ethnic race.]

The central concept expressed by Alexander follows Charles Darwin, whose preview of Chapter 3 reads: "Struggle for life most severe between individuals and varieties of the same species; often severe between species of the same genus" (Darwin 1859, p. 114). Alexander likes the group-against-group hypothesis because "it can explain any size or complexity of group; it accords with all of recorded human history; it is consistent with the fact that humans alone play competitively group-against-group on a large and complex scale; and it accords with the ecological dominance of the human species" (Lewin 1987, p. 669). Lewin explains further:

> At the core of this evolutionary explanation is the idea that intense social interaction and manipulation demands unprecedented skills in dealing within one another. Human intelligence and human reflective consciousness are therefore seen as the product of natural selection for dealing with the most challenging things in the human environment: other humans, not, as usually has been assumed, technological exigencies. (P. 669)

Speculations on the Appearance of Ethics

Writing on the appearance of ethics, Bernard Campbell considers it "one of the most important events in human evolution—the evolution of ethics. Ethics arose as a direct result of the appearance of self-awareness in the growing human consciousness" (1985, p. 359).[1] Campbell explains: "Humans could see, as a result of their self-consciousness, how many of their activities were directed to satisfy their own needs, their basic requirements for life; but they could also see that certain of their actions satisfied only social needs and led not to personal satisfaction but perhaps to frustration" (p. 360).[1] "Ethics therefore arose when humans found that they had to make conscious choices in a social context" (p. 362).[1] Campbell refers to earlier speculations that the need for values arose from anxiety, from the conscious frustration of human needs: "Internal conflict and frustration may, however, be among the most important stimuli of cultural progress, of the development of adaptive behavior and technology" (p. 362).[1]

Dobzhansky also fits the rise of ethics into human evolution: "There are two interesting sources of ethics and values—cultural and biological. The ethical standards of every individual are imparted, mainly in childhood and youth, by other members of the society for which the individual is being prepared. Ethics are acquired, not biologically inherited" (1977, p. 455).[2] Yet, although the particular moral values—the "oughts"—are cultural (artifacts), the ability to see ethical issues is a genetically controlled capability. Dobzhansky quotes from C. H.

Waddington (1960) to the effect that humans are genetically determined "ethicizing beings" and, especially in childhood and youth, "authority acceptors." Dobzhansky continues:

> Every member of a human society must become familiar with the ethical and value systems of his society. Failure to do so makes him a misfit or an outcast; it jeopardizes his and his progeny's success and survival. Therefore natural selection has exerted pressure to insure that every member of the human species comes into the world with a genetic endowment making him an "ethicizing being." (P. 455)[2]

This, of course, is a purely naturalistic hypothesis—part of the total theory of biological evolution—and you might be prompted to conclude that ethics does, indeed, have a mechanistic base that can be explained by the scientific method.

That conclusion might, in turn, seem to support the hypothesis that an ethical system is achieved wholly through natural processes. Dobzhansky addresses himself to that conclusion, for he points out that the ethics of a particular society are not genetically derived: "Anthropologists have ample evidence that different cultures demand different modes of behavior, and that these demands are usually complied with" (p. 456).[2] While some human attitudes and evaluations that seem quite instinctive (inherited) may seem to be based on value judgments, it is more reasonable to suppose that they were developed purely by survival pressures. For example, in the human family, high value is placed on motherhood, while children are cherished and loved. These values directly promote the survival of the family, and they are also displayed to some degree in animals. But, says Dobzhansky:

> This can hardly be said of many other ethics and values that are recognized in most, if not all, human societies as valid. For example, it is wrong to steal, swindle, rob, waylay, or murder other people, especially members of one's own group or society and, by extension, any human being. This is wrong even if so doing is profitable, the misdeed is undetected, and no vengeance or retribution is to be feared. On the contrary, honesty, generosity, and veracity are praiseworthy, especially if they bring hardships to persons who practice them. Human life, that of a stranger no less than that of a relative, is sacred, with the significant exception of war. Life is to be preserved at all costs (including that of incurably ill persons whose existence may be sheer misery). At the summit of ethics, we have the commandments of universal love (including one's enemies), service to others, and resistance to evil. (Pp. 456-57)[2]

It is evident in the above statement Dobzhansky has taken the quantum jump from that which is an empirical concept (survival of the fittest) to the transempirical realm, which is couched in language not amenable to evaluation by empirical science—notice the burst of emotive words, such as "wrong," "misdeed," and "praiseworthy," and even such excessively emotive phrases as "life is sacred" and "universal love."

A transempirical concept is a product of the human brain through ideation. Its naturalistic origin notwithstanding, the basic nature of the transempirical concept is such that its truth-status as a human judgment cannot be confirmed or denied by empirical (scientific) methods. As already explained in Chapters 9 and 12, "cannot be confirmed or denied" means that if an alternative judgment were to be assigned to the same object of valuation, the alternative would enjoy equal truth-status. Thus, if one individual says "action A is good," equal status applies to another individual's counter-statement that "action A is evil." Another characteristic of transempirical concepts, and perhaps a descriptive criterion as well, is that they arise from emotive mental processes and attitudes; i.e., they are always associated with an individual's feelings about what is desirable to possess or to have accomplished. Recognition of "that which is transempirical" as distinct from "that which is empirical" is the basis of a dualistic ontology that can operate in total freedom from the supernatural realm.

Returning now to Dobzhansky on ethics, he is well aware of the two ontological realms we have identified. He makes a clear distinction between two kinds of ethics:

> *Family ethics* are shared by man with the "quasi-ethics" of at least some animals; in animals as well as in men, many family ethics are genetically conditioned dispositions (although in man they may be overcome by an exercise of will). Family ethics can be envisioned as products of natural selection, which established the genetic bases of these ethics in our ancestors as well as in other animal species. *Group ethics* are products not of biological but of cultural evolution. They confer no advantage and may be disadvantageous to individuals who practice them, although they are indispensable to the maintenance of human societies. Natural selection has not made man inherently evil (as is so readily assumed by believers in original sin or proponents of territorial and other "imperatives"). Whatever proclivities to selfishness and hedonism man may have, he also has a genetically established educability that permits him to counteract these proclivities by means of culturally derived group ethics. Natural selection for educability and plasticity of behavior, rather than for genetically fixed egoism or altruism, has been the dominant directive factor in human evolution. (1977, p. 457)[2]

It seems reasonable to me that group ethics, heavily involved with transempirical concepts, would have developed concurrently with religion, and that the moral precepts would have been enforced by religious sanctions. This linkage could have developed as soon as language was capable of communicating abstract ideas, but before the invention of writing. Abstract transempirical value concepts are much easier to hand over to supernatural authority than to justify rationally. Perhaps it was simply a case of following the path of least resistance. Family ethics, being much simpler in its concepts, would have long preceded group ethics, since the former would have been much simpler to comprehend and to communicate.

The third evolutionary biologist whom I named early in this chapter, George Gaylord Simpson, has a lot to say on the subject of "biology and ethics" in a chapter of that title in his volume *Biology and Man* (Simpson 1969, pp. 130-48). He describes among biologists "the search for naturalistic ethics, that is, an ethical or moral system rationally related to the nature of things or to the material universe" (p. 130).[4] To my surprise, I found that Simpson feels quite optimistic about the prospects for achieving a naturalistic ethics.

First, however, Simpson refers to what philosophers of ethics call the naturalistic fallacy, an error in logic that takes place when one passes from pure description of nature ("it is" or "it is not") to the conclusion prescribing what should be the case ("it ought to be" or "it ought not to be"). In Chapter 12, we used Simpson's example of the naturalistic fallacy from biology, in which Alfred Kinsey claimed to have discovered that homosexuality was more prevalent than previously thought. On the grounds that the phenomenon exists in nature, he recommended that laws prohibiting the practice be modified to reflect the conclusion that it cannot be morally wrong. But Simpson has a good point to make:

> It is undoubtedly illogical to conclude that what *is* therefore *ought to be*. It is, however, equally illogical to make that the basis for a further conclusion that decision as to what ought to be cannot be based on consideration of what is—in other words, that naturalistic ethics are impossible. (P. 132)[4]

> Let us not be too dismayed if some attempts to set up naturalistic ethics turn out themselves to be fallacious, or if we cannot all agree as to either origin or criteria for naturalistic ethics. It would be enough to go on with if we could conclude that naturalistic ethics are eventually possible. In fact, I believe we can do considerably better than that. (P. 133)[4]

Naturalistic ethics, Simpson notes, is usually considered in the context of biology and evolution. Perhaps he is referring to the writings of Konrad Lorenz of the mid-1960s. As did Dobzhansky, Simpson refers to Waddington's assertion that humans have the genetic predisposition to ethicize and to accept authority. Simpson writes:

> In the evolutionary context, the problem really becomes one of why and how organic evolution produced an animal capable of cultural evolution and of the ethicizing that helps to mediate cultural progress. There is no real doubt, and neither Julian Huxley nor Waddington has doubted, that the capacity or, one can say, the necessity for ethicizing is in fact a *biological* characteristic of the human species developed by natural selection because it is adaptive for the species. In generally biological but nontechnical terms, the direction of human adaptation early became one depending on individual flexibility with mainly learned abilities, with alternatives of action, and with consequent responsibility for those actions. (P. 134)[4]

After discussing and rejecting the then-prevalent Freudian theory that involved inheritance of acquired characteristics, Simpson returns to reinforce his view of naturalistic ethics:

> It is thus plausible and indeed, I think, practically certain that ethicizing, the capacity and the necessity for some system of ethics, arose in the course of human evolution in a completely natural way. That does not mean that men early and generally recognized the natural origin of the moral sense or that they adopted specific ethical systems that were naturalistic. We all know that they did not. (P. 135)[4]

At this point, Simpson turns to the introduction of systems of revealed or inspired religion that would have taken advantage of the ethicizing and authority-accepting capability in humans. He is, as one might expect, skeptical of the value of religious ethics in terms of promoting survival. He gives two reasons. "In the first place, no given one of the systems of revealed or inspired religion and ethics is necessarily adaptive biologically or specially" (p. 135).[4] Those who promoted and manipulated those ethics may have had ends other than group survival in view. There were "bitter conflicts between rival systems of supposedly revealed ethics," and these "are obviously inadaptive even when either system alone might be sufficiently adaptive for survival" (pp. 135-36).[4] Simpson then points out that ethical systems that might have been adaptive for survival under primitive tribal or pastoral conditions would almost certainly be nonadaptive under today's social and environmental conditions. As a second reason for not placing reliance in traditional religion-based ethical systems, Simpson argues that large segments of society today (presumably atheists and secular humanists) "will not accept fiat as a substitute for reason." Unless based on reason, those systems will not be functionally adaptive today. Simpson concludes: "Therefore, continued human welfare requires ethical systems that are not supernaturalistic but are naturalistic" (p. 137).[4] I presume that what he means is that ethics derived by reason in consideration of present environmental conditions is constructed in such a manner as to promote human survival. As I see it, this does not mean that such ethics will actually pass from the category of acquired culture to that of the genetic endowment by natural selection. Shortness of time and the rapid flux of social change could not permit natural selection to do its work in the Darwinian sense.

Simpson leaves the biological area of survival ethics and enters the area of value judgments. This is a quantum jump to the transempirical realm. He reviews the theory of society as the supraorganism (Gerard; Comte), in which the "good of the state" requires sacrifice of the good of individuals. Nothing in evolutionary biology would, as I see it, point to recognition of social organization on the level of nations. The ethical concept of a nation having intrinsic value is better viewed historically as an artifact of religious origin.

Simpson refers next to "numerous biological phenomena that have been considered general tendencies of evolution and have been involved in proposals of evolutionary ethics" (p. 139).[4] These are technical points and include some of the unique features of living organisms which I discussed in Chapter 4: for example, the ability of organisms to decrease system entropy, the ability of organic systems to achieve steady states, and the general increase of complexity of organisms through time. Simpson does not think these trends or tendencies are relevant to the question of ethics: "I do not believe there is an 'ought' inherent in any of these processes" (p. 140).[4] In other words, what have often been referred to as biological laws are not in themselves ethical in content.

Taking up next the subject of ethics, Simpson makes his final encounter with the quantum step I have referred to above—the shift from empirical to transempirical ideas and constructs. Referring to Julian Huxley as an exponent of trend ethics, Simpson states: "An overall evolutionary trend is inferred or postulated. It is then concluded that the continuance of that trend is desirable and that whatever promotes it is ethical" (p. 140).[4] In a quotation from Huxley (not referenced) the word "right" is used repeatedly to describe the supposed goal of further social evolution, which is both to respect human individuality and to encourage its fullest development. Huxley defends his stated evolutionary trend on the grounds that its organic phase would constitute an "improvement" in society. Simpson does not let this assertion get by him, for he states:

> "Improvement" is an evaluating word. It means "change for the better," so that when unlimited improvement is taken as the criterion for the ethically good, what is being said is simply that it is good to become better. The argument becomes circular to the point of meaninglessness unless some external, independent criterion of "better" is found and applied. It is surprising and, at first sight, dismaying to find Huxley maintaining that the concepts of progress, of higher and lower, of better and worse, "spring automatically to the mind" and are known "as an immediate and obvious fact." If, after all, we are to take intuitive concepts as ethical principles, then purported derivation of naturalistic ethics from the facts of evolution is either irrelevant or spurious. (P. 140)[4]

Simpson continues with a review of C. H. Waddington's views on evolutionary ethics, which closely resemble those of Julian Huxley and are subject to similar criticisms. Skipping over this rather involved analysis, I turn to Simpson's conclusions. One is "that neither strictly organic nor social evolution necessarily leads to improvement in any sense of the word acceptable for the human situation. In fact either, in their ways that *in themselves* are blindly amoral, may have results extremely undesirable for mankind" (pp. 145-46).[4] Simpson sees a saving grace in the knowledge that both organic and social evolution are now to some degree, though limited, under human control, and that we can work deliberately for the desirable. Of course, in using the words "desirable" and "undesirable," Simpson stumbles into the morass of the transempirical. Whether "desirable" means "to promote survival of the human race" or anything else, it is a value judgment that is religious in nature, for it refers us to the "ulterior" questions: "Why are we here? What value have we? To whom or what are we valuable other than to ourselves?"

Although Simpson does not come right out and say what I have covered in the foregoing sentence, his concluding statement shows that he is well aware of what is derivable by reason and what is beyond either formal logic or empirical analysis:

> Man has risen, not fallen. He can choose to develop his capacities as the highest animal and to try to rise still farther, or he can choose otherwise. The choice is his responsibility, and his alone. . . . Evolution has no purpose; man must supply this for himself. . . . It is futile to search for an absolute ethical criterion retroactively in what occurred before ethics themselves evolved. (P. 148)[4]

Thus far, we have identified one underlying causative force—organic evolution—in the emergence of ethical systems within large groups of humans, but we have not really identified a viable cultural driving mechanism that can tell us how and why the specific "oughts" of an ethical system were forced into existence. In a later section, we will present an explanation to fill that need. However, we postpone consideration of that hypothesis of origin of religious/ethical systems in order to investigate further the phenomenon of "sociobiology," already introduced in concept by Simpson, but until the early 1970s, lacking that fancy name.

Two Behavioral Controls: Biological and Cultural

Before investigating further the possible biological origins of ethics and religion, we need to examine the broader topic of the possible interplay of two forms of control over basic human behavior—the biological control through genetic change on the one hand, and the cultural control through information transmission and learning, on the other.

The phenomenon on which we are focusing attention is human behavior in its broadest sense—what humans do and how they do it—but with particular emphasis upon human functions in a society of humans. We are concerned with how people act in relation to one another and to the environment and the kinds of artifacts (using the term broadly) they produce. This is in the province of the science of sociology. Most particularly, we are interested in how people act when they have available choices of two or more ways in which to act. We want to understand how the choices they make influence the likelihood that they will survive as individuals or in groups and continue to reproduce themselves.

G. Ledyard Stebbins covers this subject in a remarkably clear fashion (1982). Changes in human behavioral patterns through time, starting from, say, the earliest species of genus *Homo*, have, in general, evolved from simpler to more complex and more sophisticated modes. The driving mechanisms of observed changes—biological and cultural—are continuously interacting with a possible built-in feedback mechanism that makes it exceedingly difficult for us to separate their effects. It is to this problem that Stebbins devotes a chapter titled "The genetic and cultural heritage of humanity" (pp. 369-82).

The totality of changes observed in human behavior as time passes can be described as a form of evolution, which simply means a process of change through time. The total picture of change is commonly thought of as consisting of two kinds of evolution—genetic evolution and cultural evolution. These terms can be confusing; they must be carefully defined and their meanings closely scrutinized.

Genetic evolution is responsible for physical changes in the human body and in the capability of the human brain to perform the higher functions we have already examined. Stebbins states that genetic evolution, "established in populations by the action of natural selection modified by chance events, has provided us with a wide range of physical and mental capacities" (pp. 369-70).[3] *Cultural evolution* makes use of unique mental capabilities in the making of conscious choices and decisions leading to improved ways to act in given situations and in developing new and improved artifacts. Once invented or discovered, these improved behavioral changes are dispersed through the population by imitation and, more effectively, by language. The knowledge itself is passed on to succeeding generations by those same methods of transmission. Nothing in the innovative or inventive process is directly related in any physical sense to human genes. There is, however, an indirect connection to genetic evolution through a feedback process that is acceptable in terms of modern genetics.

The feedback mechanism between culture change and the genetic endowment is thought of as a mechanism of selection—selecting for survival, that is, of those groups whose innovations and inventions improve their chances of survival and reproduction, as compared with groups with lesser inventive capabilities. The feedback circuit is described by Stebbins in these words:

> As humans relied more and more on artisanship for survival, genes that increased the capacity for making better tools, shelters, and clothing and for cooperation through sharing and division of labor spread more and more widely because of their increased adaptive value. At the same time, societies that used their capacities more efficiently could acquire more food and defend themselves better against predators, and thus improved their chances for survival and reproduction. (P. 370)[3]

Let me try to clarify the distinction between the cultural feedback mechanism of genetic change and the basic mechanism of genetic change implied in natural selection as it functions throughout the plant and animal kingdoms. As an example, take the evolutionary change of some part of the human body. At a time when (as an older hypothesis goes) early hominids took leave of the dwindling tropical forests of Africa and occupied the expanding savanna grasslands, available foods in the new environment consisted largely of hard seeds or other tough plant tissues. Through mutations, certain individuals received a set of teeth better adapted to masticating these new food varieties, e.g., harder enamel and/or flatter molar crowns. Such individuals would be better adapted to the changed environment, and the probability for survival through adulthood and thus for propagating offspring would be increased. In due time, the genes coding for the improved dental equipment would spread through the population. The important point here, and in all similar cases, is that the individual human had no choice in selecting the improved teeth. No tray of tooth types was set before the individual with instructions to pick the one that seemed most likely to make chewing easier. Evolutionary changes in the human tooth came about in the same nonconscious framework as the evolutionary changes in the tooth of a beaver or a gopher. Noninvolvement of the thought processes is the essential point. In contrast, cultural feedback involves conscious choice, or at least the possibility of choice. Which of two hunting strategies shall we adopt? Which of two kinds of storage sites for nuts is the safer? Shall we go over to the neighboring tribe and steal some babies for adoption, or is the risk of retaliation too great? Choice in itself could have been a basic factor in inducing genetic improvements in the human brain. The better equipped one individual's brain is to conceptualize, to remember, to imagine, and to invent, the better would be the choice made by that individual in a given situation involving survival.

Philosophers point out that ethics always involves choices of action. It is easy to understand how the selection of morals at the family level of interaction could have adaptive significance through the feedback mechanism. The distinction between genetic change without choice and with choice based on the ability to conceptualize is encompassed in the term *coevolutionary circuit*; i.e., from genes to culture and back, found in sociobiology (see Edward O. Wilson and Charles Lumsden, 1981).

Stebbins continues to develop the concept of two kinds of selection in evolving humans. He notes that in time they would come to have almost equal importance for human evolution (p. 370). But, in due course, cultural evolution would have become the dominant means of behavioral or social change. As the content of the culture itself became more and more complex, and the store of information required to sustain it became vast, genetic adaptation by increase in mental capacity would have become inadequate to keep pace. This is suggested by the observation that brain size has not increased in modern humans as compared with Neanderthal Man. The most important factor, however, would be the increased rate of cultural change in contrast to the slowness of the process of genetic change. There would come a point at which genetic evolution (unguided by deliberate selective breeding) could play no significant role in social change. Cultural evolution would then be required to carry the whole burden of implementing and guiding change, using an effectively unchanging mental capacity. We must wrestle with the problem of extinction by nuclear weaponry using a genetic mental endowment that is no greater than it was in the time of Plato or Aristotle or even inhabitants of an earlier time, when civilizations first arose in Mesopotamia and the Nile Valley.

Sociobiology

At this point, we insert a short review of the development of modern sociobiology to obtain a better perspective on the problem of the origin of ethics and morality. We made brief mention of sociobiology in Chapter 3 in the context of examples of the ruling hypothesis. The popular term for the controversy we looked at is "nature/nurture," signifying the conflict between the strongly entrenched cultural school of anthropology that claims human behavior is almost entirely acquired, and the sociobiologists, reviving the thesis that behavior is strongly controlled by genes and thus in part a product of biological evolution through natural selection. Our previous section in this chapter has introduced the concept of coevolution, a central thesis of sociobology.

Sociobiology was preceded in the 1960s by the extension of modern ethology (the science of animal behavior) into the area of human behavior and human traits, especially in reference to human aggression. Konrad Lorenz was a leader in modern ethology who ventured into what are now called sociobiological areas in his 1966 work, *On Aggression*. In rereading his Chapter 13 titled "Ecce Homo," I find him saying that our understanding of the dismal track record of humanity in managing its affairs can be explained "if one assumes that human behavior, and particularly human social behavior, far from being determined by reason and cultural tradition alone, is still subject to all the laws prevailing in all phylogenetically adapted instinctive behavior" (p. 237). On the subject of aggression, Lorenz says that in 1955 he had written that human aggressive drives "simply derive from the fact that in prehistoric times intra-specific selection bred into man a measure of aggression drive for which in the social order of today he finds no adequate outlet" (p. 243). In the same chapter, however, Lorenz gives full credit to the dominance of culture evolving at a pace that greatly outstripped the snail's pace of biological evolution.

The "splash" of the 1970s is credited to two Harvard professors, Robert L. Trivers and Edward O. Wilson. Trivers, as a graduate student at Harvard in the early 1970s, wrote a paper that added a new concept to the social behavior of animals (Trivers, 1971). He was specifically dealing, not with family ethics (as we have defined it), but with group ethics—relationships of individuals to members of the social group outside the individual's immediate family. Trivers proposed the existence of a phenomenon called *reciprocal altruism*, in which the individual performs altruistic acts toward another individual outside the family. An example of an altruistic act would be food-sharing; another, the giving of warning calls (signals) of an approaching dangerous predator. This activity serves to save or protect the lives of other individuals, while involving some sacrifice or risk on the part of the giver. Bernard Campbell describes the principle of reciprocal altruism:

> This refers to behavior that is likely to be reciprocated during an individual's lifetime and where the risk to the altruist is low in relation to the risk to the victim being rescued. If conditions are reversed at some future time, the altruist may benefit greatly from

> his original act by having his own life saved. Members of social groups will tend to encourage such behavior for their mutual benefit, so that reciprocal altruism will become a part of the culture of the society. (1985, p. 361)[1]

In improving the chances of the altruist's staying alive longer than otherwise might be the case—i.e., to increase fitness for survival—the tendency to act altruistically becomes genetically fixed in the group population. The evolutionary concept has since been extended to group activities under the concept of *group selection*, in which selection operates on groups as well as individuals (Campbell 1985, p. 361). As a result there arises a genetic endowment expressed in such social attitudes as "team spirit, true altruism, bravery in battle, patriotism, and so on" (p. 361)[1]. This concept seems to fit in nicely with the "group-against-group" hypothesis (discussed on earlier pages) of the forcing mechanism of human brain development. Campbell sees in group selection for altruism a way in which "a society could develop an ethical system that would strongly channel human behavior for social ends" (p. 361)[1]. The following description of how reciprocal altruism might be seen as an empirical basis for morals is from a popular article jointly authored by Michael Ruse and Edward O. Wilson:

> Nature, therefore, has made us (via the rules) believe in a disinterested moral code, according to which we *ought* to help our fellows. Thus we are inclined to go out and work with our fellows. In short, to make us altruistic in the adaptive, biological sense, our biology makes us altruistic in the more conventionally understood sense of acting on deeply held beliefs about right and wrong. (Ruse and Wilson 1985, p. 51)

Edward O. Wilson, a zoologist and entomologist, gained in the public eye the perhaps unenviable position of "innovator" of sociobiology when, in 1975, his book, *Sociobiology: The New Synthesis*, was published by the Harvard University Press. Sometimes in the history of science a particular idea or hypothesis is evolved and discussed for a long time—even decades—before it hits the fan of public attention. What sometimes triggers this splatter phenomenon is the invention of a buzzword that catches the public fancy, not only for its intrinsic value in a special field, but also for its wide-ranging implications extending into fields of general human interest. In using "sociobiology" in his book title, and devoting a final chapter to the sociobiology of humans, Wilson brought down a storm of criticism against the "nature" side of the nature/nurture debaters.

Defined as simply "the biology of human behavior," sociobiology was nothing new. Konrad Lorenz and his collegues Karl von Frisch and Nicholas Tinbergen had seen to that in scientific research that brought them jointly a Nobel Prize in medicine and physiology in 1973. Their predecessors in ethological speculation were Herbert Spencer, Charles Darwin, and William James. The early ethologists had already touched off a storm of public indignation through the implications of Social Darwinism, bringing on a severe depression to that line of investigation among anthropologists. (Social Darwinism was a popular doctrine of the sociopolitical ideologies, justifying as a natural phenomenon the "survival of the fittest" in application to human social and political behavior.)

Most of Wilson's 1975 treatise was devoted to producing a unifying explanation of the systems of social insects. Wrote Jeffrey Saver: "Meticulously extending this approach to the many thousands of social animal species, Wilson in *Sociobiology* provided an encompassing evolutionary account of the origin and maintenance of the disparate patterns of sociality displayed by birds, primates, insects, and man" (1985, p. 15).

On the other side of the Atlantic, sociobiology was being brought to public attention by zoologist Richard Dawkins, University Lecturer in Animal Behavior at the University of Oxford. Dawkins had selected as the title of his popular 1976 book, *The Selfish Gene.* On his first page, Dawkins flatly stated: "My purpose is to examine the biology of selfishness and altruism" (1976, p. 1). Using a term we usually associate with deplorable social behavior—selfishness—to describe our genes introduced an anthropomorphism. Not only did this term liken our DNA to a thinking, planning social human being, but it clearly suggested that certain of our genes code for immoral behavior vis-à-vis moral or "good" behavior. Of course, Dawkins had also adopted Trivers's equally suggestive word "altruism," used in 1971 to describe the concept of reciprocal altruism. Altruism, usually thought of by lay persons as a desirable moral act and the antithesis of selfishness, could easily be construed as an expression of "good" (generous, loving) genes. That wasn't at all the intent of these authors. Far from it, "altruism" was intended to describe a behavior that promoted the welfare of an individual's genes (genome), and could, from the genes' standpoint, be thought of as selfish. Dawkins clarifies this point when he states: "This book will show how both individual selfishness and individual altruism are explained by the fundamental law which I am calling *gene selfishness*" (p. 7). The anthropomorphism was only exacerbated by Dawkins's use of "good genes" and "bad genes" (p. 41). Definitions of those terms are, however, impeccably correct according to evolutionary biology, and any biology teacher could only applaud such metaphorical exposition as a splendid way to get students interested in evolution. In that respect Dawkins's books (another is *The Blind Watchmaker*) have been exemplary.

Wilson's massive *Sociobiology* was followed in 1978 by a shorter work, *On Human Nature*, which won him a Pulitzer Prize. Devoted entirely to human sociobiology, and written in an appealing and generally non-technical style, *On Human Nature* drew criticism from a large audience of nonscientists who might otherwise never have heard of sociobiology. He then teamed up with physicist Charles Lumsden to produce *Genes, Mind, and Culture* (1981), in which a serious attempt was made to establish more firmly the genetic base for speculations advanced in *On Human Nature.* In that work the authors developed the concept of the coevolutionary circuit, mentioned earlier in this chapter, and in so doing introduced what we might call Stage 2 in the evolution of sociobiology.

Let us go back to what might now be called "Stage 1" in the Wilson/Trivers version of sociobiology. Charles J. Lumsden, a student of Wilson who began to participate actively in sociobiological research in the early 1980s, has called the first stage "classical sociobiology." It has been described by Lumsden and Ann C. Gushurst as follows:

> Classical sociobiology . . . has previously attempted to link genes to social phenomena in a direct manner. Using basic principles of population genetics and ecology, researchers have sought to explain and predict the environmental conditions under which dominance systems, altruism, pair bonding, parental care, homosexuality, and other forms of social behavior are most likely to arise when individuals or groups optimize various types of genetic fitness. (Lumsden and Gushurst 1985, pp. 3-4)[5]

The authors point out that this form of sociobiology has been very successful in the study of the relatively invariant, "instinctive" behavior of other animals. The dominance of instinct, seen so amazingly displayed in colonies of bees and ants, is often referred to as "hard wiring." On the other hand, classical sociobiology "has been successful only to a limited degree in the analysis of human behavior. It has, for instance, not accounted for the activities of human (or animal) thought, consciousness, advanced emotion, and decision-making" (p. 4). What was lacking in classical sociobiology was "useful treatments of entire populations that evolve by the inheritance of both genetic and cultural information" (p. 4).

The coevolutionary theory—or *gene-culture theory*—that has followed is complex by a factor of at least ten over the classical theory—or so it seems to one like myself peering in from the outside. It has a large set of precisely defined terms and concepts. First, what is meant by "culture" must be carefully defined. A unit called a *culturgen* has been defined to differentiate operationally a certain information pattern "corresponding

---

[5]Reprinted by permission of Kluwer Academic Publishers.

to a relatively homogeneous set of artifacts, behaviors, or mentifacts (mental constructions having little or no direct correspondence with reality)" (Lumsden and Gushurst 1985, p. 6).[5] A process of *epigenesis* is another essential term. It refers to "the total process of interaction between genes and the environment during development. During mental epigenesis, the substance of culture is reshaped to build the mind and its contents. Information coded in the genes guides and shapes this development" (p. 7). The logic followed in epigenesis is expressed in *epigenetic rules*.

These authors say that "much of human culture is sustained by gene-culture transmission rather than by pure cultural transmission" (p. 8). They cite numerous examples in which—given a choice of alternatives—there is "almost always revealed an innate bias favoring some culturgens over others" (p. 8). One example they give may serve to illustrate such innate bias: "the predisposition to acquire phobias against certain dangerous objects, such as heights, running water, and snakes, but not other dangerous objects, including electric sockets and guns" (p. 8). Another that particularly interests me is Noam Chomsky's theory that language "springs from an innate human faculty" and that "some fundamental principles of language may already be embedded in our brains" (excerpted from a biographical sketch by John Horgan in *Scientific American*, 1990).

One point that strikes me as significant in the application of sociobiological coevolutionary theory is that the postulated genetic influence is not justifiably applied in any specific way to culturgens on an exact deterministic basis in the value areas of ethics and religion. Rather, the genetic influence is broadly pervasive as a tendency toward, a bias toward, or a propensity for some general category of belief. For example, it seems reasonable to postulate an innate tendency toward submissiveness to authority or toward belief in the supernatural (or toward the opposites of these tendencies), whereas a societal behavior as specific as slave-holding or the belief in one God vis-à-vis many gods is beyond the direct control of a set of allelic genes.

At this point, it seems appropriate to turn to the relationship of sociobiology to ethics and morality. Lumsden and Gushurst state:

> To the sociobiologist, a striking feature of ethics is its fundamental role in the structure of human societies. Human reality is a moral reality. Details as to what, specifically, is "right" or "wrong" vary from society to society, but the obsession to regulate human behavior using ethical principles is universal.
>
> The learning of ethical principles involves both the assimilation of culturally transmitted information and the expression of individual preferences. (1985, p. 16)[5]

The actual learning of ethical principles, these authors say, has been studied in an attempt to discern the genetic predispositions that may influence ethical choices. They explain:

> According to these analyses, there is a very set way in which human children learn ethical systems, and this process can be identified in any child. While every child is capable of learning ethical systems, and even seems predisposed to do so, what each child learns varies from individual to individual and culture to culture. However, there are certain common features that can be identified in the ethical learning process, and in the content of the final belief systems. Gene-culture theories (Lumsden and Wilson 1983) together with developmental data (review in Keil 1981), point to the existence of epigenetic rules that make the culture-learner more likely to acquire certain systems of ethical knowledge rather than others. (1985, p. 18)[5]

These authors observe that because human life is so strongly ethics-centered, there will be selection pressures that favor certain kinds of moral passions, and the genetic bias thus generated will in turn influence the kinds of ethical knowledge assimilated in cultural learning (p. 18). I would add that, unaware of such genetic bias, the philosopher of ethics usually sets up in its place some a priori "self-evident" axioms about ethics. I can understand that this genetic bias may explain such concepts as "moral imperatives" and why they are so well received and without questioning by the public. Theologians set up ethical dogmas in the same manner and for the same reasons, claiming, of course, that they are God's imperatives revealed to humans.

The Philosopher's View of Sociobiology and Ethics

The 1977 Dahlem Workshop on Biology and Morals, organized and managed by molecular biologist Gunther S. Stent of the University of California—Berkeley, had as its purpose to give sociobiology a very close and critical scrutiny in regard to its application to ethics and morality. Here, we introduce some comments from two philosophers who contributed significantly to that workshop (Stent 1978).

Philosophy professors T. Nagel of Princeton University and C. Fried of Harvard University were in strong agreement that "ethics is an autonomous subject." Nagel wrote:

> Ethics is a theoretical subject developed through the collective human capacity to submit pre-reflective motivations and behavior patterns to rational criticism and thereby to modify them. The result

> is an open-ended process of development of social institutions, motives, and methods of moral reasoning. It can be best understood internally, through the methods appropriate to the subject, and its history is part of intellectual and cultural, not of biological history. (1978, p. 221)

Nagel observes: "No one, to my knowledge, has suggested a biological theory of mathematics; yet the biological approach to ethics has aroused a great deal of interest" (p. 222). He takes the view that the development of ethics must be participated in and accepted by a large number of people and that this requirement makes the pursuit of ethics a "more democratic subject than any science" (p. 224). In his discussion of methods of ethical inquiry Nagel talks about the principle of moral equality—most modern ethical positions hold that moral claims or rights of all persons are equal (p. 228). This is a conclusion that I would find difficult if not impossible to arrive at by practical reason; it concerns presuppositions about the intrinsic value of humanity that are in the transempirical or supernatural realm.

Nagel's paper closes with a firm stand on sociobiology. He realizes that humans who engage in ethical discussions are "organisms about whom we can learn a great deal from biology." He grants that "their capacity to perform the reflective and critical tasks involved is presumably somehow a function of their organic structure. But it would be as foolish to seek a biological evolutionary explanation of ethics as it would be to seek such an explanation of the development of physics" (p. 229).

Professor Fried emphasizes how close the sociobiologists come to committing the naturalistic fallacy (1978, p. 209). Being of the opinion that ethics is an autonomous subject, he comments: "If it puts me in the camp of those who hold that one cannot derive an 'ought' from an 'is', so that naturalism is indeed a fallacy, then I must comfort myself with the otherwise conflicting company of both Hume and Kant." Fried also emphasizes that ethics relates to choice, and normative ethics is concerned with guiding choice. About the latter he says: "I take it to be clear that when it is a matter of recommending a choice, no description of past or present states of the world can of itself determine that recommendation" (p. 210).

A Hypothesis of Origin of Ethical Systems and Moral Imperatives

So far, our discussion of ethics and religion has emphasized causes and mechanisms, i.e., genetic versus cultural, or nature versus nurture, but there remains to be set forth a hypothetical reconstruction of the history of these belief systems that includes a driving force or energy source. A hint as to a historical relationship lies in the idea, already stressed, that religion has served as an enforcer of morals deemed desirable for the stable operation of a society. We raise the question: "Deemed desirable" by whom, and for what self-serving motive?

The origin of moral imperatives has been sought after by moral philosophers for centuries, but up to the present this has proved a futile search. When an empirical explanation is favored, the arguments end up in tautologies and total frustration. Henry Margenau's 1964 work, *Ethics in Science*, summarized these arguments fully for the modern period prior to that publication date. His work is particularly significant because of his distinction as both a physicist and a philosopher of science. He made a serious attempt to examine science and ethics as comparable in terms of their structure and methodology, but was careful to recognize the ontological and epistemological differences between them. Near the end of his book, on the subject of obligation and purpose in ethics, he finds that science requires no reference to purposes, and that therein lies their difference:

> The reason for the difference lies in the fact that ethics needs to plead, persuade, cajole, praise, reprove, and condemn, whereas science states and rests its case on its own evidence. Such disparity has its origin in the contrast between the *est*-laws of science and the *esto*-laws of ethics, and in related differences stemming from this contrast. Both in science and in ethics one meets discordances between the laws and the behavior they regulate. In science one calls them errors, in ethics immoral acts. The measure of the competence of the scientific law to create compliance is exactness, for ethical law it is efficacy. Science seeks to enhance the exactness of its laws, ethics their efficacy. But these are wholly different tasks. (1964, pp. 289-90)

Margenau devotes one chapter to origins of imperatives and primary values in ethics. Here, one could hope for a historical explanation bearing empirical content, but we find none. He carefully examines the ethical systems of ancient religions, both Western and Eastern, but he merely states the norms of each, and these prove to be remarkably alike in many respects. He concludes that ". . . there is no general *a priori* prescription for the establishment of a workable ethical system" (p. 259). The origin of such systems, he supposes, lies in "creative intuition, which defies prediction."

Margenau, like other philosophers of ethics, have run their explanatory efforts in neutral gear, albeit racing the engine at full throttle many times, while the vehicle itself remains at rest. What I propose to do is to shift into gear and impart some forward motion toward an empirical and rational explanation that is historical. This process of formulating a dynamic historical explanation requires two steps. First is to recognize that beliefs and values expressed in statements of ethics, religion, and religious theologies are pure artifacts that lie in the belief realm within ideational knowledge. On the other hand, the propositions put forward to state or document those beliefs and values are historical records that lie in the empirical realm within perceptional knowledge. The mental process that generates such propositions is real, even though the images so created are fictions. Therefore, our reconstructed history follows an empirical track.

Our second step is to look back in time beyond recorded history to what may have happened in the prehistory of *Homo sapiens*. We need to keep in mind Gunther Stent's observation that much of human behavior is controlled by "covert deep structures" inaccessible to direct observation (1975, p. 1053). There are no records from this formative time in human evolution to give us clues as to what selection pressures may have acted to generate the inaccessible information. And thus, says Stent, "We encounter the barrier to ultimate scientific understanding of man which Descartes had recognized more than three centuries ago" (p. 1057). Because of this barrier, my hypothesis states only what is logically and physically possible, and perhaps there is where it should remain. On the other hand, like all historical science, it conforms with the axioms and/or presuppositions of science, and with laws of science.

My hypothesis paraphrases essential principles of what I propose later in this chapter to be the parallel scenario for the origin of religion; i.e., that we assume a stable primitive society at the level of the tribe consisting of (a) a small group of strong managers with low susceptibility to belief in imagined entities coupled with strong inclination to wield authority, and (b) a large number of subservient individuals with high susceptibility to believe imagined entities coupled with strong inclination to accept authority.

Assuming, then, that we will be able to present a viable naturalistic explanation of the origin of religion, we must now turn to a historically and logically earlier step, which was the origin of the ethical/moral precepts contained in the religion. I propose that those ethical/moral precepts or norms came first, with religion being subsequently (or concurrently) concocted by the managers in order to initiate and sustain what was necessary for their system to work.

My thesis may be stated as follows: All ethics and included moral imperatives were pragmatically conceived by the manager(s) or ruler(s) to gain and hold power over the managed (ruled) individuals. (These are our two classes.) As such, the imperatives were not intuitively arrived at by the managed class, but instead were purposefully invented by the manager class.

The crucial principle reads as follows: The manager (ruler) says: "Do as I say, but not as I do." An ethical duality is set up, in which the norms reserved to the manager are symmetrically opposed to the norms granted to the managed. More specifically, the most important of these norms are as follows:

- Manager's self-asserted norms: I have the right or privilege to kill, torture, punish, degrade, impoverish, and subdue all of you individuals in order to maintain my program of group survival. I also have the right to lie to you and steal from you as required for that same purpose.

- Norms conferred upon the managed: You may not kill, etc., steal, or lie. You may not do these acts to me or to any of you on your own volition or for your purposes, but I may require you to do them to members of your class or to other groups of humans if it serves my purposes and the attainment of my goals.

We see this division of imperatives in place today within every political state, including those of the Western democracies. The nation through its constitution is the manager and says to its citizens: "Thou shalt not kill." But at the same time, the nation reserves the right (a) to kill the enemy in wars, and (b) to kill any citizen who kills another citizen. The individual citizen may not steal from another citizen or from the government, but the latter can with impunity steal from the citizens, through taxation and confiscation. And, of course, the government can lie with impunity to its citizens, but prosecutes them if they lie on their tax returns or by giving false testimony in the courts (perjury). These are only samples of the many antithetical pairs of norms.

How does this dualism of opposed values derive its imperatives? They arise from purely pragmatic considerations of making the managing (governing) process capable of being carried out and sustained in a reasonably stable state. The conferred norms are designed to keep the individuals in the population safe from one another, peaceful and docile amongst themselves, and obedient to the manager. Collectively, these conferred imperatives promote what moralists have called the "greatest good," "maximum happiness," or "greatest pleasure."

We see this program in effective operation in the management of domesticated animals being produced for human consumption. Closely confined beef cattle are fed tranquilizing drugs to reduce their physical activity and aggressiveness, which in turn makes the muscle more tender and increases the production of fat. Laying hens are physically separated from each other in cages to prevent the loss of eggs and infliction of trauma on each other. In human societies, the imperatives serve to increase the numbers of healthy new individuals that will be needed to send forth to be killed in wars with enemy groups or to supply cheap labor to the managers who hold most of the wealth and natural resources. The imperatives serve to reduce interpersonal tensions that make individuals less desirable as fighters or workers, or more likely to criticize the judgments of their managers. If you were a manager, would you do it any differently? Would you not strive to evoke feelings of benevolence toward you rather than to provoke malevolence?

Merely to decree the division of moral rights could not guarantee the managers their successful dominance. Sassy individuals among the managed could so easily protest: "Who gave you this power and authority over us? What makes you any different from the rest of us? Get lost!" This is why religion had to be invented and installed. Invent a god (or gods) endowed with supernatural powers—a god who is on the manager's side. Claim direct knowledge of that god through revelation. Then turn over the enforcement of moral imperatives to that god. Tell your people: "If you don't do as I tell you, our god (or gods) will punish you in ways far worse than mine. Your soul will suffer eternal damnation and torture in a hell so terrible you can't even imagine it! See how much better it is to conform in this life and enjoy the everlasting pleasures of heaven." It may be wise to invest a high priest with power to serve as the go-between in matters of theology, but be wary and keep that agent under strict control. A key word from sociobiology applies here: manipulation; you might say "It's the name of the game."

In this cultural evolutionary succession—secular norms to supernatural beliefs—law preceded morals, the reverse order of what we usually assume. Religion allowed autonomous concepts of right/wrong and good/evil to sprout and grow in the new soil of supernaturalism. Their being today so strongly rooted in the soil of religion now poses a vexing problem for the secular humanists, who seek to transplant them into the empirical garden.

As civilization emerged, strong military and commercial/industrial structures were evolved, each mirroring the primitive manager/managed system. The officer/enlisted-men dichotomy in the military establishment and the executive/employee dichotomy of corporations fitted naturally into the nonegalitarian model—separate toilets, of course. Both military and industrial establishments supported the civil rulers, and today largely control the civil authorities of democracies through expenditures of large sums of money to control elections. Traditionally, in England for example, sons within each family of substance were distributed among leadership careers in the government/judiciary, church, military, and industrial sectors.

When recorded history arrived and began to generate a written record of what was happening, the origins from prehistory had already disappeared. Oral history had kept alive only the religious legends, although there survived also a mythology of dreadful acts of violence perpetrated by ancient rulers on their peoples. God had already become autonomous, together with the attached theologies. Now, the freethinkers who dismiss gods and theologies must try to find an autonomous but secular origin for the moral imperatives directed to the managed group. In floundering around unsuccessfully in search for the lost origins, they have settled for a priori or axiomatic status of the imperatives, or have brought up some very suspect notions of self-evident or intuitive truths.

Sociobiology and Religion

Sociobiology has much to contribute to an understanding of religion. No one has explained this relationship more thoroughly or eloquently than Edward O. Wilson in his popular 1978 book, *On Human Nature*. In Wilson's opinion: "Religion constitutes the greatest challenge to human sociobiology and its most exciting opportunity to progress as a truly original theoretical discipline" (1978, p. 175).

Of the intensity of the genetic factor in religious belief, Wilson states: "The predisposition to religious belief is the most complex and powerful force in the human mind and in all probability an ineradicable part of human nature. . . . It is one of the universals of social behavior, taking recognizable form in every society from hunter-gatherer bands to socialist republics" (p. 169). The growth of science and the attempts of nontheistic humanists, Wilson thinks, have done little or next to nothing to diminish religion (pp. 170-171). "Men," he succinctly comments, "would rather believe than know" (p. 171).

Wilson develops the coevolutionary theory in the following steps. First, religion evolves culturally in terms of its theology, dogma, and rituals, i.e., the ecclesiastical contents of religion. The ecclesiastical elements are altered by the theologians in response to social needs and thus a cultural selection—*ecclesiastical selection*—occurs (p. 176). Second, modifications that are made must "eventually be tested by the demands of the environment" (pp. 176-77). Those cultural beliefs selected might on the one hand strengthen a particular society in its ability to cope with environmental stresses, such as warfare. On the other hand, another set of changes might weaken that society. Either way, natural selection will cause changes in the gene frequencies, and thus the coevolutionary cycle is completed. Wilson summarizes the changes in the following paragraph:

> According to the hypothesis, the frequencies of the genes themselves are reciprocally altered by the descending sequence of several kinds of selection—ecclesiastic, ecological, and genetic—over many lifetimes. Religious practices that consistently enhance survival and procreation of the practitioners will propagate the physiological controls that favor acquisition of the practices during single lifetimes. The genes that prescribe the controls will also be favored. (P. 177)

Wilson uses the previously established term *religiosity* to mean susceptibility of an individual to being indoctrinated in religious beliefs (p. 186). He also refers to a genetic capacity for blind conformity and consecration (pp. 186-87). He asks: "Is the readiness to be indoctrinated a neurologically based learning rule that evolved through the selection of clans competing one against the other?" (p. 184). Of course, his answer and mine are both yes. Willing subordination must have been a potent force for survival then as now. This biological advantage, says Wilson, also operates at the highest levels of religious practice (p. 188).

Wilson says that we should make no mistake about the power of scientific materialism to defeat traditional religion (p. 192). This is his prediction:

> Most importantly, we have come to the crucial stage in the history of biology when religion itself is subject to the explanations of the natural sciences. As I have tried to show, sociobiology can account for the very origin of mythology by the principle of natural selection acting on the genetically evolving material structure of the human brain.
>
> If this interpretation is correct, the final, decisive edge enjoyed by scientific naturalism will come from its capacity to explain traditional religion, its chief competitor, as a wholly material phenomenon. Theology is not likely to survive as an independent intellectual discipline. But religion itself will endure for a long time as a vital force in society. (P. 192)

At various points in our earlier chapters the same theme has been emphasized: Traditional religion is a product of human imagination and forms a belief system within ideational knowledge. While Wilson has illuminated the sociobiology of religion as a coevolutionary process, there remains the problem of the root origins of religion, defined broadly as the human belief in supernatural entities and their alleged powers. It is to this highly speculative history of origin that we turn next.

First, however, let us review the relationship between ethics and religion in the light of both (a) my hypothesis of the origin of ethics and (b) the sociobiological hypothesis of coevolution of religion presented in the foregoing paragraphs. My thesis is that ethics came first in logical, if not temporal, order and was autonomous, i.e., was independent of religion in terms of origin. Religion, as belief in the supernatural, was also spontaneous in origin and autonomous with respect to ethics, and may have originated earlier—much earlier—in time and in stage of human intellectual development. But on the other hand, religious belief as a response to environmental forces involved no ulterior motive or purpose, and in that respect was not originally a product of conscious planning to gain an end. Instead, it is a passive emotive system that lends itself to being used and/or abused to achieve rationally-derived human goals. Thus, religion can be seen as the prime manipulative tool for enforcement of an ethics purposefully designed by managers to govern the managed.

Think of a length of weathered tree branch lying on the ground near the point where it fell from a tree—call it a "stick." A chimpanzee, frustrated in its efforts to capture termites from within their mound, picks up the stick for the purpose of using it as a tool to accomplish that goal. Poked into the opening of the mound, the stick attracts termites, which may then be withdrawn and swept into the chimp's mouth. Consider also that the stick might be modified by the chimp (by chewing on it, or rubbing it against a rock) to make it a more effective tool; or the stick might be discarded as being unsatisfactory and another stick selected in its place. By analogy, religious concepts might have been consciously and purposefully selected and modified (intensified, redirected) for greater effectiveness in enforcing the ethical imperatives. Having said this, we need to speculate on how the prime or primitive supernatural concepts might have arisen spontaneously in the first instance from within the brain in response to environmental influences.

A Naturalistic Origin of Religion

The sociobiologists seem to be agreed that the the human capacity to ethicize has become a genetic endowment, and Edward O. Wilson has effectively argued that the human capacity to accept religious concepts is now also a genetic endowment. The only essential requirement for producing a religious thought is imagination. We have already seen that imagination is uniquely a mental faculty of humans—a faculty that is a product of biological evolution. Imagination enables humans to supply explanations of phenomena, the causes of which are not immediately available. Wilson explains how the mind constructs dreams and other image forms:

> In the absence of ordinary sensory information from the outside, the cortex responds by calling up images from the memory banks and fabricating plausible stories. In an analogous manner the mind will always create morality, religion, and mythology and empower them with emotional force. (1978, p. 200)

It is not difficult to reason that imagination is of adaptive value in human evolution, when imagination is used to predict situations that, on the one hand, may be favorable to survival or, on the other hand, unfavorable to survival. For an individual to imagine a successful group hunting strategy could permit survival when the lack of that imagination could under similar circumstances result in death by starvation. The ability to imagine a dangerous encounter with a group of animals or of other humans could result in devising an evasive strategy leading to survival.

When we pass from such simple, mundane situations over which humans have some measure of control to imagination about events that seem to have no obvious cause and over which humans have no control, the adaptive role of imagination is difficult to demonstrate and perhaps should be presumed not to exist. For example, an unseen pestilence causing the rapid death of many individuals of both sexes and various ages would have sent the survivors into a state of great emotional distress. The same imagination that served a practical purpose in the circumscribed hunting situation now runs free and uninhibited in seeking an agent on which blame may be placed. Imagined supernatural agents or forces fill the void; they provide an explanation and an object of emotion—fear, anger, or hate, for example. Imagination provides further details of the agent and its relationship to humans. By analogy with human relationships, it can be imagined that the supernatural agent has a plan, purpose, or design in perpetrating the pestilence. It can be imagined that the agent is in the role of an all-powerful ruler or creator or whatever. The imagery thus produced, once accepted by individuals and standardized by a social group, has become a working mythology or a religion and, as such, is a cultural package to which morals are referred for authority and enforcement. In short, religion is an artifact.

Perhaps it is hard to see in imaginative activity about the supernatural realm any adaptive role that might become incorporated into the genetic evolution of humans. Does such activity relate to survival fitness in any meaningful way? Or is it simply a recreational activity of the brain, indulged in when there were no pressing requirements to produce images for fitness-related situations? I suppose one could argue that a satisfactory supernatural system that accounted for all ulterior questions would bring increased ability of a social group to cope with stressful situations, but on the other hand, one can deduce negative effects. For example, setting out generous helpings of scarce foods for forest spirits to appease their innate hostility, or sacrificing one's son or daughter to appease the wrath of a god could severely diminish the survival capability of the group.

I found the literature of cultural anthropology so rich and diverse on the subject of the early development of religion as to be almost unassailable. Some general principles of interest are covered by Professor Bernard Campbell and deserve review (1985, pp. 356-57). Religion would serve the important function of integration of the social group and promote its cohesion because it would direct feelings toward a symbolic center. Religion would also assist in controlling individual behavior through religious sanctions. These ideas fit well into my hypothesis that religion was adopted as a tool for the enforcement of ethical sanctions already set in place to govern a dualistic society consisting of managers and those managed. Primitive religion would have first been promoted through rituals having deep and abstract meaning. *Ritual* is seen as a condensed and compact form of social memory. Says Campbell: "Ritual is a special form of communication with a very high information content, but, unlike language, the meaning of ritual depends on its social context. A ritual act is a social act—it is society's meditation of traditional knowledge and behavior" (pp. 356-57).[1] Campbell explains further:

> In its remarkable way ritual records knowledge about social origins, about social structure, kinship, and obligations. It records behavior patterns of a fundamental nature, such as hunting, toolmaking, and food preparation. Ritual is, as it were, the DNA of society, the encoded informational basis of culture; it is the memory core of human social achievement. (P. 357)[1]

Myth, the verbal derivative of ritual, followed directly, and later, with the evolution of writing, continued to play a role of promoting social stability. Religion, which in its early stages was predominantly concerned with the supernatural realm, also served the same important social function: to bind society more tightly together. The awesome "power" of imagined supernatural forces could be harnessed in support of behavior required for social stability. Many important social activities would then take on a sacred and ritual quality, safely secured in a domain beyond reason. Campbell writes:

> Not only are traditions ritualized, but many important social activities appear to take on a sacred and ritual quality. An element in the common life of a society may come to have a special significance and in time become an act of religious observance. Especially important in this respect are the *rites de passage* (birth rites, puberty rites, marriage rites, death rites), which consolidate social roles and social structure, as well as bind members of the society together. But religion does more than that, for it directs social sentiments toward one stable and symbolic center. (P. 357)[1]

The phenomenon of death awareness, referred to earlier, may have led to the primitive religious rituals associated with burial, for which artifacts can be found. The religious concept of continuance of ancestors in a supernatural realm would have emerged and perhaps led to ancestor worship as one of the primitive forms of religion.

Laws and legal codes forming a secular structure would, I suspect, have followed the growth of an entangled web consisting of functional ethics and religion. A written legal code could be derived by one society in power from its own ethical/religious system to serve the wider function of directing the behavior of subjugated societies with different religions—conquered nations and imported slaves, for example. Campbell observes:

> These different means of binding and stabilizing society are clearly factors of overriding importance in human evolution. It remains to point out that religious and legal institutions help control human behavior at an individual level for the benefit of the society. What "is done" and "is not done" is a powerful determinant of human behavior; it is dictated by tradition, religion, and law, through each individual's peers and his elders. Religious and legal sanctions control human behavior at its most fundamental level, and the two kinds of sanctions are often so closely allied as to be indistinguishable, as in the Ten Commandments. Among these vital functions of religious and legal institutions falls the control of the way people express their biological needs, by means of sanctions against theft, aggression, adultery, and so on. (Pp. 358-59)[1]

If, as I have proposed in an earlier section, fitness for survival was enhanced by accepting authority on ethical matters, could it not also have been enhanced by accepting authority on religious statements? As before, let us assume a stable primitive society at the level of the tribe consisting of (a) a small group of strong managers with low susceptibility to belief in imagined entities coupled with strong inclination to wield authority, and (b) a large number of subservient individuals with high susceptibility to believe imagined entities coupled with strong inclination to accept authority. I have claimed that this bipolar or dualistic arrangement, seen everywhere today in political, military, industrial, and religious establishments on various scales, is remarkably successful, as compared with egalitarian arrangements generally. Perhaps at a time when culture was extremely primitive and its evolution extremely slow, religiosity became a genetic endowment with alleles coding for a range of intensities from weak to strong. In terms of what modern sociobiology asks us to take seriously, this hypothesis is neither extreme or unreasonable.

In this chapter I have argued the case that religion is, at its rock-bottom base, a phenomenon of completely naturalistic origin; i.e., it is an artifact. Thus, a naturalistic hypothesis accounts for all belief systems composed of supernatural phenomena. This is not a self-contradictory statement, as it might seem to be at first reading. The relationship between natural and supernatural is fully and precisely accounted for in a monistic ontology—that of mechanistic materialism (see Chapter 11). Imagination, a physical/chemical functioning of the brain (a neurophysiological activity), produces religious imagery, which is a form of reality amenable to inquiry by empirical science. This view accomplishes an ontological reduction from a dualistic to a monistic system. When religion is seen as naturalistic in origin, then both ethics and religion are fully tractable to empirical science for investigation as real phenomena. If this general hypothesis is viable, it must accommodate all ethical and religious systems existing in the past, present, and future. The range of imaginative constructs is surely unlimited in the human population. Therefore, we can expect to find limitless variation in religious systems the world over through human history.

Credits

1. Reprinted with permission from: Campbell, Bernard G. *Human Evolution: An Introduction to Man's Adaptations*. 3rd Edition. (New York: Aldine de Gruyter) Copyright © 1985 Bernard Campbell.

2. From Theodosius Dobzhansky, Evolution of Mankind, Chapter 14, pp. 438-63, in T. Dobzhansky, F. J. Ayala, G. L. Stebbins, and J. W. Valentine, *Evolution*, W. H. Freeman and Company, San Francisco. Copyright © 1977 by W.H. Freeman and Company. Used by permission.

3. *From Darwin to DNA, Molecules to Humanity*, by G. Ledyard Stebbins, W. H. Freeman and Company, New York. Copyright © 1982 by W. H. Freeman and Company. Used by permission.

4. From George Gaylord Simpson, 1969, *Biology and Man*, Chapter 10, Harcourt Brace World, New York. Originally published under the title "Naturalistic Ethics in the Social Sciences" in *American Psychologist*, vol. 21, pp. 27-36. Copyright 1966 by the American Psychological Association. Reprinted by permission of the American Psychological Association.

5. From Charles J. Lumsden and Ann C. Gushurst, "Gene-culture Co-evolution: Humankind in the Making." Pp. 4-28 in James H. Fetzer, ed., 1985. *Sociobiology and Epistemology*. D. Reidel Publishing Co., Dordrecht. Reprinted by permission of Kluwer Academic Publishers.

CHAPTER 14

Creationism—Is It Religion or Pseudoscience?

In Chapter 11, in our sections on the dualistic ontologies of theism and deism, we observed that many scientists are also Christians; i.e., believers in the standard theologies and dogmas of Christianity. These believers have adopted various personal strategies to reconcile supernatural doctrines of their religion with the presuppositions of a materialistic (naturalistic) science. Here, we examine these strategies and their consequences.

It is the divine intervention of God in what is otherwise a naturalistic system that should interest us here. If, at intervals, God creates a new object or structure out of nothing (*ex nihilo*) or by a mysterious process of reorganization of matter and energy that is not lawful in science, God in effect subverts nature for at least a brief moment in cosmic history. This process we call *divine creation* (usually, simply "creation"), and the belief that the process occurs is called *creationism*. (Remember that the suffix "ism" implies a belief system.) Creationism spans a wide range in duration and intensity of the asserted creation events. Any individual who believes in an act or process of divine creation, whatever may be its extent in space and time, is therefore correctly named a *creationist*.

A creationist stands in opposition to a scientist; these two individuals stand on opposite sides of a fence or wall that separates their unlike and mutually exclusive ontologies and epistemic fields. It is, however, possible for a single individual to be both a creationist and a scientist; i.e., a *creationist/scientist*, but only with the fiat that both roles cannot be played simultaneously in interpreting the unbroken chain of natural history. We can see this as a split-personality (Jekyll-and-Hyde) relationship. A specific event of history in a specific time segment must fall into either (a) divine causation or (b) natural causation. Our logic is as follows: "If *a*, then not *b*. If *b*, then not *a*." To follow with the proposal "Both *a* and *b*" is therefore not logically possible. Moreover, one cannot get out of this bind by proposing that God is the sole causative agent of all natural causes, which in turn are the causative agents of the observed event. This "First Cause/Secondary Causes" model, long a standby of the eighteenth-century school of natural theology, is obviously a cop-out, since it adds up to 100 percent supernatural creation.

Consider the analogy of cosmic history as an unbroken chain made from all possible combinations of two kinds of links, *a* and *b*. The chain shown below has ten links, comprising the set.

All *a:*	o-o-o-o-o-o-o-o-o-o	Only religion (pure creationism)
a or *b*:	o-x-o-x-o-x-x-x-o-x	Alternating religion and science
All *b*:	x-x-x-x-x-x-x-x-x-x	Only science (pure naturalism)

All permutations and combinations of the two pure models run into difficulty when we ask what preceded the first link, because there ensues an infinite regression back in time, whether we start with science or with religion. For science it reads: "What existed before the Big Bang?" For religion it is the "Who-created-God?" regression. We can get out of this difficulty by postulating that the chains extend to infinity back in time; i.e., some form of universe has always existed, or God has always existed, or both have always existed.

The final point we will reemphasize at the close of this chapter is that when a theist declares any link in the chain to be an *a*-link (whereas all the others are *b*-links), an element of the science set has been replaced by an element of the religion set. When this substitution has been accomplished, the entire ensuing sequence is flawed by that single antecedent event of divine creation and must be viewed as false science, or pseudoscience. The reason that replacement of a single link changes the character of all ensuing links is that each successor link is dependent upon its predecessor in a cause-effect relationship. In the purely naturalistic mode, the entire chain is generated without a teleology (i.e.,without being directed to a goal or purpose). Any insertion of an act of God (a miracle) injects a direction toward an ultimate goal that can never be discerned by scientific observation. Moreover, that divine act itself can never be detected by the scientist because, by definition, it is a supernatural act. There exists only the claim that such an act occurred, and science cannot deal in such claims. By the same token, science must reject revelation as a means of obtaining empirical knowledge.

We can now turn to a review of several scenarios of divine creation that have been developed over past centuries and perpetuated to the present; these provide a never-ending source of doctrinal infighting by the theologians.

The Literalist Creationist Scenarios

From the viewpoint of science, and particularly of those scientists who are also Christians, there is a recognized breakdown of creationism into two major classes: (A) *Young-age* (or *young-earth*) scenarios; (B) *Great-age* (or *old-earth*) scenarios. Adherents of the first class believe that the earth was created only a few thousand years ago, the most common date quoted being that of Bishop Ussher at 4004 B.C., but with a limit on the order of, say, 10,000 years. Their "year" is the same time unit we use today, namely, the tropical year. Adherents of the great-age scenarios accept the going ages assigned to the universe, solar system, and Planet Earth, as established by scientists on the basis of radiometric age determination and other absolute dating methods.

The young-agers are now most closely associated with present-day Christian fundamentalism, which holds the Bible (in particular, the King James Version) to be inerrant. Their belief system is usually labeled *fundamentalist creationism.* This program is also called *recent creation.* It relies upon a strict literal interpretation of the Book of Genesis and includes two major events: the six-day creation and the one-year Flood of Noah. The intervening Antediluvial period of 1350 years also contains essential details of divine creation.

Is "Creation Science" Pseudoscience?

Nearly everyone in the United States is familiar with the attempts of the fundamentalist creationists to bring the teaching of their young-age scenario into the science classrooms of the public schools as an alternative theory to modern neo-Darwinian evolution. To facilitate this insinuation into science education, they have have labeled their system *creation science.* They have attempted to find empirical evidence of recent creation and the Flood, but that proffered "evidence" has been discredited and their arguments vitiated by numerous scientists in a large number of published articles and books. (See Strahler, 1987, for a full treatment and bibliography.)

It is appropriate now to reflect on this question: Does "creation science" qualify for inclusion with mainstream science as one and the same field of human knowledge, or is it, on the other hand, pseudoscience? Chapters 7 and 8 laid the foundation for assessing this question. We looked closely at pseudoscience, using examples that did not include "creation science," but which enabled us to recognize specific features of pseudoscience. Chapter 8, to which we now refer for guidance, set down the characteristics of the scientific community, its practitioners, and its self-imposed guidelines. Most important was the establishment of the essential components of a cognitive field and the comparative referencing of both science and pseudoscience to those components, following the guidance of philosopher Mario Bunge.

Recall, first, Bunge's contention that, whereas there exist numerous important and respected cognitive fields that are nonscientific—the "belief fields," he calls them—the crucial point is that "any cognitive field that, though nonscientific, is advertised as scientific will be said to be *pseudoscientific*" (Bunge 1984, p. 39).[1] To repeat a few lines from Chapter 8, a claim put forward by its adherents that a belief field is science is a fraudulent claim; it is a misrepresentation of the real nature of that belief field. Our charge, then, is that the fundamentalist creationist view of the universe, based on the literal interpretation of the book of Genesis and clearly a belief field, when presented by its adherents as if it were, instead, a research field under the name of "creation science," constitutes pseudoscience. Under our revised classification of knowledge fields, presented in Chapter 9, we would restate this distinction as follows: When a field of ideational knowledge (religion) is offered to the public as perceptional knowledge (empirical science), that claim is false or fraudulent. Whether the false claim arises from ignorance or an intent to defraud makes no difference in the ontological/epistemological distinction.

As explained in Chapter 8, Bunge describes and defines science in terms of twelve conditions, all of which must be satisfied. He further stipulates that if any cognitive field fails to satisfy all of these conditions, it should be judged as *nonscientific.* Bunge follows with a second list; it gives those conditions that will identify a cognitive field as pseudoscience. (At this point, you may wish to refer to the two lists of conditions given in Chapter 8.) It is to the second list that I refer here by number. Whereas in Chapter 8 each point was directed at the two historical scenarios of earth history used as examples (Velikovsky's *Worlds in Collision* and von Däniken's *Chariots of the Gods*), we now direct those points at "creation science."

Point 1. [Little change occurs in the components in the course of time.] Keeping in mind that the single unique component of "creation science" is its set of religious tenets, that component has not changed in the slightest through the modern period of an organized body of "creation scientists." Rooted in a single book of the Old Testament and strictly limited to one specific English version of that religious document, the basic hypothesis of "creation science" (recent and rapid creation followed by a catastrophic Flood) cannot be modified in the light of any external observation of the real world. No change in the tenets of "creation science" can be induced from external (empirical) sources, because to make any change would be, in effect, an admission that the tenets were previously faulty, i.e., in error. If changed, a revised tenet of "creation science" would, in turn, become vulnerable to further change; hence, no tenet currently held could be accepted as eternally true. To avoid this logical self-destruct process, no change can be tolerated, no matter what empirical science comes up with in the way of new information.

Contrast this inflexible position of "creation science" with that of mainstream science, which has a set of tentative theories and hypotheses, all of them vulnerable to the impact of new information based on observation. Not only is forced change in those tentative models to be expected, it is to be diligently sought after. This single, glaring contrast between mainstream science and "creation science" is by itself sufficient reason to declare the latter to be pseudoscience.

Point 2. [There is no research community as such. Instead, the cognitive community consists of believers, who, although calling themselves scientists, conduct no scientific research.] Without question, "creation scientists" form a community of believers. First and foremost in their outlook is the inerrancy of the Scriptures. To become enrolled in its membership, all inductees of that community take a pledge of belief in the religious tenets of "creation science." All this is freely stated in the literature disseminated by the Institute for Creation Research and allied organizations. Turning to the matter of research community and scientific research, members of the ICR stoutly maintain that they are, indeed, a scientific research community and that they do, in fact, carry on scientific research and publish the results of such research. Mainstream scientists will counter with the charge that the actual quantity of scientific research produced by all of the "creation scientists" is minuscule, that it fails to meet standards of the mainstream scientific community through peer review and open journal publication. The charge has merit. Serious research articles supporting the hypothesis of recent and sudden creation are almost totally lacking in the journals and treatises of mainstream science.

In this connection, consider the following statement that appeared in the September 1985 issue of *Science 85*:

> A three-year data-base search of 4,000 scientific publications—focusing on the names of people associated with the Institute for Creation Research and on phrases and keywords such as "creationism"—didn't turn up a single paper. A follow-up study of 68 journals found that only 18 of 135,000 total manuscript submissions concerned scientific creationism, and all 18 were rejected. Reasons cited included "flawed arguments," "ramblings," and a "high-school theme quality." (vol. 6, no. 7, p. 11)

The published literature of "creation science," almost without exception, exists apart from the literature of mainstream science. Publication of articles and technical monographs on "creation science" appear almost exclusively in publications owned and controlled by creationist organizations sponsored by fundamentalist religious groups. Here the publications are sequestered from peer review by mainstream science. The articles often contain within the body of their texts references to religious dogma and religious goals. A single issue of the Australian journal *Ex Nihilo*, for example, is a mix of articles on religion and science, spanning a range in level from lessons for children to so-called technical papers. This mixture of overtly religious matter with scientific topics is a hallmark of the "creation science" educational publications offered to the general public.

As to the quality of alleged "scientific" research conducted by "creation scientists," it differs strikingly from that seen in mainstream science. First, it is directed entirely at the two functions: (a) attacking the so-called evolutionary view of the cosmos, and (b) supporting their alleged recency and rapidity of universal creation and the Flood catastrophe. The first function is entirely negative or destructive, one in which the "research" activity is limited almost entirely to a search of the published work of mainstream science. Gleaned from this activity is a long list of alleged inconsistencies in the published or oral statements and of disagreements among the mainstream scientists. These supposed "flaws" of science are typically documented by out-of-context quotations—often using citations of statements long since discredited and superseded by new findings. Propositions and principles alleged to be held by mainstream scientists are shown to be untenable, using the fallacious straw-person argument. Article after article in the "creation science" literature consists of this abuse of second-hand material in an attempt to discredit mainstream science. To call the production of such a literature "scientific research" is a travesty of the meaning of that term.

"Creation scientists" will claim that they have a positive program of field research looking for new evidence to support recent and rapid formation of geologic materials, such as petroleum and coal. In comparison with field research programs in mainstream geology and geophysics, these creationist projects are trivial in scope and almost devoid of substantive results. "Creation scientists" simply do not participate in mainstream science research, whether it relates to controversial questions of origin or not; nor do they produce independently a comparable body of scientific results. It seems fair to say that in terms of results, there is no research community as such in the cognitive field of "creation science." Thus, on the second point alone, "creation science" is clearly recognized as pseudoscience.

[Note: Points 3, 4, 5, 6, 8, 9, 10, and 11 are omitted here for the sake of brevity. See Strahler 1987, pp. 526-527 for a full discussion.]

Point 7. [The specific background is small or nil; a pseudoscience learns little or nothing from other cognitive fields. Likewise, it contributes little or nothing to the development of other cognitive fields.] Bunge's reference here is to borrowed specific scientific laws, principles, and methods. Because the Creator had his own unique and mysterious ways of creating time, matter, and energy, the specific background of recent creation in terms of mainstream science is simply nil. By the same token, science cannot contribute to an understanding of the ways in which God created everything; that knowledge is forbidden to humans. Chalk up another clear indicator that "creation science" is pseudoscience.

Point 12. [No other field of knowledge, except possibly another pseudoscience, overlaps the stated cognitive field. This means that the field is isolated and free from control of other cognitive fields.] Certainly mainstream science does not overlap the religious doctrine of special recent creation by a supernatural being. In that respect the creationist tenet is completely isolated from all research fields.

Any one of the points we have reviewed can by itself serve to identify "creation science" as pseudoscience. Points 1 and 2, singly or together, provide positive identification of "creation science" as pseudoscience.

We shall not further examine young-age fundamentalist creationism here, but turn instead to other and more interesting creationist programs that mix science and religion in an attempt to accommodate or merge the two fields.

Religious Alternatives to Recent Creation

In this review, the alternatives to recent creation are arranged in order from most conservative to most liberal. We begin with attempts to reinterpret the words of Genesis 1: 1-2, to give some accommodation to the geologists' needs for more time.

The *gap theory* of biblical creation separates the first and second verses of Genesis 1 by a large time gap. According to Bible scholar Bernard Ramm, a minor wording change in the second verse, suggested as early as 1791, made possible the insertion of a great time gap between an initial creation and a later creation (1954, p. 135). Time was what the stratigraphers of England and France of that period needed most, and this was where they found it. They could fit into the time gap their entire program of successive catastrophic faunal extinctions, each ending a long period of deposition of fossiliferous strata.

Those who subscribe to the gap theory assume that the original creation of the universe and earth, described in verse 1, was complete and perfect. In contrast, the earth described in verse 2 was in terrible condition. Biblical scholars can interpret the Hebrew words for "without form, and void" as also connoting "ruined and empty." This interpretation leads to the supposition that the Creator's perfect world was subjected to a gigantic cataclysm, leaving a shattered, uninhabited earth. As a result, it was necessary for God to restore the earth by means of a six-day "re-creation." For this reason, the gap theory is also known as the "creation-ruination-re-creation theory," or "restitution theory" (Ramm 1954, p. 135).

Today, the gap theory pleases no large group, least of all the fundamentalist creationists. For them, the gap has been trumped up merely to reconcile Bible and evolutionary geology. They are puzzled as to why God would go through the entire process of guided evolution, resulting in plants and animals closely resembling living species, only to destroy that life entirely and "restock" the biosphere with a closely similar fauna and flora. The modern fundamentalists regard evolution through extended geologic time as an evil and cruel process causing much suffering and death before culminating in emergence of humans. God would not want to engage in such activity. In contrast, God's six-day creation was humane and free of evil. Fundamentalists just do not want any "geologic ages" in the scenario, whether they conflict with the Bible or not.

The *day-age theory*, which has a number of variations, is based on the expansion of each "day" of the six creation days into a much longer time interval, making possible the evolutionary sequence and the great accumulations of strata that go with it. Authority for stretching the mean solar day as required is often found in 2 Peter 3:8, which reads: "But, beloved, be not ignorant of this one thing, that one day *is* with the Lord as a thousand years, and a thousand years as one day." The basis of this verse is in Psalm 90:4: "For a thousand years in thy sight are but as yesterday when it is past, and as a watch in the night." It is obvious to anyone reading the epistle that the verse is taken out of a very different context.

One variant of the day-age theory separates each of the six days of creation by a large time gap sufficient for big blocks of geologic time and for organic evolution. This can be called the *isolated-day theory*. The fundamentalists nevertheless insist on creation being accomplished entirely within each of the separated six days of mean solar time, so the intervening time spans are unnecessary and simply represent stagnation. Both the day-age scenario and its variant are rejected by the fundamentalists, and neither one is of any scientific interest.

Gosse's *Omphalos* and Indeterminate Creation

Before turning to the serious business of contemporary "theistic evolution," we should take time out for something in a lighter vein. It is from the pen of Philip Henry Gosse (1810-1888), an experimental zoologist who was a contemporary of Charles Darwin, with whom he was personally acquainted and exchanged correspondence during the 1850s (Morowitz 1982, p. 20). As a lay preacher for a Protestant sect with extreme fundamentalist views (The Plymouth Brethren), Gosse was obsessed with the need to reinforce the biblical doctrine of special creation against the rising mountain of geologic evidence favoring transmutation and a continuous evolutionary tree of life. In a book bearing the title *Omphalos: An Attempt to Untie the Geological Knot*, 1857, Gosse successfully defended the Creator's entire Creation from all possible attack, rendering it completely invulnerable to any and every objection that might be brought against it. And how did he do this?

Omphalos describes a stone of religious significance, shaped in the form of a navel; it was used in cultist rites in religious practices of the Greeks and Romans. The Greek word means "navel" or "umbilicus." The *Omphalos* residing in the Temple of Delphi was the most famous of all: it represented the center of the earth. But it was another navel—Adam's navel in particular—that raised embarrassing questions for the fundamentalists from the first time it was seen—seen, that is, in the eyes of religious artists. As all of you know, Michelangelo's Creation-of-Adam scene on the Sistine Chapel ceiling shows a half-reclining Adam, in the buff and displaying a magnificent umbilicus, no doubt about it, especially since the ceiling frescoes were given a severe cleaning. If Adam was created from scratch, out of nothing, how come he had a navel? The longstanding argument of theologians was that Adam had a navel because God wanted him to look as if he had developed *in utero*, like all other humans (except Eve) to follow. Eve is also shown with a navel, as for example in the expulsion scene as rendered by Jacopo della Quercia (ca. 1425) and Masaccio (ca. 1427). In the navel-art competition, consider also Jan van Eyck's rendition of both Adam and Eve in the Ghent Altarpiece (ca. 1430). It was a good decade for navels! Martin Gardner develops this idea further:

> This is not as ridiculous as it may seem at first. Consider, for example, the difficulties which face any believer in a six-day creation. Although it is possible to imagine Adam with a navel, it is difficult to imagine him without bones, hair, teeth, and fingernails. Yet all these features bear in them the evidence of past accretions of growth. In fact there is not an organ or tissue of the body which does not presuppose a previous growth history. (1957, p. 125)

The number of examples is as great as the objects in nature that reveal growth or change through time. A stately tree in the Garden of Eden must have shown numerous annual growth rings appropriate to its species and the prevailing climate, all placed in the trunk at the instant of its creation. We can suppose (as Gosse did) that Adam had partially digested food and its residue in his alimentary canal, food that he had never eaten. With impeccable logic, Gosse carried his principle to its logical conclusion: The Creator created everything in the universe to look as if it had an antecedent existence that flowed smoothly into the time stream of the new universe. Fossils, for example, were created in place in rock strata, designed to look as if they were once living forms, which they never were. The Creator thus built into the universe a complete line of things that never were.

All findings of science that relate to geologic age of the earth and the universe are completely explained by such "false records of a non-existent past" (Morowitz 1982, p. 20). Take, for example, the ratio of uranium-238 to lead-206, the latter an isotope produced by the spontaneous decay of the former at a constant rate. When the geochemist today analyzes the ratio of these parent and daughter isotopes in a particular rock, the age of the rock is revealed, and it may extend back 3 or 4 billion years in the past. For Gosse the answer would have been obvious: the Creator determined the isotopic ratio when He created the rock and did so with the intent to lead geologists into assigning a spurious age to the rock, but one fully consistent with all other indications of its age (also placed there by the Creator). And what of the red-shifts in the spectra of galaxies, a phenomenon that is interpreted to assign ages of several billion years to a single galaxy and an enormous radial velocity in an expanding universe? The Creator took care of that, too, with an appropriate Doppler effect. One of the nicest things about the *Omphalos* scenario is that scientists can search for centuries to turn up new kinds of fossils, minerals, and rocks containing evidences of their ages, and in all cases the results will be perfectly consistent with the table of geologic time as it is so far constructed—or will it? Suppose that, somewhere, the Creator has arranged for dinosaur bones to be intermixed with human bones—what then? Knowing that the Creator is always consistent and humane, and would never stoop to such trickery just for his amusement, we do not need to worry about that ever happening.

And, of course, there is no reason why the event of Creation should be assigned to the year 4004 B.C., give or take a couple of thousand years. Why not in the year 1 A.D., or in the year 1900, precisely at Universal Time 0001 hours? Gosse, himself, pointed out this logical extension of his *Omphalos* view of Creation (Gardner 1957, p. 126). He suggested that Creation may have occurred only a few minutes ago and we would be unaware of it! We realize that the Creator would have also provided all humans with a complete memory of all events of life prior to Creation, and there would exist newspaper files complete through that date, along with libraries and all other necessary artifacts. Because the instant between nothing and something was just that—an infinitesimal of time—there would be no way it could have been detected or recorded. Don't forget, the Creator would have also set up the complete line of ancient documents and inscriptions—Dead Sea Scrolls, included—to place the false creation date at 4004 B.C., but how could one detect that? The entire Bible would have been created by God, complete with English and other language translations; family bibles would have been created complete with handwritten insertions of records of family births and deaths. Thus it would be impossible for humans to know when Creation occurred, and it could have occurred at any time within the past five thousand to six thousand years, as the fundamentalists require. The total scheme of Divine Creation based on the *Omphalos* principle might be called *indeterminate creation*, setting it apart from theories we have already reviewed, such as the gap and day-age theories.

For naturalistic scientists, the *Omphalos* theory can simply be ignored (or accepted) as religion. As for scientists holding fundamentalist religious views, *Omphalos* allows them to continue research into the geologic past, secure in the belief that they are seeing God's work as He intended it to be. If there be seeming deception in God's creative work it is not malicious in any sense, but a means of promoting harmony and peace within all humanity. The creation of a relict past universe as it could have been, but never was, presents us with a marvelous continuity of past with present, ameliorating the psychologically disturbing trauma of sudden catastrophic creation.

Contemporary Views of the Theism/Science Coupling

Leaving behind with no regrets the spate of interpretive theological stories invented to give modern science the vast span of time it needs to explain the universe, but perhaps with regret that so perfect a theory as Gosse's has to be relegated to the status of a delightful fairy tale, we turn to the seething cauldron of modern Christian theology. In this pot theologians brew what are euphemistically called "liberal" scenarios that they think can wed theology to modern science. In biology, these scenarios go under the general heading of *theistic evolution*, an egregious example of an oxymoron, when we realize that "theistic" refers to a belief system in the ideational realm, whereas "evolution" rests firmly grounded in the empirical science of the perceptional realm. We can easily remedy this contradiction of terms by changing the adjective "theistic" to the noun "theism," to give us the noun-couple *theism/evolution*. Broadening the scope of inquiry to the whole universe, *theistic science* (equally oxymoronic) can be altered to *theism/science*. The slash keeps the two realms well separated, but they remain butted together like two stage actors competing for the same role. Perhaps a hyphen would be preferable. You may, if you prefer, substitute "versus" for the slash, a move that emphasizes the combat between the two. A scientist who espouses theism/science can be referred to as a *theist/scientist*, a shortened version of "a theist who is also a scientist," or vice versa.

One model, or program, of theism/science was been sketched in general terms in Chapter 11 as a dualistic ontological model (Model A in Figure 11.2). It is given the title Theistic-Teleological Dualism. It shows divine intervention by an omnipotent God. Whereas the vertical arrows, pointed down at four boxes, suggest a series of great creative acts, the label reading "Pervasive divine guidance, control, purpose" suggests some continuous divine activity. Evidently, some different versions of the scenario are possible, but all are firmly rooted in two beliefs: (a) God is a potent creative agent, and (b) God supplies a plan or purpose (a teleology), which he follows unerringly from start to present, with the human as the supreme achievement.

The human motive for theism/science seems to lie with the theologians, who seek to give back to God part of the job he lost when Darwin's theory dispensed with a creator. Unemployment is a terrible thing, particularly to anyone endowed with good health and a creative mind. Protestants, especially, look upon being idle as a sin of the first order, and that unacceptable condition must not be allowed to persist.

Another feature that all versions of contemporary theism/science share is a freer interpretation of the Old Testament. As an example of one position of a Christian theologist, not a "creation scientist," I quote from a paper written by a distinguished Jesuit scientist, Rev. James W. Skehan, S.J.: "The creation stories in Genesis 1-2 are theological reflections on the origin of the Universe and on humanity as the apex of creation" (1983, pp. 307-308). Further on, Skehan writes: "Among the biblical theologicians there is wide agreement that the story of the creation of the earth and of mankind in the first chapters of Genesis, is presented to recount the beginnings of the religious history of the people of Israel, and is not a scientific analysis to establish either the age or mode of origin of the earth."

The fundamentalist creationists' outspoken denunciation of old-earth, evolution-accepting scientists, and a return flow of hostility by the latter to the former, have prompted a number of protesting views from the scientific group consisting of theist/scientists. Their statements are worth quoting to develop further the position of theism/evolution. W. H. Hildemann, a biologist in the Department of Microbiology and Immunology of the University of California, Los Angeles, in a letter to *Science* (Journal of the American Association for the Advancement of Science) explains:

> Many evolutionary biologists appear to be responding in an uncompromisingly hostile manner as if no compromise were conceivable in teaching about the origins of life. Overlooked is the fact that many of us teaching life sciences in the universities and high schools are *both* Christians and evolutionists. The view has long been held among many, if not most, educated Christians that evolution is God's awesome method for achieving the creative process—in other words, adaptive diversity of species. One need only look at the relatively rapid appearance of new variants of

> animals and plants or pesticide-resistant and antibiotic-resistant strains of organisms to realize that this process continues unabated. The sadness of the rigid reasoning of spokesmen for the Institute for Creation Research is in considering creation and evolution as irreconcilable. Many biologists who also believe in a supreme being governing an orderly universe of marvelous design deplore the efforts of these "creationists" to force their literal religious views into the curriculum. One may also object to the attitude of intellectual arrogance among certain evolutionists who push their view that the original forms of life appeared entirely by accident or that matter itself sprang from nothing. The evidence of evolution does not and cannot reveal the source of the basic chemical elements or the primal source of life. (1982, p. 1182)

Of course, all versions of theism/evolution (theism/science) are thoroughly unacceptable and repugnant to the fundamentalist "creation scientists." They argue that theism/evolution is a theological compromise with which to placate theists while allowing the false doctrine of naturalistic evolution to be upheld. Randy L. Wysong, a "creation scientist," refers to theism/evolution as a "hybrid," a "baptized evolution" (1976, p. 63). Whatever role in evolution a more liberal theologian attempts to ascribe to the Creator, it is declared patently false if it departs in the slightest from the literal interpretation of Genesis.

Just how many, or what proportion, of the members of the science community hold to some version of the general theme of theism/evolution (theism/science) is very difficult to determine or even to estimate. The position itself is a satisfactory working philosophy to those who embrace it and, as such, it is not likely to produce a strong evangelistic movement. Nontheistic (atheistic, agnostic) scientists seem to feel little urge to argue with the theists/scientists, perhaps feeling that the mixture of theism and naturalism it requires could be something of a contradiction, but of no consequence if it remains a personal matter.

For myself, I wonder about the ability of a theist/scientist to "serve two masters" while carrying out research in the historical areas of biology, geology, and astronomy; but as long as no mention of miracles appears in published research in the scientific journals, I see nothing to worry about. An important practical point is that many theist/scientists want no intrusion of religion into the science classroom and this view unites the theist/scientist with the nontheistic (materialistic) scientist in opposing the fundamentalist creationists' efforts to get their anti-evolutionary doctrine into the classroom.

Is there any inner conflict in holding this dualistic position? We ask them: Does such belief inhibit your scientific activity in any way? I feel concerned that some level of conflict within the individual may exist, but not necessarily for those in all branches of pure science, and perhaps not at all for those in the applied or normative sciences.

For most theists/scientists there seems to be no problem with God's role in initiating the Big Bang, postulated by many cosmologists to have started the universe. Science has no means of probing farther back in time than that instant. There seems to have been no problem in gaining acceptance by the religious establishment (fundamentalist creationists excluded) of the scientific estimates of the age of the universe and its components, such as galaxies and stars. There has been general acceptance of the geochemists' determinations of the age of the earth as a planet, of rocks that form the earth's crust, and of fossils contained in those rocks. A large number of well-educated persons within the Catholic, Protestant, and Jewish communities, including their clergy and leaders, give broad support to the scientists' time scale and to the descent with modification that is shown in the fossil record.

Although a relatively liberal theological view of the significance of the Genesis account may seem to enable the scientist to practice without inhibition, there actually remains a large body of religious belief that attributes an active and directive role to God. Members of the Christian faith believe that God is a living force, for they pray to God for strength and for the enactment of beneficial events, such as an end to poverty, cure of a disease, or for peace on earth. They believe in miracles—events that cannot be explained by science and must therefore be acts of God. Moreover, God has always had a purpose or plan, directing all changes and events toward a goal. Study of such divine direction and its revelation in a design or pattern is teleological, but much more far-ranging in time than the natural teleology we have already referred to in biology (see Chapter 4).

Thus, in the historical sciences, the theist/scientist is faced with the problem of deciding which events or portions of the record represent God's work and which remain as naturalistic phenomena. To conduct research as if God had played no part is a rejection of one's own religious belief; surely this would be traumatic. On the other hand, if one proceeds on the assumption that God's hand was at all times into everything, historical science becomes nothing more than the documentation of God's work. Because God also created all laws of science, reductionist explanations invoking underlying laws are also attempts to understand God's plan. That was the thrust of natural theology. Indeed, documenting God's glorious and perfect work was held by the early naturalists to be the purpose and justification of natural science. That same view is expressed in one form or another by many modern theists/scientists; perhaps for them it successfully bridges the two realms of reality.

An illustration of a highly thoughtful version of theism/evolution is the philosophy of Pierre Teilhard de Chardin (1881-1955). A French paleontologist and Jesuit, Teilhard seems to have concentrated upon the purposefulness of evolution throughout all the cosmos, culminating in the human species, *Homo sapiens*, as the supreme goal. Based on one of Teilhard's posthumously published works, *Man's Place in Nature* (1966), I judge that he accepted a largely naturalistic origin of life but considered that supernatural teleological forces became progressively more important through organic evolution and finally dominated in the rise of humans. Thus, concerning evolution he states: "We must assume the existence and influence, underlying it, of some powerful dynamism" (p. 32). Teilhard regards the phenomenon of humans and their civilization as too complex to be accommodated in deterministic natural selection (p. 72). The phenomenal growth of human brain power, which he calls "hominisation," is considered unique: "Hominisation, a mutation that, in its development, differs from all the others" (heading on p. 72). A direct reference to a divine driving force seems limited to his final sentence:

> And it is at this point, if I am not mistaken, in the science of evolution . . . that the problem of God comes in—the Prime Mover, Gatherer and Consolidator, ahead of us, of evolution. (p. 121)

Teilhard's text is strong on teleology but vague on the details of how this direction was accomplished. We really don't know if it was envisioned as continuous or punctuated direction. Teilhard's worldview appeals to many because of its mystical overtones, invoking a numinous aura. Most persons among the general public don't want too many details. Perhaps many mainstream scientists who espouse theism feel the same way. For them, probing for details of God's modus operandi, as if God were a fellow scientist, might uncover too many difficult problems, inconsistencies, and contradictions. The analytical task is for philosophers, preferably those whose point of view is nontheistic and assumes that all theologies are constructs of the human mind.

We next consider two forms of theism/science, both in vogue today. The difference between the two lies in the timing of God's creative activity: (a) punctuated (progressive) creation, in which divine activity is episodic, occurring in bursts clearly separated from periods of natural activity; (b) continuous creation, in which divine activity is continuous and pervasive.

Punctuated Creation (Progressive Creation)

Progressive creation is an acceptance of the scheme of organic evolution through geologic time, but with the provision that God intervened at intervals to create new lines of life forms. Essentially we have an extension of pure deism that calls for God to drop entirely out of the picture, leaving the universe to a fully naturalistic mode of operation. In progressive creation, however, this one-two sequence is repeated over and over again (see Chapter 11 and Figure 11.1, B, Deistic-Mechanistic Dualism). Note that this program is essentially the same as in the logic-model introduced in opening paragraphs of this chapter. Between episodes of God's intervention, the universe goes on automatic pilot, so to speak, with God's laws taking over while God rests. Perhaps God remains in the pilot's seat at all times, eyeing the instrument panel and listening to the traffic controllers talking on the radio, and always on the lookout for trouble. In that event, God manually takes over the controls and makes a change of course.

Examples cited by the creationists are God's intervention to create *Hyracotherium*, the first genus of the line of horses, which then evolved through the Cenozoic Era, and in like manner to create the early hominoid (ancestral ape) that then evolved into the hominids and modern humans (Morris 1974, p. 220). The idea seems to be that evolution would not generate new lines such as new orders, classes, or genera, so that to keep the evolutionary tree growing, God needed to step in and graft on (as it were) a new branch from time to time. Generally, divine intervention is postulated for a time gap in the historical record preceding the seemingly abrupt appearance of a new species or higher taxon.

Fundamentalist young-earth creationists reject progressive creation both because it does not agree with the literal meaning of the six-day creation and it limits God's intervention to a sporadic activity. On their side of the fence, evolutionary biologists resent the idea that their theory is not permitted to account for new branches on the tree of life. Also, the complex dualistic program runs contrary to the basic principle of parsimony.

Progressive creation seems to be a relative newcomer to contemporary creationism, as compared with the gap and day-age theories, although it derives from similar scenarios put forth in earlier centuries. The modern version was promoted in the 1950s as one of the alternatives seriously considered by the American Scientific Affiliation. In line with the modern evolutionary theory of punctuated equilibrium, this version might better be called *punctuated creation*. My conception of this idea is that God becomes bored with evolutionary stasis and zaps the system at intervals with just the right mutations needed to create a new species, genus, or higher taxon. Before that, of course, God had to zap a lot of complex organic molecules in order to get them to organize into DNA, RNA, and the necessary enzymes to start the chain of life. This zapping he did in an attempt to accelerate evolution to the point that he could introduce the human species, which was to give God a great deal more to think about and many interesting problems of evil to resolve.

To avoid the tedious details of accounting for each of the vast numbers of galaxies, stars, and planets—to say nothing of the enormous numbers of new species of life on earth through geologic time—the punctuational creationists usually confine their efforts to three great creation events: (1)

creation of the universe, (2) origin of life (biopoesis) on our planet, and (3) the creation of the human species. They will gladly settle for those three, if only they can arrange that the topics are aired in the public school science classrooms. That this three-event approach may be the current instrument (a trident) of the American Scientific Affiliation's power structure is suggested by their 1986 brochure, *Teaching Science in a Climate of Controversy*. It asks that students be encouraged to question intensively all that evolutionary science has to say about these three key events, playing up the argument from ignorance, and leaving the door wide open to divine creation. Teachers are instructed to discuss the creationistic versions freely. It is suggested that teachers emphasize that many scientists accept divine creation along with mainstream science. With surprising candor, the text reveals that all Western theists—Christians, Jews, and Moslems—are creationists in the broad sense. Both "progressive creation" and "theistic evolution" are specifically named as being within the spectrum of acceptable Christian belief (A.S.A. 1986, p. 17). The authors point out that an erroneous impression to be avoided is "that all creationists are united against all evolutionists" (p. 17). One strategy they emphasize is to keep up the pressure against the recent-creation/flood dogma as being unscientific and a "dangerous pseudoscience" (pp. 15-16). Progressive (punctuated) creation, they insist, lies in a "broad middle ground" in which one can have his or her Christian beliefs along with good science. Having one's cake and eating it is a formula hard to beat.

A recent version of punctuated creation has been promoted by Davis A. Young, a professor of geology at Calvin College in Grand Rapids, Michigan (Young 1977, 1982). An avowed creationist, so far as belief in the biblical record is concerned, Young is also convinced of the great length of geologic time required by geologic evidence, both from the stratigraphic record and from radiometric age dating of ancient rocks. He also accepts organic evolution but with the belief that God can and does perform miracles, one of which is the creation of humans (Young 1982). He accepts the naturalistic evolution of the hominoids and earlier hominids and the similarity of *Homo sapiens* to those older species. It does not seem to bother Young that modern humans may appear to have evolved from more primitive species of the genus *Homo*, as the mainstream evolutionists theorize. He reasons that God would simply have created *H. sapiens* in such a way as to have suggested this evolutionary connection. (Good old Gosse—always there when you need him!)

Origin Science Versus Operation Science—A Creationist Ploy

Secularization of one large area of science by creationists who favor punctuated (progressive) creation has been attempted through the device of dividing empirical science into two categories: (a) *operation science* and

(b) *origin science*. "Operation science" consists of what we have already recognized in Chapter 4 as timeless scientific knowledge. A good example of the use of this term is found in *The Mystery of Life's Origin*, by creationists Charles B. Thaxton, Walter L. Bradley, and Roger L. Olsen (1985, pp. 8, 202-204). These authors explain that all theories of science (e.g., relativity theory, atomic theory, quantum theory) "deal with various facets of the operation of the universe, let us call them operation theories of science. Our point of clarification notes the difference between operation theories and origin theories, such as theories about the origin of life" (p. 8).

Physics and chemistry fall into the category of operation science, as do the applied sciences—engineering and technology; these fields emphasize a view of nature that is atomistic, reductionistic, and universal. Operation science, for a creationist (broadly defined) requires only the belief that God initially created the laws of science and keeps them intact. (You will need to forget about miracles in this discussion, sticking firmly to a pure deism.) Teaching the principles of physical science in this framework of laws of science (its proponents say) involves no controversy and can be pursued quite independently by theists, deists, and atheists alike. Operation science is thus made a "sanitized science," segregated by a sort of *cordon sanitaire* from the messy, sticky residue of origin science.

Origin science is composed of natural events the "creation scientists" designate as "origins." They seem to be referring to the initial appearance of a unique structure of matter such as the universe, a star, a planet, a living cell, or a species. Of course, Christian doctrine, as expressed in its creeds, is that all such initial appearances are the work of the Creator. Many, and perhaps most Christians who are also scientists simply keep silent on this requirement when doing science teaching or research—it is dogma aired in the sequestered space of the church or the home.

Fundamentalist "creation scientists," in contrast, speak out strongly and publicly on this point, stating flatly that scientific proof of such divinely wrought "origins" is impossible. This topic is presented early in the content of the ICR textbook, *Scientific Creationism* (Morris 1974, pp. 4-5). Both naturalistic evolution and supernatural creation are stated to deal with "origins," and it is considered of vital importance to study the subject of "origins" (p. 4). "At the same time," they state, "it must be emphasized that it is impossible to *prove* scientifically any particular concept of origins to be true" (p. 4).

Let us start with the assertion that it is impossible to prove scientifically the truth of any concept of "origins." Our discussion of the nature of science in Part One placed strong emphasis on a fundamental concept in empirical science: no proposition based on observation can be proved to be true or false. The most that can be done is to reduce to a small value (or increase to a large value) the probability that the proposition is false when declared true. As I explained earlier, the words "proof" and "true" apply only in the area of logic and mathematics, in which the initial propositions or axioms are accepted a priori as true. Thus, for the creationists to say that no hypothesis of "origins" can be proven to be true is simply not not an issue for the empirical sciences. When applied to recent creation by a Creator, which is a religious statement invoking the supernatural, science and the scientific method are, again, not an issue.

Science does concern itself with causes (i.e., with explanations) and that concern is legitimate. Historical geology and biology search for causes (explanations) of everything and anything material that can be examined today, especially explanations of the record of the past preserved as fossils and of the enclosing rock. That the study of events in past geologic time can be carried out in strict observance of the requirements and criteria of the scientific method is agreed on by most geologists and biologists. Even Sir Karl Popper, who in his earlier writings seemed to say that evolution was not a scientific hypothesis, later reversed his position to accept the prevailing modern view (see below). Historical geology and biology must consider the origin of everything observable. Every geologist who examines a particular rock asks: "What is the origin of this rock?" And so do the "scientific creationists." When they look at the same rocks they search for the origins of those rocks—they postulate that the rocks originated in a great biblical Flood. They even attempt to prove that postulate by interpreting in their own way the features of the rocks and the fossils they contain. Creationists cannot have it both ways. They cannot first assert that origins cannot be proven and then go right out into the field in search of geologic evidence by which to prove the asserted origin. Such self-contradiction is an essential part of the modus operandi of "creation science." Moreover, assertion simultaneously of contradictory statements can be a form of self-deception.

Popper's change of position needs to be documented here, so that it cannot be said I offer it as mere hearsay. John R. Cole, writing on the subject of the use of misquotations by creationists, refers to this incident in detail (1981, pp. 43-44). He repeats an oft-quoted statement by Popper: "Darwinism is not a testable scientific theory, but a metaphysical research programme—a possible framework for testable scientific theories" (Popper 1976, p. 168). In evaluating Popper's statement, you need to know that the word "metaphysical" has two meanings in philosophy, and both are in common use. On our side of the Atlantic, "metaphysical" is commonly associated with supernatural phenomena and beliefs. In Europe (where Popper did his lecturing and writing), it is rigorously used by philosophers to refer to basic philosophical questions "such as ontology, the body-mind problem, the problem of origins . . . the problem of natural versus supernatural," and "other questions of ultimate significance" (Ferm 1936, p. 92).

I think it would be a serious mistake to assume that Popper intended to mean that Darwin's theory lies in the supernatural realm. Creationists would like to use that meaning to support their claim that "evolution is religion." Popper's change of position was given in published form: "I have changed my mind about the testability and logical status of the theory of natural selection; and I am glad to have an opportunity to make a recantation" (Popper 1978, p. 344). Even the best-informed scientists and science-news reporters have confused and abused the meanings of "logical prediction" and "temporal prediction," as have the "creation scientists." (An egregious example is in the book by Thaxton, Bradley, and Olsen, referred to in an earlier paragraph; see 1984, p. 204.) Historical hypotheses, such as organic evolution, are capable of producing logical predictions, whereas temporal predictions (predictions of events far in the future) are, as a general rule, limited at best to guesses. Logic forbids the extension of the universal proposition "All x's are . . ." to "All x's will always be . . . ." It is in the sense of logical prediction (deduction of a consequence of a proposition) that Popper accepts the testability of Darwin's hypothesis.

Continuous Creation

Suppose we postulate that divine creation is a continuous happening keeping perfect pace with real time. Assume that the universal store of matter (mass/energy) is held constant through all time, and every thing in the universe either remains unchanged or undergoes change with the passage of time (or undergoes an alternation of those states). All change is the work of God and nothing changes unless it is changed by God. Any thing that remains unchanged (in stasis) does so at the will of God. Then, "creation" broadly means all change of any things in the universe. To this we add that that change follows laws created and sustained by God, who may also modify or suspend these laws at any time without limit. This is *continuous creation*. It is pure creationism—"all *a*" (all religion) in our opening logic model—because nothing happens unless God causes it to happen.

Much of theism/evolution equates to continuous creation, as here defined. Continuous creation, in turn, is an application of the theological doctrine or principle of *immanentism*, which considers that God's presence is diffused throughout all the physical universe, so that God is continuously active in all change, including, of course, biological evolution. God is immanent and has the quality of immanence. At the same time, God can be *transcendent* in the sense of being above and apart from the physical realm; just don't ask how he does this!

Immanentism has been recently emphasized by Arthur Peacocke, whose career in research on physical biochemistry in England gave way in later years to an interest in theology, which he studied, and led to his ordination as an Anglican priest. My description here is borrowed from a review of Peacocke's 1986 work, *God and the New Biology*, by J. W. Haas, Jr. (1989). (Haas, a professor of chemistry at Gordon College in Massachusetts, is currently editor of *Perspectives on Science and the Christian Faith*, journal of the American Scientific Affiliation.) Haas describes Peacocke's view of the universe as follows:

> Since matter has a "continuous, inbuilt creativity," creation is going on all the time. God's relationship to the world is perennially and perpetually that of Creator and we can view his actions in the processes of nature by scientific study. For Peacocke, nature is a mode of God in action, not just a stage for His action. "In the actual processes of the world, and supremely in human self-consciousness, God is involving himself and expressing himself as Creator" (p. 97). Peaocke focuses on God's *creatio continua* over against *creatio ex nihilo*. (Haas 1989, p. 164. Quotes are from Peacocke 1986.)

Peacocke sees the interplay of law and chance in all natural process as part of the order of creation. Because God penetrates as well as includes the whole universe, he can keep chance from getting out of control. Obviously the entire scenario is one of purposeful design—totally teleological. Perhaps "guided chance" would be the descriptive term for this extreme form of theistic evolution, but that term is not only oxymoronic—"chance" means "not guided"—but totally antithetic to the main thrust of naturalistic evolution as a pure trial-and-error process. To accept Peacocke's total immanentism means that the role of science is entirely revelatory, as in natural theology; i.e., to expose God's work and his methods of doing it.

Haas describes Peacocke's approach as *panentheistic*. *Panentheism* is defined by Peter Angeles as follows: "All things are imbued with God's being in the sense that all things are *in* God. God is more than all that there is. He is a consciousness and the highest unity possible" (1981, p 111). The suggestion behind panentheism is that supernatural and natural realms are fused into one homogeneous substance, like homogenized milk. The ingredients that go into the product are clearly different at the outset, but lose their identity in the product. Creationists are prone to homogenize supernatural and natural realms. That would be a logical fallacy because homogenization is forbidden by the definitions of the two terms.

One might think that a researcher in physical biochemistry would be curious about the nature of the God-substance that permeates the universe. Looking at the universe from quantum mechanics, it consists of particles. Surely the God-substance would fill all of space between all of the particles and one might think of it as a perfectly homogeneous and nonviscous fluid. If so, how could God's instructions for planned creative change be transmitted through this fluid? In the absence of a central nervous system with neuronal transmission of information, we might speculate that information is transmitted by waves similar to seismic or sound waves. A much more intriguing question is whether the divine substance merely surrounds the elementary particles or actually permeates them with the same uniform density as the enclosing medium. Theists who are also scientists approach theology with a very different mindset than they approach their science. They are reluctant to hypothesize about supernatural processes, perhaps considering this area off-limits to humans, and that such speculations might earn God's disfavor. Perhaps it's just a case of lack of interest in analytical theology on the part of the scientists.

Pantheism and Panpsychism

At this point, we may need to consider two additional models suggested by panentheism. One is *pantheism*, defined by Peter Angeles as follows: "The belief that God is identical with the universe. All is God and God is All. God and nature . . . are synonymous, or two words for the same thing" (1981, p. 111). Pantheism is a purely monistic model in which there is no distinction between supernatural and natural. It is beyond the fold of the Western theistic religions, which require a dualism of supernatural and natural. In any case, pantheism seems to have played no role in the science/creation issue, whether at the literalist/fundamentalist level or that of theism/science.

An interesting variation of panentheism is *panpsychism*, "the belief that God is completely immanent *in* all things in the universe as a psychic force (mind, consciousness, spirit, soul)" (Angeles, 1981, p. 111). The designation of God as a "force" immediately gives us a mechanism by which God might be able to accomplish change in the physical universe. Thus panpsychism sounds like a possible candidate as a model for Peacocke's version of continuous creation. All that needs to be added is a strong dose of teleology. The psychic force can perhaps be identified with what we described in Chapter 4 as "entelechy," "élan vital," or "radial energy." Perhaps it equates to the Cartesian "soul," or the guiding "consciousness" existing "outside the parameters of space and time," proposed by Nobel Laureate, George Wald (see Chapter 1). A mysterious psychic force is also what the parapsychologists are looking for as the mechanism of psychokinesis.

The Anthropic Cosmological Principle—A Creationist's Dream?

Who would have guessed that the most empirically positivistic of all scientists—those physicists who investigate the highly mathematicized field of quantum mechanics—would become strangely attracted to creationism as a direct result of their great successes in nearly completing a grand unifying theory of cosmic reality? Could it be that scientists searching under the glaring light that distinguishes science from religion can be led straight to a creationistic scenario for the origin and evolution of our universe? Perhaps the argument from design is simply too attractive for any human to resist, no matter how it is packaged.

Quantum physicists always knew they were exceptional humans, but only in recent years has this ego trip taken them to the point that they believe the universe was specially created for them, mind you, not just for humans as a species (for the Christian theists have been saying that right along) and not just for scientists in general. We have long been subjected by the logical positivists to the thought that only physics is science, but this concept goes much further.

"Bizarre" may best describe the *Anthropic Cosmological Principle*. It has joined a long line of cosmic principles of science, starting with Aristotle's Principle of Continuity that pictured nature as passing gradually upward from a least perfect earthly shell to the outermost and most perfect shell that is God. This Ptolemaic geocentric model of the universe was to serve the Church through the Dark Ages, after which it was replaced by the Copernican heliocentric model. Tony Rothman, a physicist who holds a Ph.D. from the Center for Relativity at the University of Texas, describes the principle in an intriguing essay in *Discover* magazine; its title: "A 'What You See Is What You Beget' Theory." He observes that the new reigning paradigm became the *Copernican Principle*, which claims that no one part of the universe is more privileged than any other part (Rothman 1987, p. 90).[2]

The Copernican Principle gave rise to the cornerstone of modern cosmology, the *Cosmological Principle*, to the effect that the universe is homogeneous as well as isotropic. Cosmologist Joseph Silk, who holds the Ph.D. from Harvard University and is professor of astronomy at the University of California, Berkeley, tells us that it was extended to the *Perfect Cosmological Principle*, under which "the universe should present a similar aspect when viewed from any point in space *and* time" (Silk 1980, p. 3). It also provided that temperature be constant through time. Silk points out that the inclusion of time in this definition led to the formulation of the steady-state theory, now ruled out as a viable contender and leaving the Big Bang as the favored theory. This exclusion of steady-state also requires exclusion of the constancy of both time and temperature from the Cosmological Principle. Included, however, is the assumption that physical laws are the same for all parts of the universe and have been so through all past time, at least as far back as the singularity.

The Anthropic Cosmological Principle has been lucidly described by Silk. The following excerpt is from his 1980 book, *The Big Bang: The Creation and Evolution of the Universe*:

> This assumption takes precisely the opposing view to that of the perfect cosmological principle by asserting that we are viewing the universe at a privileged time, although the present universe would appear the same when viewed from any point in space. This privileged era is assumed because of the necessity for special conditions to arise that are favorable for the evolution of life. For example, if the universe were much hotter or denser than it is now, galaxies could not form. If the force of gravity were very different from its observed strength, planetary systems either would not form or would not be congenial to life as we know it. It is, after all, a remarkable coincidence that the age of the earth turns out to be similar (within a factor of four) to the ages of the oldest stars or galaxies that astronomers have found. The anthropic cosmological principle explains this similarity by fiat. The universe could be vastly more irregular and disordered than it is. The anthropic cosmological principle asserts that if it were, conditions would be uncongenial to life. Thus as observers, we inhabit a very special universe, and only this universe is isotropic and homogeneous. The anthropic argument is a very fundamental one, as it purports to explain the Copernican cosmological principle, which is central to practically all viable cosmologies. (Silk 1980, pp. 3-4)

Silk says that we can appeal to the Anthropic Cosmological Principle in selecting the Big Bang model for the origin of the universe:

> For, were the universe to evolve in a highly irregular fashion, we would not likely be present to bear witness. All these chaotic cosmologies, given enough time, are likely to develop in a way that is hostile to life. Only the standard Big Bang model, from an infinity of choices, is destined to provide an environment congenial to the evolution of life. (Silk, p. 5)

"Anthropic" connotes only "relating to the human species." Silk's words in his first paragraph do not necessarily say that the universe was designed by a designer from its outset to be the home of the human species. He implies only that the probability that the terrestrial environment should have turned out to be as we find it is extremely minute when stacked against an almost infinite number of alternative possible universal laws, physical constants, and evolutionary histories that the human mind can conceive of as logically possible.

Under Darwinian evolution by natural selection the most obvious conclusion should be that life on our planet is what it is because life processes and structures could only derive from the available chemical and physical resources and environmental circumstances—i.e., the givens. Adaptation of life forms to the given environment simply prevailed, there being no alternative. Considerations of hypothetical other universes are simply specious, irrelevant, and uncalled for.

In Silk's second statement, some suspiciously teleological and religious overtones creep in. The phrase "to bear witness" raises the query: Witness to what? To God's planned creation? The words "destined to provide an environment congenial to the evolution of life" is nothing if not teleological. My response is that the concept of destiny has no place in naturalistic evolution.

Tony Rothman tells us that the term "Anthropic Principle" was coined in 1974 by Brandon Carter, a Cambridge University cosmologist (Rothman 1987, p. 91). His idea was that rules of quantum physics require physicists to include an accounting of the properties of their experimental apparatus. Of course, physicists are an integral part of the apparatus. Carter had said: "What we can expect to observe must be restricted by the conditions necessary for our presence as observers" (p. 91). The general idea is simply that what we observe about the universe must contain all those conditions necessary to finally produce intelligent humans.

Rothman also tells us that this "biological selection effect" was most explicitly developed by quantum physicist Robert Dicke of Princeton University, who in 1961 presented the first full-blown anthropic argument. Dicke's reasoning included the observation that our universe must exceed the minimum age that would have allowed the element carbon to be produced, because, wrote Dicke, "it is well known that carbon is required to make physicists"—at least physicists as we know them (Rothman, p. 91).[2]

Dicke's observation that carbon is needed for all earthly life is satisfactory, albeit specious. What follows is a non sequitur. If, as we all know, physicists need carbon in their bodies, the stellar synthesis of carbon had to come first. Evolution of life made use of whatever elements were then and now are available, including carbon. The suggestion that carbon was synthesized so that intelligent life might exist on earth is preposterous, especially so when coming from anyone who is not a fundamentalist creationist. Who planned it that way? Dicke said that the observed age of the universe "is limited by the criteria for the existence of physicists" (Rothman, p. 92). Dicke seems to have got it "bass-ackwards," and we need to correct it to read: "The existence of physicists (who came last) was limited by the criteria for the observed age of the universe." Dicke's statement of the Anthropic Principle has since been specified as the "Weak Anthropic Principle" (WAP).

Coming down to basics, Dicke's Weak Anthropic Principle carries the inescapable, but unspoken, assumption that physicists (and perhaps even their mothers) possess instrinsic positive value or worth. Value for what and for whom? Value to society? Value to God? All of these are religious concepts; they are meaningless and irrelevant in empirical science in explaining the functioning and history of observed cosmic phenomena.

An extension of the Weak Anthropic Principle has been made to include interpretation of the fixity of the fundamental physical constants and other problems related to quantum mechanics. This extension goes by the name of the Strong Anthropic Principle (SAP). It was formulated by Brandon Carter to read: "The universe must be such as to admit the creation of observers within it at some stage" (quoted by Rothman 1987, p. 94). Rothman adds: "Most scientists interpret the SAP to mean that the universe must be nearly as we know it or life wouldn't exist; conversely if life didn't exist, neither would the universe" (p. 94).[2] The first sentence (along with Carter's) is acceptable to the degree that it means nothing more than that life on our planet has adapted to the universe as it is. The "converse" is essentially a statement that if there is no reality in some particular object we usually consider real (i.e., to exist) then it follows that there is no reality in any and all other objects we consider real. Has anything new and profound been included in this description of the SAP?

A subject worthy of attention within the aegis of the SAP is the observation that our present universe reflects the particular values of a number of fundamental constants of nature, and that if even one of these had been sufficiently different from what it is, the universe would have been profoundly different than it is. Among those constants named by Rothman are Newton's gravitational constant G, Planck's constant h, the speed of light c, and the weak interaction coupling constant a_w that governs how fast neutrons decay radioactively (p. 94). Those physicists who have been swept up in the anthropic fever love to play the game of "What if?" What if such-and-such a constant had been different by such-and-such a factor? What then? In this way, they can imagine many different universes that might have been or perhaps now exist elsewhere.

As an example given by Rothman, "Carter has proposed that if G were slightly larger, all stars would be blue giants, which probably don't have planets; if it were much smaller, they would be red dwarfs, which don't produce supernovae" (p. 95). Rothman comments: "An objection to this sort of thinking is that perhaps life doesn't require planets or supernovae, and cosmologists have been unduly anthropocentric" (p. 95). A sound objection, I think it should be sustained. "Nevertheless," Rothman continues, "anthropocists have filled the literature with literally hundreds of such arguments, claiming that all the fundamental constants and many of the elementary particle masses are fixed by the SAP."[2] How, I ask, can a theory originating in the human brain have served as a physical agent to fix the constants of nature at or before the singularity of the Big Bang, some 10 to 20 billion years ago?

For those of you wishing to go even farther and deeper into the "What if?" game, I recommend Rothman's description of the "Many Worlds Interpretation," proposed by Hugh Everett in 1957 (Rothman, p. 95). It will be helpful if you have studied Schrödinger's cat paradox and the principles of radioactive decay. The analysis leads to the conclusion that there would have emerged a nearly infinite number of universes, each with its unique set of fundamental constants.

An even further extension of WAP and SAP is the Final Anthropic Principle (FAP), which Rothman describes as "the notion that the universe not only must give rise to life, but that, once life is created, it will endure forever, become infinitely knowledgeable, and ultimately mold the universe to its will" (p. 96).[2] Rothman likens this goal to that final state described by paleontologist/Jesuit Teilhard de Chardin and named by him the Omega Point. "Thus," says Rothman, "man—or Life—will be not only the measure of all things but their creator as well."

Criticism of the Anthropic Principle has come from among the physicists. One is Heinz Pagels, who wrote:

> Carter's worldview is the product of an anthropocentricism as profound as that which underlay the pre-Copernican view of the universe; the anthropic principle is born of a most provincial outlook on what life is. Its adherents assume that all life must resemble, in broad form at least, life on this planet. (As quoted by Rothman, p. 96, from Pagels 1985.)

More serious, says Rothman, is the following charge by Pagel:

> Unlike other principles of physics, the anthropic principle is not testable. It is all well and good to imagine universes with various gravitational constants and estimate the prevailing physical properties, but there is no way we can actually go to an imaginary universe and check for life. We are stuck with our universe, and powerless to alter its fundamental constants." (As quoted by Rothman, p. 96, from, Pagels 1985.)

Rothman counters that it is not necessary that all theories be testable to be judged scientific in quality. He cites the Copernican Principle, the Principle of Simplicity, and the Principle of Beauty as examples of propositions that cannot be verified directly (p. 96). Nevertheless, he points out, these principles proved enormously fruitful in advancing quantum mechanics by directing investigators toward hypotheses of interactions of matter that could be tested in the laboratory.

Popular interest in WAP/SAP/FAP was heightened by the appearance in 1986 of a massive tome, titled *The Anthropic Cosmological Principle*, by John D. Barrow and Frank J. Tipler. It has been favorably cited for its thoroughness of treatment of the entire subject, and for its detailed account of the history of teleology. In his review of the work, Martin Gardner, whose books and articles have so successfully deflated many pseudoscientific propositions and cults, wrote in the *New York Review of*

Books that "The FAP should properly be called the CRAP—Completely Ridiculous Anthropic Principle" (as related in Rothman 1987, p. 96).

The Anthropic Principle attracts teleological and religious interpretations from all quarters. As with the problem of origin of the universe, physicists and astronomers are strongly tempted to opt for a creator to have designed the entire program of cosmic evolution. Physicist and popular author Paul Davies seems to have succumbed to that temptation. His bottom line reads: "The laws which enable the universe to come into being spontaneously seem themselves to be the product of exceedingly ingenious design. If physics is the product of design, the universe must have a purpose, and the evidence of modern physics suggests strongly to me that the purpose includes us" (Davies 1984, p. 243). Rothman's bottom line says it a bit differently, but much more to my liking: "When confronted with the order and beauty of the universe and the strange coincidences of nature, it's very tempting to take the leap of faith from science into religion. I am sure many physicists want to. I only wish they would admit it" (1987, p. 99).[2]

Philosopher of science and physicist Mario Bunge gives us a fitting paragraph with which to close this section. After disposing of creationist cosmology as "fantasy," he writes:

> A related fantasy is the so-called anthropic principle, according to which the universe would have evolved precisely in such a manner that human beings could emerge as they did. This hypothesis may be interpreted in either of two manners. According to one reading it would be nothing but a tautology, namely the statement that man emerged when he did. According to the alternative construal it would be a theological dogma amounting to the assertion that God created the universe in such a manner that, at the right time, it would become inhabitable by human beings—preferably the gullible kind. In either version the anthropic principle is alien to science. And in its second version it must face the objection that, if God was so smart in timing the phases of the universe, why did He create a creature so dumb that it is fouling up his own God-given nest. (1991a, pp. 140-141.)

Are All Forms of Theism/Science Partly Pseudoscience?

We almost never think that Christians who espouse the contemporary varieties of theism/science—punctuated creation and continuous creation—are practicing pseudoscience. Is that claimed immunity derived from the acceptance of a great age of the universe and earth combined with at least lip service to a naturalistic form of organic evolution?

We now refer back to the introductory paragraphs of this chapter, in which a logical analysis is made of the separation of divine causation and natural causation. Consider first punctuated creation, in which events of divine creation in a supernatural manner alternate with periods of lawful naturalistic process. Think of a ticker tape (if you can remember what that was) issuing from a machine that produces a single continuous color band, showing either divine creation (red) or naturalistic process (black). The history of the universe thus consists of red bands alternating with black bands (never overlapping and never parallel). We have labeled as pseudoscience the fundamentalist creationists' Genesis scenario of recent six-day Creation and one-year Flood, which they claim is bona fide empirical science. It can, instead, be shown to be a religious belief. A believer in punctuated creation will plead not guilty to promoting pseudoscience with the following argument: Granted that the creation events (red sections of tape) are religious belief, they are always temporally isolated from the naturalistic events (black sections of tape); i.e., these contrasting kinds of events never overlap.

The analog of this relationship is that of a double-throw (DT) electrical switch. Unlike the off/on light switch with which you are familiar in the home, the DT switch can be used to switch from one power source to another to feed the same appliance. When you throw the switch, you break one power source just prior to engaging the other source. Let biological evolution be the appliance. Let one power source be God; the other the naturalistic lawful process. The theist's argument is based on which power source is used, but ignores the power source that has been relegated to disuse. I ask that you focus on the latter act. Switching off the naturalistic process is a denial that empirical science functions within certain segments of time; it is religious denial of empirical science for some finite period of time. Moreover, this religious denial is repeated innumerable times in the history of the universe. Thus an important part of all universal history is claimed to have been nonempirical, and instead to have been religious. As we demonstrated in our chapter introduction, the suspension of naturalistic process is not permitted by empirical science; therefore to claim that such suspension repeatedly occurs is false science. To the degree that suspensions of the naturalistic process are required, punctuated creation is indeed pseudoscience. Punctuated evolution is thus contaminated throughout by pseudoscience. Regardless of the degree of contamination, the total product cannot be labeled as genuine science, but rather as bogus science.

Try this criterion out on continuous creation. Here, the divine component is immanent and ever-present in contact with all matter. There are no temporal alternations. Contamination of the empirical content with the religious content is total. Thus continuous creation is pseudoscience from beginning to end. Bottom line: All forms of creationism are pseudoscience to some degree.

Is Religion Also Pseudoscience?

Does this mean that religion must be called "pseudoscience"? We suggest the answer: "No, or not necessarily." Religion is secure in its own ontology and epistemology, a position we have gone to great length to establish in earlier chapters. The question would only arise when theologians claim that God can physically influence or alter the states of matter and space-time that can be observed by science or the natural laws that describe the behavior of that universe. Such claims are strictly forbidden by the very definition of "supernatural" as a realm separate from the natural realm. To conform to this definition based on separatism does, however, require the abandonment of the miracle. As we have just concluded, a miracle requires a substitution of the supernatural for the natural, a mental act that is pseudoscientific.

If theistic religion, Western-style, is to keep completely free and clear from pseudoscientific utterances, its significance and value to humans must be sought in nonempirical realms other than itself. Teleology, the belief that physical and biological evolution of the universe—any change, that is—is directed to a value-laden goal, is perhaps the most fertile concept that theism can cultivate. Another and closely related transempirical concept of enormous consequence treatable by theism is that *Homo sapiens* is the supreme and ultimate product of universal evolution. Perhaps the concept of the immortal human soul as consisting of an everlasting supernatural essence can be cultivated without conflict with empirical knowledge. Most aspects of ethics and morality can be viewed as a priori knowledge in the form of supernatural moral imperatives. These are all value concepts generally off limits to empirical science. Through such legitimate religious concepts, when used by theologians to sustain rather than to degrade humanity, theism can greatly enhance and enrich the lives of many susceptible persons without impinging negatively upon the lives of others.

Chapter 11 concluded with a short section labeled "Is Naturalistic Monism the Answer?" Yes, it can be seen as one answer (and perhaps the only answer) to the kinds of problems we have discussed here. Naturalistic (mechanistic) monism dissolves the existential claim for an autonomous supernatural realm that can be pitted against the natural realm. Instead, it transfers the ideational concepts of religion to a neurophysiological base within empirical science. The ideational concepts themselves remain in the ideational category as products of the human imagination, but claims of divine creation have been defused and reduced to myths of historical interest only; as such, they pose no challenge to empirical science.

Credits

1. From Mario Bunge, "What Is Pseudoscience?" *Skeptical Inquirer*, vol. 9, no. 1, pp. 36-46. Copyright © 1984 by the Committee for the Scientific Investigation of Claims of the Paranormal. Used by permission of the author and publisher.

2. From Tony Rothman, *Discover*, vol. 8, no. 5, May 1987, pp. 90-99. Tony Rothman/© 1987 Discover Magazine. Used by permission.

EPILOGUE

The Skeptical Scientist

An epilogue gives an author free rein to go prescriptive, shamelessly indulging in a succession of oughts and ought-nots. You, my readers, are in turn granted full license to contradict or disregard any or all of my suggestions.

First, I urge each of you would-be scientists to take as a guiding principle the philosophical position of moderate skepticism, not only in your scientific research, but also in shaping your personal worldview. Philosopher of science and physicist Mario Bunge, whose published works I have relied upon heavily and quoted from liberally, tells us: "The scientist's skepticism is methodical and partial, not systematic and total. The serious researcher is neither gullible nor nihilistic: he does not embrace beliefs uncritically, but he does admit, at least until new notice, a host of data and theories. His skepticism is constructive, not just critical" (1991a, p. 146).

Professor Bunge goes on to say that our moderate skeptic keeps in mind a set of presuppositions that can be viewed as the philosophical requirements of the pursuit of a strong, healthy science. One is *materialism*: everything in the physical universe is concrete or material and behaves lawfully. A second is *realism*: the world exists "independently of those who study it." A third is *rationalism*: "our ideas ought to be internally consistent and they should cohere with one another." A fourth is *empiricism*, meaning that every scientific proposition we propose should be testable. A final presupposition is not so obvious, and not always adhered to: "The data and hypotheses of science are not stray but constitute a system"—i.e., they are *systematic* (1992a, p. 146). If you fail to fulfill all of these presuppositions, Bunge warns, you risk doing pseudoscience.

The skeptical scientist is ever on the alert to recognize and expose errors and fallacies of logic, tautologies that introduce no increase in information content, and those self-contradictory terms known as oxymorons. Of course, the same observation applies to all serious, rational discourse, whatever the subject, but it applies with particular force to the recognition and demolition of pseudoscience. From what you would-be scientists have read in my book, do you feel confident in your ability to cope with the panderers of pseudoscience?

The skeptical scientist finds no place for the verb "to believe" and the noun "belief" in scientific writing, scrupulously consigning all emotive expressions of faith to discourse in the belief fields of knowledge. This prohibition arises from the skeptical scientist's acceptance of a probabilitic view of the limits of scientific understanding—a paradigm that excludes any possibility of establishing either the truth (an absolute) of any scientific proposition or the perfection in measurement of any dimensional constant. Those of you who have read and digested the contents of my book should have no trouble with this formula. My complaint is not trivial; it arises from an attack that relativists of various ilks are currently directing at scientists in an attempt to denigrate the products of science. These detractors of science say: "After all, your science is just another belief system resting in faith, just like our systems of belief in religion and ethics." So I ask you to join me in a global campaign to convince the science writers and the journalists of all news media never again to write or say: "*Scientists believe* that such-and-such is the case (or is true)." Many probabilistic expressions implying degrees of tentativeness are available to replace the forbidden "B-word." To be sure, that requires a couple of extra words, but well worth the effort. Will you help me in this campaign, futile as it may seem?

Much of what I have written in the second part of this book is about the six transempirical (ideational) knowledge fields that surround and interface with science and related forms of empirical (perceptional) knowledge. I therefore suggest that you reexamine and evaluate your entire worldview, taking always the position of moderate skepticism. Use the seven hexagons of Figure 9.1 as a notebook page to jot down your personal beliefs, particularly in the hexagons of religion, ethics and morality, and sociopolitical ideologies. For each ideational proposition you compose, ask youself: "What, if anything, has science to say about this particular statement of my belief?" Does each proposition about your religion or your ethics or your sociopolitical affiliation conform with all five of the presuppositions of science laid out above? For example, is your belief (if you have one) that God exists within a supernatural realm accurately described as materialistic, realistic, rational, empirical, and systematic? Perhaps your answer would be: "No on all counts, but that's acceptable to me." Try ethics next. Suppose you jot down: "The right to life is one of a long list of inalienable human rights." Can your knowledge of science support that statement, applying information gained by observation? Is it a rational statement? Is it testable? Apply these same questions to the category of sociopolitical ideologies. Suppose you have jotted down: "Governments are instituted among humans, deriving their just powers from the consent of the governed." Can science in any meaningful way guide humans in judging the "correctness" or "validity" of such propositions?

In the area of sociology of science there is plenty of room for large doses of skepticism. This is especially evident in the question of what role science plays or should play in society. Since the time of Abraham Lincoln's presidency, when the National Academy of Sciences was formed to apply scientific knowledge to benefit the American nation and public, we have heard oratorical cries for "Science in the Service of Society." Does that motto specifically designate science as a servant of industry, or as a servant of the military establishment? Only in the past few years has this motto been altered to read: "Science in the Service of our Planet." Suppose you plan to specialize in a branch of science that offers little prospect of ever being of tangible utility to humankind—for example, you want to study Cambrian fossils or supernovae or the speciation of finches on the Galapagos Islands. Should (ought) you feel guilty of wasting your talent and/or the funds given you to do your research? Would it be "wasteful" to give research funds to the paleontologist, instead of to a biologist searching for a cure for AIDS? Do you think you would be making a positive contribution to humankind by doing a Ph.D. dissertation on the origin of petroleum and following that by a career of searching for oil pools as the employee of a major oil exploration company? All these questions involve ethical values with focus upon which of two alternatives is "the better" or "the more desirable": ergo, which *ought* to be implemented.

Apropos of my above questioning of the role of science in society, perhaps you, as a scientist, should (ought to) be just a bit skeptical of grandiose plans proposed by politicians (kept in office by the PACs) to improve science and math education in the public schools, using standard tests to establish criteria of improvement in comparison with students of other nations. What sector of our society is this rather naive program intended to benefit? Is it intended to furnish better trained employees for industry (to improve corporate profit margins) and better qualified military recruits (possibly an ethically desirable aim in itself), or is it designed to increase the rate of realization of new theories and discoveries on the frontiers of science? Do we need a greater total flux in millions of young brains of the rote memorization, regurgitation, and subsequent loss of recall of innumerable terms, laws, and formulas of elementary science and math? Or should we be seeking out and sponsoring those few gifted persons who can swiftly rise to the cutting edge where theories are being devised and evaluated? Are you for populism, or elitism, or some of both?

Here's a related topic: Through programs such as the Reserve Officer Training Corps (ROTC), the armed services support students through college, asking only that they remain available and fit for future duty in those services. A similar military subsidy carries postgraduate students through medical school, asking that in turn they serve a term of duty in a military medical facility. Through the National Science Foundation (NSF) college graduates with exceptionally strong academic records may be awarded NSF fellowships, carrying them through to a science Ph.D. For these special few, no specific form of service to the nation is required, nor is any form of monetary repayment asked. Is this an indication that the NSF encourages careers in pure science research that promise at best only very modest incomes and may generate no direct benefits to society?

This brings me to what you may see as a rather strange list of oughts and ought-nots in the widely approved content of our public school curricula. Turning to the Bill of Rights (the Bill of Oughts?) in the First Amendment, those phrases referred to by constitutional lawyers as the Establishment Clause and the Free Exercise Clause have so far effectively barred the promotion of religion from the classroom, and even put the kibosh on prayers in school. So far as science teaching is concerned, this doctrine excludes the teaching of religious creationism in the science classroom. On the other hand, the teaching (inculcation) of a particular ethical system and its morals is mandatory and even assumed to be a national obligation, since those moral propositions have been transformed into our laws of conduct; i.e., the "oughts" have been forged into "musts." However, these morals which we all recognize as both "good" and "necessary" for the harmonious working of our democracy are clearly of Western religious origin, for we find many of them in the ancient holy scriptures of the Judeo-Christian faiths or in the New Testament. My first question is this: If the teaching of religion is excluded from the school curriculum, should (ought) not the teaching of morals and moral values also be excluded? We would have to presume that the task of teaching of morals must then fall to the private sector, through religious and private secular agencies.

Two additional questions can be appended to the above, one specific and one general: (1) Should the students be taught the legal code to which they are obliged to conform on pain of punishment or death, whereas at the same time morals may not be taught? (2) Should our public schools teach *about* religions in general, using only purely descriptive and historical statements? At least a few prominent scientists favor this second program. These questions bring us back to scientists and scientific knowledge for the following reasons: We have concluded in Part Two of this book that the major category of perceptional knowledge includes both science and history; i.e., everything that is external reality (concrete things). Everything observable to the senses is encompassed in this knowledge realm, and that includes the written and oral statements that have stored the information content of history. Written legal codes have been a part of human history since the time of Hammurabi, along with human history in general (i.e., accounts of what happened). The final tie into science comes through the sociology of science and the "scientific ethos" claimed for it. Scientists must not lie, steal, or cheat in their investigative activities and their journal publications. Deliberate falsifying and tampering with data are considered among the deadly sins against science. So my final question is: Would it be possible to generate a purely empirical, rational, and plausible scientific ethos solely from the "is"es contained in the purely empirical body of science and history? If so, science would be freed from a dependence upon all moralistic ideational constructs, rendering science invulnerable to destructive relativistic attacks by fundamentalist religious groups, the left-wing factions of the New Social Science, the New Agers, and the panderers of pseudoscience. Secular humanists claim they can derive a succcesful ethical system based on science, but unless science decontaminates itself from a set of ideational morals, no general secular ethical system is possible.

In summation, most of the questions I have posed above are about transempirical values; they ask us to judge "ought to be" against "ought not to be." In contrast, science internally poses questions of empirical content in which scientists are asked to judge "is" against "is not." Keep all your apples in one box and all your golf balls separate in another box. The skeptical scientist will always check to be sure that the "oughts" stay in one ontological box and the "is"es in another. Those who believe and preach a pseudoscience, a particular religion, a moral prohibition, or a particular political or social ideology are prone to blur or erase the ontological distinctions, whether out of ignorance or in the practice of deliberate deception. The skeptical scientist will recognize these relativistic tactics that only serve to demean and degrade science. Young (and old) scientists, keep your eyes and ears open at all times or you may be conned before you realize what is happening.

REFERENCES CITED

Albritton, Claude C., Jr. 1963. *The Fabric of Geology*. Reading, Mass.: Addison-Wesley Publ. Co.

American Scientific Affiliation. 1986. *Teaching Science in a Climate of Controversy*. Ipswich, Mass.: American Scientific Affiliation.

Anderson, Philip W. 1990. "On the Nature of Physical Laws." *Physics Today* 43: 9.

Angeles, Peter A. 1976. *Critiques of God*. Buffalo, N.Y.: Prometheus Books.

———. 1980. *The Problem of God: A Short Introduction*. Buffalo, N.Y.: Prometheus Books.

———. 1981. *Dictionary of Philosophy*. New York.: Barnes & Noble Books, Division of Harper & Row.

Atkins, Kenneth R. 1972. *Physics—Once Over Lightly*. New York: John Wiley & Sons.

Atkins, K. R., Holum, J. R., and Strahler, A. N. 1978. *Essentials of Physical Science*. New York: John Wiley & Sons.

Ayala, Franscisco J. 1977. "Philosophical Issues." Chap.16, pp. 474-516 in Dobzhansky, et al. 1977.

Bahm, Archie J. 1971. *The World's Living Religions*. Carbondale, Ill.: Southern Illinois University Press.

———. 1980. "Humanistic Ethics As the Science of Oughtness." Pp. 210-26 in Storer. 1980.

———. 1984. "Humanism Is a Religion." *Free Inquiry* 5, no. 1: 44.

Bainbridge, William S., and Stark, Rodney. 1981, "Superstitions: Old and new." Pp. 46-59 in Frazier. 1981.

Barrow, John D., and Tipler, Frank J. 1986. *The Anthropic Cosmological Principle. New York: Oxford Univ. Press.*

Beckett, Chris. 1990. "The Great Chain of Being." *New Scientist* 125, no. 1709: 60-61.

Berger, James O., and Berry, Donald A. 1988. "Statistical Analysis and the Illusion of Objectivity." *American Scientist* 76: 159-165.

Bischof, N. 1978. "On the Phylogeny of Human Morality." Pp. 53-33 in Stent. 1978.

Boulding, Kenneth E. 1984. "Toward an Evolutionary Theology." Pp. 142-158 in Ashley Montagu, ed. *Science and Creationism*. New York: Oxford Univ. Press.

Bowles, Norma, and Hinds, Fran. 1978. *Psi Research*. New York: Harper & Row.

Bridgman, Percy W. 1936. *The Nature of Physical Theory*. Princeton, N.J.: Princeton University Press.

Broad, William J. 1981. "Creationists Limit Scope of Evolution Case." *Science* 211: 1331-1332.

Brown, Harold I. 1977. *Perception, Theory and Commitment: The New Philosophy of Science.* Chicago, Ill.: Precedent Publishing Co.

Bucher, Walter H. 1941. "The Nature of Geological Inquiry and the Training Required for It." Tech. Publ. No. 1377. New York: Amer. Inst. of Mining & Metallurgical Engineers.

Bunge, Mario. 1977, 1983. *Treatise on Basic Philosophy.* vol. 3, Ontology I, Furniture of the world, 1977; vol. 5, Epistemology & Methodology I, Exploring the world, 1983a; vol 6, Epistemology & Methodology II; Understanding the world, 1983b. Dordrecht and Boston: D. Reidel.

———. 1984. "What is Pseudoscience?" *The Skeptical Inquirer* 9, no. 1: 36-46.

——— 1991a. "A Critical Examination of the New Sociology of Science." Part I, *Philosophy of the Social Sciences* 21: 524-60.

——— 1991b. "A Skeptic's Beliefs and Disbeliefs." *New Ideas in Psychology* 9: 131-141.

Campbell, Bernard. 1985. *Human Evolution: An Introduction to Man's Adaptations.* 3rd ed. Hawthorne, N.Y.: Aldine Publ. Co.

Carnap, Rudolph. 1936-1937. "Testability and Meaning." *Philosophy of Science* 3: 419-21; 4: 1-40.

Chamberlin, Thomas C. 1904. "The Methods of the Earth-sciences." *Popular Science Monthly* 66: 66-75.

Clements, Tad S. 1985. "Religion vs. Science." Pp. 548-554 in Stein. 1985.

Cloud, Preston E. 1980. "Beyond Plate Tectonics." *American Scientist* 68: 381-87.

Cohen, I. Bernard. 1985. *Revolution in Science.* Cambridge, Mass.: Belknap Press of Harvard Univ. Press.

Cohen, Morris R., and Nagel, Ernest. 1934. *An Introduction to Logic and the Scientific Method.* New York: Harcourt, Brace and Co.

Cole, John R. 1981. "Misquoted Scientists Respond." *Creation/Evolution* 2, no. 4: 34-44.

Condon, Edward U. 1969, *Scientific Study of Unidentified Flying Objects.* New York: Bantam Books.

Crutchfield, James P., Farmer, J. Doyne, Packard, Norman N., and Shaw, Robert S. 1986. "Chaos." *Scientific American:* 255, no. 6: 46-57.

Dantzig, Tobias. 1954. *Number: The Language of Science.* 4th ed. New York: The Free Press, Division of Macmillan Publishing Co.

Darwin, Charles. 1859. *On the Origin of Species by Means of Natural Selection.* London: John Murray. Reprint 1986 by Penguin Books, Hammondsworth, Middlesex, England, with an "Introduction and Bibliography" by J. W. Burrow.

Davies, J. T. 1973. *The Scientific Approach.* London and New York: Academic Press.

Davies, Paul. 1984. *Superforce: The Search for a Grand Unified Theory of Nature.* New York: Simon & Schuster.

Dawkins, Richard. 1976, *The Selfish Gene.* New York and Oxford: Oxford Univ. Press.

Diamond, Jared. 1987. "Soft Sciences Are Often Harder Than Hard Science." *Discover* 8, no. 8: 34-39.

Dobzhansky, Theodosius. 1962. *Mankind Evolving.* New Haven: Yale Univ. Press.

———. 1967. *The Biology of Ultimate Concern.* New York: New American Library.

———. 1977. "Evolution of Mankind," Chap. 14, pp. 438-63 in Dobzhansky, et al., *Evolution.* 1977.

Dobzhansky, T., Ayala, F. J., Stebbins, G. L., and Valentine, J. W. 1977. *Evolution.* San Francisco: W. H. Freeman and Co.

Doolittle, Russell F. 1983. "Probability and the Origin of Life." Pp. 85-97 in Laurie R. Godfrey, ed. 1983. *Scientists Confront Creationism.* New York: W. W. Norton and Co.

Douglas, Alfred. 1977. *Extra-Sensory Powers: A Century of Psychical Research.* Woodstock, N.Y.: The Overlook Press.

Durbin, Paul T. 1988. *Dictionary of Concepts in the Philosophy of Science.* Westport, N.Y.: Greenwood Press.

Dutch, Steven I. 1982. "Notes on the Nature of Fringe Science." *Jour. of Geological Education* 30: 6-13.

Fair, Charles. 1974. *The New Nonsense: The End of Rational Consensus.* New York: Simon and Schuster.

Faul, Henry, and Faul, Carol. 1983. *It Began with a Stone.* New York: John Wiley & Sons.

Feigl, Herbert. 1953. "The Scientific Outlook: Naturalism and Humanism." Pp. 8-18 in Feigl and Brodbeck. 1953. (First published in 1949 in *American Quarterly* 1.)

Feigl, H., and Brodbeck, M. 1953. *Readings in Philosophy of Science.* New York: Appleton-Century-Crofts.

Ferm, Vergilius. 1936. *First Adventures in Philosophy.* New York: Charles Scribner's Sons.

———. 1937. *First Chapters in Religious Philosophy*, New York: Round Table Press.

Feyerabend, Paul. 1980. "How to Defend Society Against Science." Pp. 55-65 in Klemke, et al., 1980.

Flew, Antony, G. N. "Hume." Chapter 15, pp. 253-74, in D. J. O'Connor, ed. 1964.

——— . 1980. *Philosophy; An Introduction.* Buffalo, N.Y.: Prometheus Books.

———. 1985. "Unbelief in Miracles." Pp. 452-58 in Stein. 1985.

Francis, Mark. 1985. "Herbert Spencer." Pp. 650-52 in Stein. 1985.

Frank, Philipp. 1946. *Foundations of Physics*. International Encyclopedia of Unified Science 1, no. 7. Chicago, ILL.: Univ. of Chicago Press.

Frazier, Kendrick, ed. 1981. *Paranormal Borderlands of Science*. Buffalo, N.Y.: Prometheus Books.

Fried, C. 1978. "Biology and Ethics: Normative Implications." Pp. 209-20 in Stent. 1978.

Gardner, Martin. 1957. *Fads and Fallacies in the Name of Science*. New York: Dover Publications.

———. 1981. *Science: Good, Bad and Bogus*. Buffalo, N.Y.: Prometheus Books.

Geisler, Norman L. 1983. "A Scientific Basis for Creation; The Principle of Uniformity." *Creation/Evolution* 4, no. 3: 1-6.

Gleick, James. 1987. *Chaos: Making a New Science*. New York: Viking Penguin Inc.

Goldsmith, D., ed. 1977. *Scientists Confront Velikovsky*. Ithaca, N.Y.: Cornell Univ. Press.

Gould, Stephen J. 1982. "Darwinism and the Expansion of Evolutionary Theory." *Science* 216: 380-87.

Grene, Marjorie. 1985. "Perception, Interpretation, and the Sciences." Pp. 1-20 in David J. Depew and Bruce H. Weber. 1985. *Evolution at the Crossroads: The New Biology and the New Philosophy of Science*. Cambridge, Mass.: The MIT Press.

Haas, J. W., Jr. 1989. "Arthur Peacocke's *New Biology*: New Wine in Old Bottles." *Science and Christian Belief* 1, no. 2: 161-66.

Hansel, C. E. M. 1980. *ESP and Parapsychology: A Critical Reevaluation*. Buffalo, N.Y.: Prometheus Books.

Hardin, Garrett. 1984. "'Scientific Creationism'—Marketing Deception As Truth." Pp. 159-166 in Montagu. 1984.

Hawking, Stephen W. 1988. *A Brief History of Time*. New York: Bantam Books.

Hempel, C. G. and P. Oppenheim. 1953. "The Logic of Explanation," Pp. 319-352 in Feigl and Brodbeck. 1953.

Hendry, Allan. 1979. *The UFO Handbook: A Guide to Investigating, Evaluating and Reporting UFO Sightings*. Garden City, N.Y.: Doubleday & Co.

Hildemann, W. H. 1982. "Creative Evolution." Letters to Editor. *Science* 215: 1182.

Hooke, Robert. 1665. *Micrographia*. London: Martyn & Allestry. (Reprint 1961. New York: Dover Books.)

Hospers, John. 1980. "Law." Pp. 104-111 in Klemke, Hollinger, and Kline. 1980.

Howson, Colin, and Urbach, Peter. 1989. *Scientific Reasoning: The Bayesian Approach*. La Salle, Ill.: Open Court Publishing Co.

Hull, David L. 1988. *Science As a Process: An Evolutionary Account of the Social and Conceptual Development of Science.* Chicago, Ill.: University of Chicago Press.

Hume, David. 1748 (1988). *An Enquiry Concerning Human Understanding.* Buffalo, N.Y.: Prometheus Books.

Johnson, J. G. 1990. "Method of Multiple Working Hypotheses: A Chimera." *Geology* 18: 44-45.

Kaufmann, William, J. 1985. *Universe.* New York: W. H. Freeman & Co.

Kitcher, Philip. 1982. *Abusing Science: The Case Against Creationism.* Cambridge, Mass.: The M.I.T. Press.

Klass, Philip J. 1968. *UFOs—Unidentified.* New York: Random House.

Klemke, E. D., Hollinger, Robert, and Kline, A. David, eds. 1980. *Introductory Readings in the Philosophy of Science.* Buffalo, N.Y.: Prometheus Books.

Kofahl, Robert E. 1981. "Letters to Editor." *Science* 212: 873.

Kramer, Edna E. 1970. *The Nature and Growth of Modern Mathematics.* Princeton, N.J.: Princeton Univ. Press.

Kuhn, Thomas S. 1962. *The Structure of Scientific Revolutions.* Chicago, Ill.: Univ. of Chicago Press.

Kulp, J. Laurence. 1950. "Flood Geology." *Jour., Amer. Scientific Affiliation* 2: 1-15.

Kurtz, Paul. 1981. "Is Parapsychology a Science?" Pp. 5-23 in Frazier. 1981.

———. 1983. *In Defense of Secular Humanism.* Buffalo, N.Y.: Prometheus Books.

Lewin, Roger. 1980. "Evolutionary Theory Under Fire." *Science* 210: 883-87.

———. 1982. "Biology Is Not Postage Stamp Collecting." *Science* 216: 718-20.

———. 1987. "The Origin of the Modern Human Mind." *Science* 236: 668-70.

Lightman, Alan P. 1983. "Weighing the Odds." *Science 83* 4, no. 10: 21-22.

Lorenz, Konrad. 1966. *On Agression.* New York: Harcourt Brace & World.

Lotka, A. J. 1922. "Contribution to the Energetics of Evolution." *Proc. Nat. Acad. Sci.* 8: 147-55.

Ludwig, Jan, ed. 1978. *Philosophy and Parapsychology.* Buffalo, N.Y.: Prometheus Books.

Lumsden, Charles J., and Gushurst, Ann C. 1985. "Gene-culture Coevolution: Humankind in the Making." Pp. 4-28 in James H. Fetzer, ed. 1985. *Sociobiology and Epistemology.* Dordrecht, Holland: D. Reidel Publishing Co.

Margenau, Henry. 1964. *Ethics and Science.* Princeton, N.J.: D. Van Nostrand. Reprint. 1979. Huntington, N.Y.: Krieger Publ. Co.

Marshall, Eliot. 1983. "A Controversy on Samoa Comes of Age." *Science* 219: 1042-1045.

Marty, Martin E. 1985. "Unbelief within Christianity." Pp. 98-103 in Stein. 1985.

Mayr, Ernst. 1972. "The Nature of the Darwinian Revolution." *Science* 176: 981-89.

———. 1977. "Darwin and Natural Selection." *American Scientist* 65: 321-27.

———. 1982. *The Growth of Biological Thought.* Cambridge, Mass.: Belknap Press of Harvard University Press.

McAlester, A. Lee. 1968. *The History of Life.* Englewood Cliffs, N.J.: Prentice-Hall.

Menninger, Karl. 1969. *Number Words and Number Symbols: A Cultural History of Numbers.* Cambridge, Mass.: The M.I.T. Press.

Merton, Robert K. 1973. *The Sociology of Science: Theoretical and Empirical Investigations.* Chicago, Ill.: Univ. of Chicago Press.

Montagu, Ashley, ed. 1984. *Science and Creationism.* New York: Oxford Univ. Press.

Moore, John N., and Slusher, Harold S., eds. 1970, 1974. *Biology: A Search for Order in Complexity.* Grand Rapids, Mich.: Zondervan Publishing House.

Morowitz, Harold J. 1968. *Energy Flow in Biology.* New York: Academic Press.

———. "Navels of Eden." *Science 82* 3, no. 2: 20-22.

Morris, Henry M. 1946, 1978. *That You Might Believe.* (Revised edition. 1978) San Diego, Calif.: Creation-Life Publishers.

———. 1974. *Scientific Creationism.* (Public School Edition, General Edition) San Diego, Calif.: Creation-Life Publishers.

———. 1984. *A History of Modern Creationism.* San Diego, Calif.: Master Book Publishers.

———. 1988. "The Compromise Road." *ICR Impact Series,* no. 177.

Nagel, Ernest. 1961. *The Structure of Science: Problems in the Logic of Scientific Explanation.* New York: Harcourt, Brace & World. (Reprint. 1979. Indianapolis, Ind.: Hackett Publishing Company.)

Nagel, T. 1978. "Ethics As an Autonomous Theoretical Subject." Pp. 221-32 in Stent. 1978.

Nash, Carroll B. 1978. *Science of Psi: ESP and PK.* Springfield, Ill.: Charles C. Thomas, Publ.

Nielsen, Kai. 1984. "Reason." Pp. 533-44 in Stein. 1985.

Numbers, Ronald. 1982. "Creationism in 20th-Century America." *Science* 218: 538-44.

O'Connor, D. J. ed. *A Critical History of Western Philosophy.* New York: The Free Press; A Division of Macmillan, Inc.

Odum, H. T., and Pinkerton, R. C. 1955. "Time's Speed Regulator: The Optimum Efficiency for Maximum Power Output in Physical and Biological Systems." *American Scientist* 43: 331-43.

Omohundro, John T. 1981. "Von Däniken's Chariots: A Primer in the Art of Cooked Science." Pp. 307-331 in Frazier. 1981.

Overton, William R. 1982. "Creationism in the Schools; The Decision in McLean Versus the Arkansas Board of Education." *Science* 215: 934-43.

Parsons, Keith M. 1989. *God and the Burden of Proof.* Buffalo, N.Y.: Prometheus Books.

Peacocke, Arthur. 1986. *God and the New Biology*. London: Dent and Son.

Peano, Giuseppi. 1973. *Selected Works of Giuseppi Peano*. Trans. by H. C. Kennedy. Toronto: Univ. of Toronto Press.

Pensée Editors. 1976. *Velikovsky Reconsidered*. Garden City, N.Y.: Doubleday and Co.

Pickert, G., and Görke, L. "Construction of the System of Real Numbers." Pp. 93-153 in Behnke, H., Bachmann, F., Fladt, K. and Süss, W, eds. *Fundamentals of Mathematics.* Cambridge, Mass.: The MIT Press.

Pirie, N. W. 1953. "Ideas and Assumptions About the Origin of Life." *Discovery* 14: 238-42.

Polkinghorne, John. 1989. *Science and Creation: The Search for Understanding*. Boston, Mass.: New Science Library, Shambhala Publications, Inc.

———. 1991. "The Nature of Physical Reality." *Zygon* 26: 221-236.

Pool, Robert. 1989. "Chaos Theory: How Big an Advance?" *Science* 245: 26-28.

Popper, Karl R. 1959, 1968. *The Logic of Scientific Discovery.* London: Hutchinson. New York: Basic Books, Harper & Row.

———. 1963. *Conjectures and Refutations: The Growth of Scientific Knowledge*. New York: Basic Books. (Harper Torchbook ed. 1968. New York: Harper & Row.)

———. 1972, 1979. *Objective Knowledge: An Evolutionary Approach.* Revised ed. Oxford: Clarendon Press, Oxford University Press.

———. 1976. *Unended Quest: An Intellectual Autobiography*. La Salle, Ill.: Open Court Publ. Co.

Price, George M. 1923. *The New Geology*. Mountain View, Calif.: Pacific Press.

Ramm, Bernard. 1954. *The Christian View of Science and Scripture.* Grand Rapids, Mich.: W. B. Erdmans Publ. Co.

Ransom, C. J. 1976. *The Age of Velikovsky*. Glassboro, N.J.: Kronos Press.

Raup, David M., and Stanley, Steven M. 1971. *Principles of Paleontology*. San Francisco: W. H. Freeman & Co.

Rensberger, Boyce. 1983. "Margaret Mead on Becoming Human: The Nature-Nurture Debate." *Science 83* 4, no. 1: 28-46.

Rhine, J. B., and Pratt, J. G. 1957, *Parapsychology: Frontier Science of the Mind.* Springfield, Ill.: Charles C. Thomas, Publ.

Riepe, Dale. 1985. "Ludwig Feuerbach." Pp. 221-24 in Stein. 1985.

Root-Bernstein, Robert. 1981. "Letters to Editor." *Science* 212: 1446.

Rothman, Milton A. 1988. *A Physicist's Guide to Skepticism.* Buffalo, N.Y.: Prometheus Books.

Rothman, Tony. 1987. "A 'What You See Is What You Beget' Theory." *Discover* 8, no. 5: 90-99.

Ruse, Michael, ed.. 1988. *But Is It Science? The Philosophical Question in the Creation/Evolution Controversy.* Buffalo, N.Y.: Prometheus Books.

Ruse, Michael, and Wilson, Edward O. 1985. "The Evolution of Ethics." *New Scientist* 108, no. 1487: 50-52.

Russell, Bertrand. 1920. *Introduction to Mathematical Philosophy*. 2d ed. London: Allen & Unwin. Reprint. New York: Simon & Schuster.

———. 1937. *The Principles of Mathematics.* 2nd ed. New York: W. W. Norton & Co. (First ed. 1907.)

——— . 1945. *A History of Western Philosophy*. New York: Simon & Schuster, Inc.

Ryzl, Milan. 1970. *Parapsychology: A Scientific Approach*. New York: Hawthorn Books.

Sagan, C., and Page, T., eds. 1972. *UFOs, A Scientific Debate.* Ithaca, N.Y.: Cornell Univ. Press.

Saver, Jeffrey. 1985. "An Interview with E. O. Wilson on Sociobiology and Religion." *Free Inquiry* 5, no. 2: 15-22.

Schuchert, Charles, and Dunbar, Carl. O. 1933. *A Textbook of Geology; Part II—Historical Geology.* New York: John Wiley & Sons.

Scriven, Michael. 1959. "Explanation and Prediction in Evolutionary Theory." *Science* 130: 477-482.

———. 1980. "The Exact Role of Value Judgments in Science." Pp. 369-291 in Klemke, et al. 1980.

Shea, James H. 1982. "Twelve Fallacies of Uniformitarianism." *Geology* 10: 455-460.

Shore, Lys Ann. 1989. "New Light on the New Age." *The Skeptical Inquirer* 13: 226-40.

Silk, Joseph. 1980. *The Big Bang: The Creation and Evolution of the Universe.* San Francisco: W. H. Freeman and Co.

Simpson, George Gaylord. 1944. *Tempo and Mode in Evolution.* New York: Columbia Univ. Press.

———. 1963. "Historical Science." Pp. 24-48 in Albritton. 1963.

———. 1964. *This View of Life.* New York: Harcourt Brace & World.

———. 1969. *Biology and Man.* New York: Harcourt Brace & World.

Skehan, James W. 1983 "Theological Basis for a Judeo-Christian Position on Creationism." *Jour. of Geologic Education* 31: 307-314.

Smart, J. S. 1979. "Determinism and Randomness in Fluvial Geomorphology." *EOS, Jour., Amer. Geophysical Union* 60, no. 36: 651-55.

Smith, John Maynard. 1978a. "The Evolution of Human Behavior." *Scientific American* 239, no. 3: 176-91.

Stebbins, G. Ledyard. 1982. *Darwin to DNA, Molecules to Humanity*. New York: W. H. Freeman and Co.

Stein, Gordon, ed. 1985. *The Encyclopedia of Unbelief*. Buffalo, N.Y.: Prometheus Books.

Stenger, Victor J. 1990. *Physics and Psychics: The Search for a World Beyond the Senses*. Buffalo, N.Y.: Prometheus Books.

Stent, Gunther S. 1975. "Limits to the Scientific Understanding of Man." *Science* 187: 1052-1057.

Stent, Gunther S., ed. 1978. *Morality As a Biological Phenomenon*, Report of the Dahlem Workshop, Berlin, 1977. Berlin: Abakon Verlagsgesellschaft.

Storer, Morris B., ed. 1980. *Humanist Ethics: Dialogue on Basics*. Buffalo, N.Y.: Prometheus Books.

Storer, Norman W. 1977. "The Sociological Context of the Velikovsky Controversy." Pp. 29-39 in Goldsmith, D., ed. 1977. *Scientists Confront Velikovsky*, Ithaca, N.Y.: Cornell Univ. Press.

Story, Ronald. 1976. *The Space Gods Revealed*. New York: Harper & Row.

Strahler, Arthur N. 1954 "Historical Geology, Dynamic Geology, and Recoverability." Abstracts of Papers, National Academy of Science, Autumn Meeting, Columbia Univ., New York. *Science* Nov. 12: 15.

———. 1980. "Systems Theory in Physical Geography." *Physical Geography* 1, no. 1: 1-27.

———. 1983. "Toward a Broader Perspective in the Evolutionism-Creationism Debate." *Jour. of Geologic Education* 31: 87-94.

———. 1987, *Science and Earth History—The Evolution/Creation Controversy*. Buffalo, N.Y.: Prometheus Books.

Strahler, A. N. and Strahler, A. H. 1974. *Introduction to Environmental Science*. Santa Barbara, Calif.: Hamilton Publishing Co.

Teilhard de Chardin. 1966. *Man's PLace in Nature*. (Tr. by Réné Hague.) New York.: Harper & Row.

Thaxton, Charles B., Bradley, Walter L., and Olsen, Roger L. 1984. *The Mystery of Life's Origin: Reassessing Current Theories*. New York: Philosophical Library.

Thomsen, Dietrick E. 1983. "A Knowing Universe Seeking to be Known." *Science News* 123: 124.

Trefil, J. S. 1978. "A Consumer's Guide to Pseudoscience." *Saturday*

Trivers, Robert. 1971. "The Evolution of Reciprocal Altruism." *Quarterly Review of Biology* 46: 35-37.

Tsonis, A. A,. and Elsner, J. B. 1989. "Chaos, Strange Attractors, and Weather." *Bull., Amer. Meteorological Soc.* 70: 14-23.

Unger, Tom. 1984. "The Sociobiology Debate: What Is It Really All About?" *American Rationalist* 29, no. 4: 57-59.

Velikovsky, Immanuel. 1950. *Worlds in Collision.* Garden City, N.Y.: Doubleday and Co.

von Bertalanffy, Ludwig. 1950. "The Theory of Open Systems in Physics and Biology." *Science* 111: 23-29.

von Däniken, Erich. 1969. *Chariots of the Gods? Unsolved Mysteries of the Past.* (Tr. by Michael Heron.). New York: G. P. Putnam's Sons.

———. 1971. *Gods from Outer Space.* New York: G. P. Putnam's Sons.

———. 1973. *The Gold of the Gods.* New York: G. P. Putnam's Sons.

Waddington, Conrad H. 1960. *The Ethical Animal.* London: Allen & Unwin.

Wartofsky, Marx W. 1968. *Conceptual Foundations of Scientific Thought: An Introduction to the Philosophy of Science.* New York: Macmillan Co.

Wheatley, James M. O., and Edge, Hoyt L., eds. 1976. *Philosophical Dimensions of Parapsychology.* Springfield, Ill.: Charles C. Thomas, Publ.

Whitcomb, John C., and Morris, Henry M. 1961. *The Genesis Flood.* Philadelphia: Presbyterian and Reformed Publishing Co.

Wilson, Clifford. 1972. *Crash Go the Chariots.* New York: Lancer Books.

Wilson, Edward O. 1975, *Sociobiology: The New Synthesis.* Cambridge, Mass.: Harvard Univ. Press.

———. 1978. *On Human Nature.* Cambridge, Mass.: Harvard Univ. Press.

Wilson, Edward O., and Lumsden, Charles. 1981. *Genes, Mind, and Culture: The Evolutionary Process.* Cambridge, Mass.: Harvard Univ. Press.

Wittgenstein, Ludwig. 1922. *Tracatus Logico-Philosophicus.* (Tr. by C. K. Ogden.) London: Routledge and Kegan Paul.

Wysong, Randy L. 1976. *The Creation-Evolution Controversy.* Midland, Mich.: Inquiry Press.

Young, Davis A. 1977. *Creation and Flood.* Grand Rapids Mich.: Baker Book House.

———. 1982. *Christianity and the Age of the Earth.* Grand Rapids, Mich.: Zondervan Publishing House.

Zeisel, Hans. 1981. "Letters to Editor." *Science* 212: 873.

Ziman, John. 1980. "What is Science?" Pp. 35-54 in Klemke, Hollinger, and Kline. 1980.

Name Index

Subject Index